D0161411

The Ecology of Atlantic Shorelines

The Ecology of
Atlantic Shorelines

Mark D. Bertness

Brown University

Illustrations by Kelly Benoit Bird

Sinauer Associates, Inc. *Publishers*
Sunderland, Massachusetts U.S.A.

Sinauer Associates Inc.
23 Plumtree Road
Sunderland, MA USA 01375

FAX : 413-549-1118
Internet: publish@sinauer.com
www.sinauer.com

Library of Congress Cataloging-in-Publication Data

Bertness, Mark D.
        The ecology of Atlantic shorelines / Mark D. Bertness.
            p.   cm.
        Includes bibliographical references and index.
        ISBN 0-87893-056-6 (pbk.)
        1. Coastal ecology—Atlantic coast (U.S.) 2. Animal
communities—Atlantic coast (U.S.)   I. Title.
QH104.5.A84B47   1999   98-39306
577.69′9′0974—dc21   CIP

Printed in U.S.A.

5 4 3 2 1

For Janette and Sarah—all that really matters.

Contents

Preface

I have always been drawn to seashores. I grew up on the shores of Puget Sound, collecting shells and crab molts, and as an adult I have explored Atlantic shorelines with the same enthusiasm. This book is for those who share my curiosity about the plant and animal communities of shoreline habitats.

I wrote this book as a general introduction to the ecology of shoreline communities of the Atlantic Coast of North America. I undertook this project because no introduction was available to stimulate my own students to think about the organization and patterning of Atlantic shoreline communities. As a student I was introduced to west coast shoreline ecology by reading Tom Carefoot's *Pacific Seashores* (1977). Even though Carefoot's book was about Pacific rather than Atlantic shorelines, dealt almost exclusively with rocky shores, and was 20 years old, I still found myself having students read it to get a feel for intertidal ecology. I wrote this book to provide an Atlantic Coast perspective.

My book is aimed at interested readers of all persuasions and backgrounds. While I had undergraduate budding ecologists primarily in mind, I have made the book accessible to general readers with interests in resource management, ecology, and conservation by minimizing jargon and stressing ideas rather than details. I have also tried to balance an accessible style with enough content to hold the attention of scientists from other disciplines as well as lay naturalists interested in understanding how shoreline communities work.

It was not my intention to write a detailed, comprehensive treatment of Atlantic shoreline ecology. A number of more general introductions to marine ecology are already available. Jeff Levinton's introductory marine ecology text and Ivan Valiela's graduate-level marine ecology text (both 1995) are broader, more complete treatments of general marine ecology. Two excellent general intertidal ecology texts have also recently been published, both with a British perspective (Raffaelli and Hawkins 1996; Little and Kitching 1996). My book overlaps little with these books; it is intended rather as a field guide to the ecology of Atlantic shoreline communities—a primer on how these communities work and are organized. It presents my personal perspective on shoreline communities, and I make no excuses for its biased perspective. My interests in shorelines have always been focused on the forces that generate repeated patterns in natural communities, and this book primarily examines how predictable processes generate patterns on Atlantic shorelines. When possible, I have tried to fit my community perspective into a broader systems framework. Those with a systems view of shoreline ecosystems, however, may find my efforts in this regard entirely lacking. I also have not given Canadian shorelines as much attention as they deserve.

The book is organized into two sections. The first four chapters are a general introduction to shoreline ecology. In these chapters I draw on general examples to introduce readers to the processes that shape shoreline communities everywhere. The last three chapters specifically examine the dominant shoreline habitats on the Atlantic coast of North America: rocky shores, soft-sediment beaches, and salt marshes.

Chapter 1 provides the background information on geological history, tides, introduced species, and hydrodynamics that is necessary to begin studying the ecology of Atlantic shorelines. The next chapter, on shoreline economy, examines the primary productivity of shorelines, introduces the major plant and animal groups found in shoreline environments, and ends by discussing the ecological and evolutionary balance between producers and consumers. Chapter 3 examines the often unusual reproduction problems faced by most shoreline organisms and the consequences of larval dispersal for the recruitment and population structure of shoreline organisms. The final introductory chapter discusses benthic processes that affect the distribution and abundance patterns of shoreline organisms once they have settled.

Chapters 5, 6, and 7 comprise the book's second section, on the organization and dynamics of major habitat types. These chapters were explicitly written to introduce particular Atlantic shoreline communities. Chapter 5 introduces the plants and animals of rocky shorelines, examines the physical and biotic forces that give these communities their distinctive structure, and describes the zonation patterns that characterize rocky shores. Chapter 6 examines the unique inhabitants, dynamics, and patterns of soft-sediment or sandy beach communities. The final chapter in the book discusses the structure and dynamics of the salt marsh plant communities that are so characteristic of Atlantic shorelines.

To enhance the value of this book as a teaching tool and field guide, I have made each chapter as self-contained as practical. The habitat chapters in particular are designed to be "stand-alone" general introductions to these habitats. This has resulted in some reiteration in the habitat chapters of general information that is covered more fully in the introductory chapters. I have tried to keep these redundancies to a minimum. Throughout the text I cite references I feel are the most appropriate for readers who wish to delve further into any of the topics discussed. I also end each chapter with a brief summary and a list of suggestions for further reading, and end the habitat chapters with an additional list of key papers that will lead students into the primary literature. The short glossary is included to enhance the accessibility of the book for novice readers.

At a time when shorelines worldwide are being threatened at an unprecedented pace by development, human population growth, and eutrophication, I hope this book will stimulate a new generation of students to appreciate and study shoreline ecology, as well as to learn how to apply our understanding of these important systems to conserving and managing them.

Mark Bertness
September 1998
Providence, Rhode Island

Acknowledgments

Many people contributed to the writing of this book. Michael Tate, Bob Paine, and Gary Vermeij are most responsible for my consuming interest in shoreline ecology. Michael Tate's enthusiasm and questioning mind first attracted me into science. Years ago Bob Paine's work initially captured my imagination and got me hooked on marine ecology. Gary Vermeij has at different times been my mentor, role model, and hero.

A number of colleagues at Brown also deserve special thanks. Doug Morse has been a perfect colleague, nudging me when nudging was needed and always encouraging me. He read an entire initial draft of this book before it was suitable for anyone to see. Steve Gaines encouraged me to write this book in the first place. The late Dick Goss and Frank Rothman helped more than they realized. My interest in salt marshes grew out of Dick's invitation to see if I could find something interesting to do in the salt marsh in his backyard. Frank made Brown a place I was proud to call home.

Numerous students and friends helped make this book a reality and definitely made it more readable. Steve Pennings, Scott Shumway, and Jonathan Levine helped in the early stages with needed criticism. My "book club" of

John Bruno, Tatyana Rand, George Leonard, Pat Ewanchuk, Coleman Kennedy, and Nancy Emery read the book cover-to-cover and helped much more than they realized. Bill Niering, Jon Witman, Ron Etter, Herb Wilson, and Phil Yund helped by critiquing chapters. Pete Peterson and Tony Chapman reviewed the final product and made numerous valuable suggestions. George Leonard, Todd Minchinton, Steve Pennings, and Herb Wilson provided valuable help in the final stages.

Kelly Benoit Bird illustrated the entire book and put up with my sometimes irrational ranting and raving the whole way. Her efforts enhance the book immeasurably and are evident on every page. I am grateful for her devotion to this project. My field work over the past number of years has benefited from the support of Jack Skerry, Bill Robertson, Tom Suchanek, Tim Miller, Al Beck, and Michelle Dionne. I thank them for the large role they played in making this book possible. It is also a pleasure to acknowledge the exceptional job everyone at Sinauer Associates has done in helping me bring this project to closure. Their help and suggestions have dramatically improved the quality of this book.

My work on Atlantic shorelines would not have been possible without the generous support of the National Science Foundation, the National Institute of Global Environmental Change of the Department of Energy, and the Andrew Mellon Foundation.

Finally, I am most grateful for the friendship and support that my wife Janette and daughter Sarah have given me while I worked on this project.

Chapter 1

The Setting

An understanding of Atlantic shoreline communities must begin with an appreciation of the historical and hydrodynamic forces that have shaped and will continue to shape them. Intertidal habitats are dynamic over a wide range of temporal and spatial scales. At large scales, shoreline features of entire continents have been shaped by geological events that occurred over hundreds of millions of years. Continental drift and the positions of continental margins relative to the shifting plates that move over the earth's semi-liquid interior have played a large role in establishing the types of intertidal habitats that dominate particular continental margins. Equally important in dictating the nature of current shoreline habitats at high latitudes, however, have been the glacial ice sheets that have scoured shorelines, deposited rocks and sediments, and changed sea levels by binding large amounts of the earth's water. The effects of global climatic cycles occur on temporal scales of thousands to tens of thousands of years. Superimposed on the effects of continental drift, climatic effects set the stage for modern shoreline habitats.

At smaller scales, shorelines are equally dynamic. Driven by gravitational interactions between the earth's water mass and the moon and sun, daily tides subject shorelines to predictable cycles of submersion in seawater and exposure to atmospheric conditions. Wind and tidal forces acting on water masses also generate waves and currents that sculpt intertidal habitats by eroding and depositing sediments on shorelines, while also influencing the delivery of food to organisms and the dispersal of their larvae.

This chapter examines the physical forces that have shaped and continue to shape the shoreline habitats of the Atlantic coast of North America. I begin by considering geological history as it relates to today's shoreline habitats and the kinds of organisms that occupy them. I then examine the importance of introduced species on western Atlantic shorelines and the influence of water movement on intertidal habitats and organisms.

CONTINENTAL DRIFT AND THE AGE OF THE EAST COAST

The east and west coasts of North America are strikingly different. The west coast is characterized by young, rugged coastal mountain ranges, wave-exposed rocky outcrops, and wave-swept, high-energy sand beaches. In contrast, the east coast is characterized by ancient, highly eroded mountain ranges and extensive marshes and sedimentary shorelines at lower latitudes. Understanding these differences requires a historical perspective and an understanding of continental drift, the dynamic movement of continents over the earth's surface.

The discovery of the fluid movement of continents over the earth's surface was one of the greatest scientific advances of the twentieth century. Prior to this time, the continents were thought to be fixed in position, even though early explorers and cartographers noted that the continents fit together like pieces of a puzzle. Alfred Wegener proposed the first serious theory of continental drift in 1912. He theorized that all the modern continents were initially part of one large supercontinent, Pangea. In Wegener's initial scheme, North America and Eurasia were at first joined to form Laurasia, or the northern portion of Pangea, while present-day South America, Africa, India, Antarctica, and Australia were joined to form the southern portion, Gondwanaland. Laurasia and Gondwanaland were separated by the Tethys (mother) Sea. [🐚1] Wegener hypothesized that the modern distribution of continents resulted from the separation and migration of these huge land masses over the earth's surface. Wegener amassed volumes of biological and geological evidence for his theory, but was unaware of a plausible mechanism for movement of the continents. Incorrectly, he proposed that gravitational pull from the moon was responsible for continental movement.

After nearly 50 years of ridicule, Wegener's ideas were confirmed and given a believable mechanism. In the 1950s and 1960s, geologists discovered that the crust of the earth is divided into a number of plates, both with and without continents. [🐚2] These plates move over the fluid inner mantle (aesthenosphere) of the earth as molten material flows from the earth's core. As this fluid mantle material comes to the surface at the seams between plates, their spreading edges grow, and their leading edges are pushed

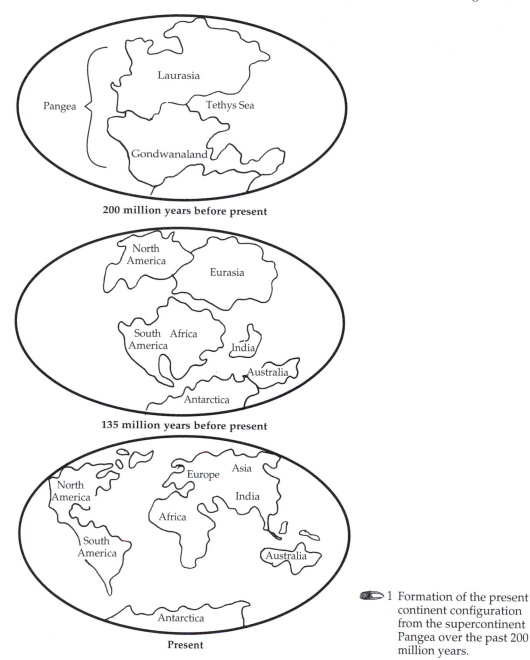

200 million years before present

135 million years before present

Present

1 Formation of the present
continent configuration
from the supercontinent
Pangea over the past 200
million years.

together. This movement causes violent seismic activity, the buckling of continental margins, and the birth and growth of mountain ranges.

Continental drift largely explains the large-scale geomorphology of the
east and west coasts of North America, as well the rest of the world. The
North American continental plate has its margins in the middle of the

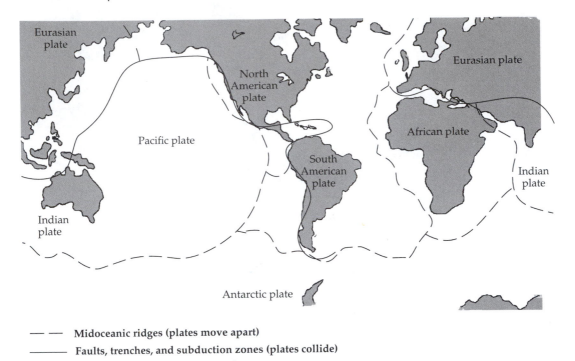

— — Midoceanic ridges (plates move apart)

——— Faults, trenches, and subduction zones (plates collide)

🜚 2 Arrangement of the continental plates, ridges where plates are moving apart, and faults, trenches, and subduction zones where plates are colliding.

Atlantic Ocean and along the west coast of North America.[🜚3] Sea-floor spreading at the mid-Atlantic ridge pushes the North American plate westward, where its western edge collides with the Pacific plate. The collision of these plates results in high seismic activity, subduction of the Pacific plate, and the uplifting of the North American plate to form coastal mountain ranges. Conversely, the position of the east coast, set far back from the spreading edge of the North American plate, explains many of its unique features. Like the east coasts of South America and Africa, the east coast of North America is geologically ancient, having been stable for more than 100 million years. Over this time, the coastline has eroded, wearing down mountain ranges such as the Appalachian and Smoky Mountains. This erosion has left wide continental margins of accumulated sediment, often bound by extensive marshes south of New England. North of New England, however, ice sheets have scoured the coast of accumulated sediment.

SEA LEVEL CHANGE

In addition to the history of continental movement, changes in global sea level and the scars left by glaciers are important determinants of the geomor-

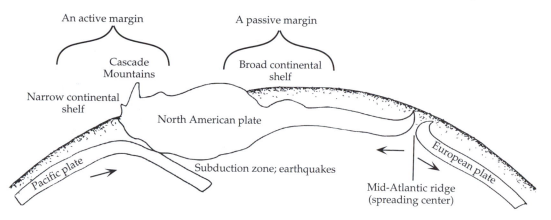

3 Movement of the North American plate by sea-floor spreading at the Mid-Atlantic ridge. Collision with the Pacific plate has given rise to the Cascade and Rocky Mountains. The Atlantic coast of the plate, however, has been stable for over 100 million years.

phology of modern shorelines. Sea level has fluctuated markedly in the past, largely as a consequence of variation in climate. [🌀4] Global climate is sensitive to the distance and angle between the sun and the earth, variation in solar radiation, and variation in the earth's atmosphere. Cooler temperatures lower sea level by shrinking the volume of water on the globe as large proportions of the earth's oceans become bound in polar ice sheets. [🌀5]

During the last major Ice Age, which started 120,000 years ago and lasted 100,000 years, over 30 percent of the North American land mass was covered by ice sheets over 2 kilometers thick. These ice sheets lowered the global sea level 120 meters, exposed the coastal margins of North America, and scoured northern latitudes down to bare rock. They also deposited massive amounts of sediment and glacial debris at their bases and compacted land masses by their sheer weight.

Over the last 20,000 years, temperatures have increased, the ice sheets have retreated, and the sea level has risen. The effect of this recent sea level change on modern landscapes and shorelines is dramatic. The Pleistocene ice sheet scraped bare the northern portion of the east coast of North America, leaving New England with a rugged rocky shoreline stripped of sediments and coarse rock debris. Narragansett Bay and Long Island Sound, just south of the glaciers' farthest reaches, are remnants of large river basins filled with coarse glacial debris and sediments deposited over nearly 100,000 years. Similarly, the Great Lakes formed at the foot of the ice sheet as the ice melted into large drainage basins. Long Island and Cape Cod are glacial moraines, essentially piles of debris that were pushed and deposited by the Pleistocene glaciers.

The entire east coast of North America has been strongly affected by the 120-meter sea level rise over the last 20,000 years. Coastal islands such as Long

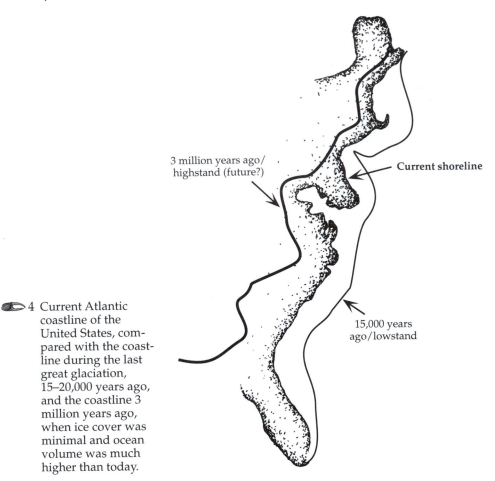

3 million years ago/
highstand (future?)

Current shoreline

15,000 years
ago/lowstand

🐚 4 Current Atlantic
coastline of the
United States, com-
pared with the coast-
line during the last
great glaciation,
15–20,000 years ago,
and the coastline 3
million years ago,
when ice cover was
minimal and ocean
volume was much
higher than today.

Island and Martha's Vineyard, once part of the mainland, were cut off from
the coast during this period. River beds created by the melting of glaciers were
flooded, creating shallow soft-sediment bays like Narragansett Bay and Long
Island Sound. Southward from where the Pleistocene glaciers had direct consequences, accumulated sediments on the continental margin were flooded,
leaving a broad, shallow continental shelf and offshore barrier islands.

Winter Ice

Whereas severe Ice Age conditions were largely responsible for eroding and
sculpting the rocky shores of the east coast of North America, ice continues
to be a powerful influence on northern Atlantic shorelines (Stephenson and
Stephenson 1971). The damaging effects of ice on shoreline organisms
include freezing, crushing, abrasion, and ripping organisms from the sub-

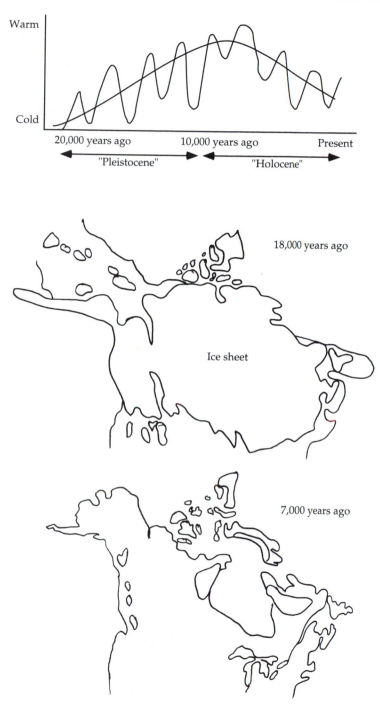

5 Variation in global temperature over the past 20,000 years and its consequences for North American ice cover.

strate. While ice damage is most common in polar and subpolar regions (Neushul 1960; Zaneveld 1996), ice may also have important effects on north temperate shorelines and the organisms that live there (Mathieson et al. 1982). Canadian shorelines are particularly affected by ice.

Along the coast, ice can be either frozen to the shore or free-floating in the sea. The fringe of ice attached to the shore, often referred to as the **ice foot**, can be over a meter thick and can completely cover the intertidal habitat (Dinsmore 1972). While fringe ice can sometimes protect shoreline organisms from ice scour, when it thaws, it can rip plants and sessile invertebrates from the substrate (Mathieson et al. 1982) and abrade shorelines as it is moved up and down by tides. This scouring can leave a characteristic barren belt of rock in shoreline habitats north of Nova Scotia (Stephenson and Stephenson 1971).

Free-floating ice in the sea originates when ice from glaciers falls into the sea (**icebergs**), the sea freezes (called **sea ice**), or rivers or estuaries freeze (called **freshwater ice**) (Dinsmore 1972). The main effect of free-floating ice on shoreline organisms occurs when ice contacts the shore and scours the substrate, removing plants and animals from intertidal habitats (Bergeron and Bourget 1986; Minchinton et al. 1997).

The extent of the damage to shoreline habitats and organisms from ice is variable and depends on latitude, exposure, and a variety of other local factors (see Stephenson and Stephenson 1971; Minchinton et al. 1997). In the northwestern Atlantic from the Arctic to 45°N latitude (Halifax, Nova Scotia), conditions are conducive to the annual formation of ice (Dinsmore 1972). [🐚6] In the Arctic, scouring by sea ice is often severe enough to limit the colonization of shoreline organisms to only a few months of the year (Wilce 1959). At the southern limit of the formation of sea ice, near Nova Scotia, Canada, sea ice is relatively thin and restricted to sheltered inlets and bays. Rare ice scouring events can occur at these latitudes when arctic sea ice drifts southward (see Stephenson and Stephenson 1971; Minchinton et al. 1997). Farther south, ice effects are limited to inland bays and estuaries, but can be substantial there. In New Hampshire, for example, fringe ice can be responsible for killing up to 50 percent of the shoreline seaweed in protected bays (Mathieson et al. 1982), and ice often damages inland salt marshes as far south as Rhode Island (Bertness 1984b).

BARRIER ISLANDS

South of Canada and New England, barrier islands are one of the most characteristic features of the east coast of North America. Barrier islands are a product of the age and stability of the east coast and its sea level history. They are common from the coast of Long Island to the Atlantic coast of Florida. They are not common north of Long Island because Pleistocene glaciers scoured sediments from these shores, leaving only bare rock or glacial debris. South of Long Island, however, the position of the shore far from a tectonic plate border, combined with over a hundred million years of

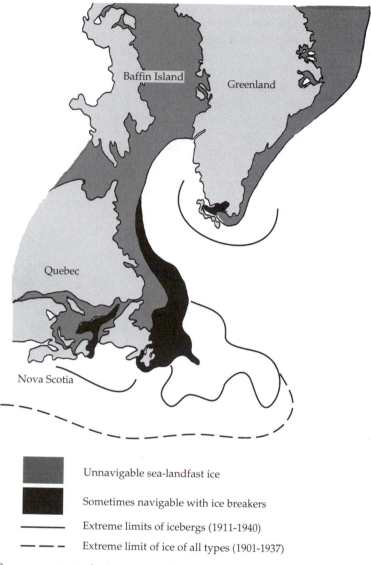

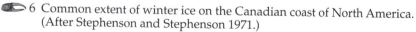

	Unnavigable sea-landfast ice
	Sometimes navigable with ice breakers
——	Extreme limits of icebergs (1911-1940)
– – –	Extreme limit of ice of all types (1901-1937)

6 Common extent of winter ice on the Canadian coast of North America. (After Stephenson and Stephenson 1971.)

sediment erosion from the continental margin, has left a massive accumulation of sediments. Barrier islands form when ocean waves, tides, and wind act together to pile up sediments on shorelines, creating large dunes stabilized by vegetation. When the sea level rises, these dunes are partly submerged and cut off from the shore. Some remain as long, uninterrupted barrier islands, whereas others break into smaller pieces, fragmented by storms and rip currents that erode channels through the obstructive barriers.

Because sea level has fluctuated markedly over the past 3 million years, and the east coast of North America has a large sediment load, dynamic barrier island formation, migration, and loss is a characteristic feature of the history of the east coast. Ancient barrier islands that formed when the sea level was at its highest 3 million years ago are now located 60 kilometers inland. The current barrier islands off the Atlantic coast are largely the result of melting global ice sheets and increasing sea level over the past 18,000 years. Typically, contemporary barrier islands on the east coast result from a single thin barrier complex on the coast of Holocene (Recent) origin. Off the coast of Georgia, however, amplified tides have led to the development of larger, wider barriers, built by both Holocene and earlier Pleistocene events. [🐚7]

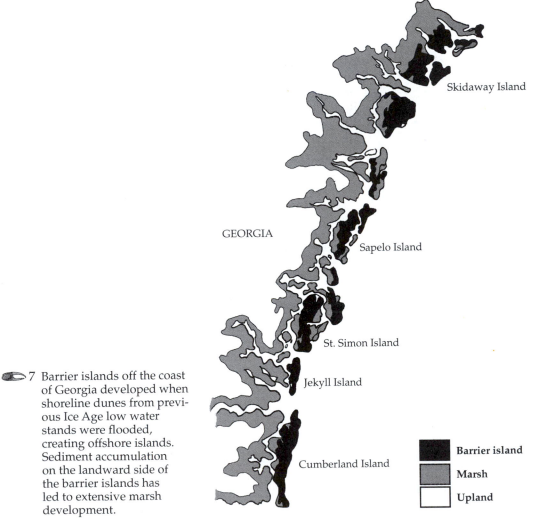

7 Barrier islands off the coast of Georgia developed when shoreline dunes from previous Ice Age low water stands were flooded, creating offshore islands. Sediment accumulation on the landward side of the barrier islands has led to extensive marsh development.

GEORGIA

Skidaway Island

Sapelo Island

St. Simon Island

Jekyll Island

Cumberland Island

■ Barrier island

■ Marsh

□ Upland

Barrier islands are dynamic structures that affect local water movement patterns and are constantly being eroded and reshaped by water movement and sediment deposition. The lagoons that form between barrier islands and the mainland fill with sediments and become shallow-water habitats that are invaded by salt-tolerant plants, promoting the formation of salt marshes. In contrast, the seaward edges of barrier islands are wave-swept, high-energy habitats characterized by constantly shifting sand dunes and spits. The shapes of barrier islands are continually changing. On the east coast of North America, coastlines are exposed to currents moving north to south, parallel to the coast, as a result of global ocean circulation patterns. The effect of these currents on barrier islands is usually to erode the northern ends while building up sediment on the southern ends. The southern ends of barrier islands often have a characteristic drumstick shape caused by these predictable accretion patterns. As long-shore currents reach barrier islands, they bend inward toward the land masses (see below for explanation) and curl around the islands. This leads to sediment piling up on and around the southern end of the barrier island. The southern end of the island expands, and soft sediment conducive to marsh development accumulates on its landward edge. Longshore development of barrier islands can also lead to complicated patterns of islands running into one another, merging, and forming high sediment deposition areas between them, which grow together to form marsh habitats.[🐚8]

SPECIES ORIGINS AND INVASIONS

The geological history of North America has also left its signature on the species composition of shoreline communities on the east coast. Over the past 3 million years, changes in sea level and climate have molded the communities of these shores. Changes in sea level have dictated the movement of species within the region as well as between the east coast and other biogeographical provinces. Past climatic extremes, especially Ice Age events, have driven many species to extinction and limited the distributions of others.

Ice Age events over the past 3 million years decimated the fauna and flora of the western Atlantic and appear to have caused more species extinctions and displacements there than on many other coastlines (e.g., the eastern Pacific or Atlantic; see Vermeij 1991). The Pleistocene ice sheets displaced organisms dependent on rocky shores, as they left no suitable hard-substrate habitats to the south. Pleistocene Ice Age conditions also strongly affected shorelines far south of New England. Steven Stanley (1986) has shown that massive species extinctions due to cold Ice Age conditions occurred as far south as Florida and the Bahamas.

Following these Ice Age extinctions, the east coast of North America was invaded by species from other biogeographic regions. Geerat Vermeij (1989, 1991) has shown that much of this reinvasion came from the trans-Arctic

Time

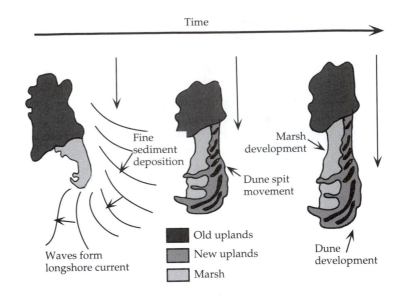

Fine
sediment
deposition

Marsh
development

Dune spit
movement

🦐 8 Barrier islands are
typically exposed
to longshore cur-
rents that lead to
characteristic sedi-
ment deposition
patterns and
island growth
and morphology.

Waves form
longshore current

■ Old uplands
■ New uplands
□ Marsh

Dune
development

migration of taxa with Pacific origins. He has suggested that Pacific taxa successfully colonized the Atlantic because of the ecological opportunities created by the Ice Age decimation of North Atlantic fauna and flora. Relatively
few Atlantic taxa successfully invaded the Pacific when geological opportunities occurred, possibly because the Pacific coast biota experienced fewer
Ice Age extinctions, and thus offered fewer ecological opportunities for
invaders.[🦐 9] Following Ice Age extinctions and subsequent cimatic
warming, migrants from eastern and southern waters also colonized the
northwestern Atlantic Ocean (Franz and Merrill 1980a,b).

Changes in sea level have also dramatically influenced the shape of
shorelines and opportunities for dispersal. For example, when little water
was frozen in ice caps during the Pliocene highstand, 3 million years ago,
the peninsula of modern Florida was flooded, Cuba was a set of islands
rather than a single island, and the Isthmus of Panama was submerged. This
situation allowed the exchange of organisms between the Atlantic and
Pacific Oceans and movement of organisms between the Gulf of Mexico and
the east coast of North America. Conversely, during the extensive Pleistocene glaciations 20,000 years ago, biotic exchange between the Atlantic and
Pacific was precluded, and the continental barrier between the east and west
coasts of Florida was more formidable than it is currently. The establishment
of these barriers to dispersal isolated populations and led to speciation
events. The Atlantic stone crab *Menippe mercenaria*, for example, is one of
many species shown to consist of two subspecies that diverged 3 million
years ago when the sea level rose and established the Florida peninsula (Bert
1986).[🦐 10]

Littorina history

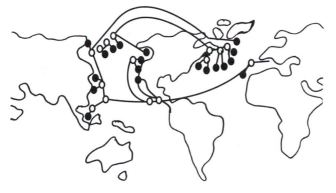

Littorina phylogeny

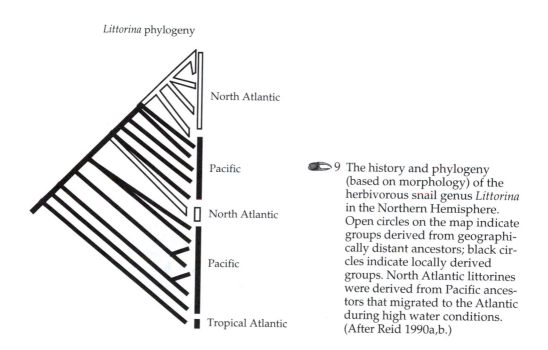

North Atlantic

Pacific

North Atlantic

Pacific

Tropical Atlantic

9 The history and phylogeny (based on morphology) of the herbivorous snail genus *Littorina* in the Northern Hemisphere. Open circles on the map indicate groups derived from geographically distant ancestors; black circles indicate locally derived groups. North Atlantic littorines were derived from Pacific ancestors that migrated to the Atlantic during high water conditions. (After Reid 1990a,b.)

INTRODUCED SPECIES

An important and often overlooked component of shoreline communities in the western Atlantic is the recently introduced species that have come to North America due to human activities. Since Charles Elton's seminal book on the biology of invasions in 1958, ecologists have been aware of the effects that intentional and inadvertent introductions of species have had on terres-

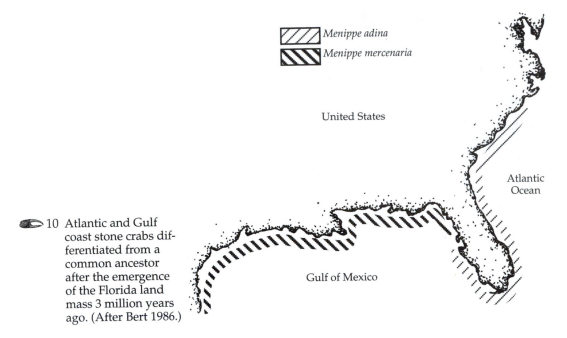

Menippe adina
Menippe mercenaria

United States

Atlantic
Ocean

10 Atlantic and Gulf
coast stone crabs dif-
ferentiated from a
common ancestor
after the emergence
of the Florida land
mass 3 million years
ago. (After Bert 1986.)

Gulf of Mexico

trial communities. It is only in the last few decades, however, that ecologists
have come to realize the pervasiveness of introduced species in marine habitats. Ironically, the failure of marine ecologists to recognize the importance
of introduced species was partially because the exotic nature of many common taxa had not been suspected.

Global introduction of exotic species has a long history. The past 500
years have been particularly important as Europeans have colonized the
New World and engaged in long-distance commerce. The resulting spread
of non-native species has led to the breakdown of traditional faunal boundaries and the homogenization of the world's fauna and flora in what Elton
has called the largest biological crisis in history. Whereas the effects of introduced species have received the most attention in terrestrial habitats, particularly on islands such as Hawaii and Australia (which until European
contact had relatively unique fauna and flora), shallow-water coastal habitats are also greatly affected by human-assisted invaders. James Carlton
(1987, 1992) has conservatively estimated that over 1000 conspicuous marine
invader species have been documented worldwide, and suggests that this
number would be much larger if it included less conspicuous and poorly
described organisms.

The widespread, human-assisted invasion of shallow coastal waters
began in the fifteenth century and continues to this day. The large wooden
ships used by early explorers in the fifteenth through the nineteenth centuries were ideally equipped to move shallow-water species around the

globe. The hulls of these ships have been described as "floating biological islands." Ships sat in ports, where they were colonized by local fauna and flora. They then set sail, and at the end of a journey that could take as long as 3 months, the surviving organisms would colonize new shores. Seaweeds, barnacles, tunicates, bryozoans, and other **fouling organisms** often covered the bottoms of these ships in dense assemblages up to a meter thick (Carlton 1992). Moreover, wood-boring clams (teredos) and isopods (woodlice) typically riddled wooden ships, often forcing them out of service. These borers left ship hulls honeycombed with passages that collected sediment and sediment-inhabiting organisms, and also provided living space for crabs and other mobile organisms.[🦪11] In addition to these formidable fouling communities on their hulls, pre-twentieth-century vessels also carried **ballast** to weigh down their hulls for rough oceanic voyages. This ballast was typically beach rocks, sand, or scrap iron. Ships would take on ballast for the trip at one port and then dump it at the journey's end to reduce their weight for maneuvering in shallow coastal waters. Cobbles and sand used as ballast may have transported entire assemblages from one port to another.

Since the early twentieth century, fast-moving, metal-hulled ships coated with toxic antifouling paint have replaced wooden ships, and are not used by boring organisms. They also move fast enough to limit the development of large fouling communities, so that the pre-twentieth-century movement of "islands of shoreline" from coast to coast has ceased. However,

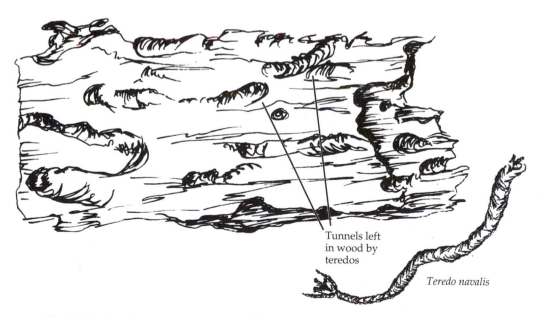

Tunnels left
in wood by
teredos

Teredo navalis

🦪11 Teredos (shipworms) were a serious problem for shipping in the sixteenth through nineteenth centuries. These wood-boring clams can riddle wooden ships with shell-lined tunnels and turn them into habitats for many fouling organisms.

contemporary ships use water as ballast, which is taken into their hulls at one port and jettisoned at the next. James Carlton (1985) and his colleagues have shown that ballast water often teems with planktonic larvae, and that these larvae can survive movement across the Atlantic and Pacific Oceans. They have suggested that ballast water movement may result in a constant inoculation of the shallow-water harbors of the world with non-native propagules. The larvae of segmented worms, crustaceans, flatworms, and molluscs are all commonly found in ballast water, suggesting that ballast water may be a relatively nonselective transport vector for moving entire coastal plankton assemblages across oceanic basins into similar habitats.

This five-century process of transporting shallow-water fauna and flora around the globe has led to considerable homogenization of the world's shallow-water assemblages. The success of invaders has been particularly high in geologically young areas such as New England, where the fauna and flora have not fully recovered from recent climatic extinction events, presenting ecological opportunities for potential invaders (Vermeij 1991). [🐚12] The most striking aspect of the species that have successfully

Codium fragile
spp. *tomentosoides*
Deadman's fingers
from Europe, 1950s

Littorina littorea
Northern periwinkle
from Europe, 1860s

Carcinus maenus
Green crab
from Europe, late 1700s

Membranipora membanacea
White lace bryozoan
from Europe, 1987

Styela clava
Asian tunicate
1973

Botrylloides diagensis
California tunicate
1972

🐚 12 Some successful invaders that have become dominant species on the east coast of North America.

invaded shallow-water habitats on the east coast of North America is not their number, but how dominant some of them have become. For example, the common rough periwinkle, *Littorina littorea*, was introduced to North America either in rock ballast or as food in Nova Scotia in the mid-nineteenth century, and is now the dominant intertidal herbivore on rocky shores from New England to Chesapeake Bay (Carlton 1982). This snail is responsible for limiting the abundance of ephemeral seaweeds throughout this range as well as modifying many habitats by keeping them clear of algae and sediment (Bertness 1984a; see Chapter 5). The introduced tunicates *Botrylloides diagensis* and *Styela clava* are also conspicuous invaders in New England, where they can dominate shallow subtidal habitats and displace native organisms (Mathieson et al. 1991). In the past decade, the white lace bryozoan *Membranipora* has also invaded New England waters, with immediate repercussions. *Membranipora* settles on large seaweeds, and without natural predators, this encrusting bryozoan covers and kills large portions of shallow-water kelp forests.

One of the dominant predators on western Atlantic shorelines, the green crab (*Carcinus maenus*) is itself an introduced species. It was probably transported from Europe to southern New England in the late eighteenth century with rock ballast, and in 200 years has become one of the most dominant omnivorous consumers on hard- and soft-substrate shorelines from southern Canada to the Chesapeake (Glude 1955).[🐚13] There are also examples of

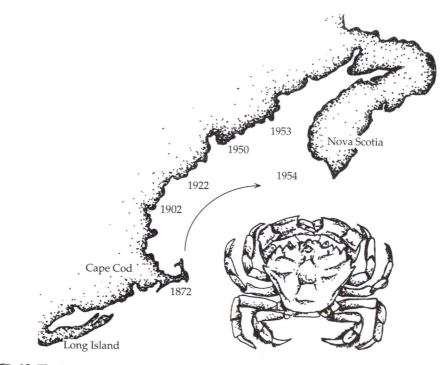

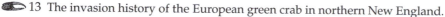 13 The invasion history of the European green crab in northern New England.

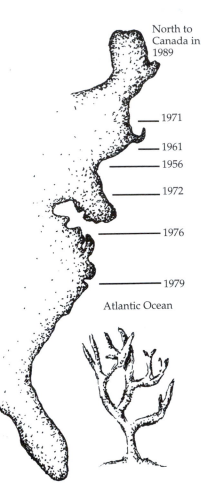

North to Canada in 1989

1971

1961
1956

1972

1976

1979

Atlantic Ocean

🦪 14 The invasion history of *Codium fragile* spp. *tomentosoides* on the east coast of North America.

invasions by seaweeds. The Japanese seaweed *Codium fragile* spp. *tomentosoides*, or dead man's fingers, was initially introduced to Europe in the early twentieth century, either as a fouling organism or on oyster spat. In the mid-1950s it was introduced to New York Harbor, probably as a hitchhiker on fouled boats (Carlton and Scanlon 1985). In the past 40 years, *Codium* has rapidly moved north to Maine and Canada, and has moved more slowly against prevailing coastal currents south to the Carolinas.[🦪14] Introduced species are a feature of western Atlantic shorelines that need to be taken into account in considering any aspect of the ecology, history, or evolution of these assemblages.

TIDES

The rise and fall of the tides is one of the most characteristic features of all shoreline habitats. Tides (from the Old English word for time) are complicated but highly predictable consequences of the gravitational pull of the moon and sun on the earth's water masses. The tidal heights at a particular location, however, can be modified by a wide variety of local, regional, and global factors. Tides play an extremely important role in shaping the abundance and distribution patterns of shoreline organisms.

Tides are caused by the gravitational forces of the moon and sun as well as the centripetal force of the earth as it spins around its axis. The moon exerts the strongest gravitational pull on the earth. While the moon is smaller than the sun, the sun is much farther from the earth. Consequently, the gravitational pull of the moon on the earth's oceans is twice that of the sun. The moon's pull on the earth's oceans causes the oceans to bulge toward the moon. Opposing this gravitational pull, however, are centripetal forces caused by the earth's rotation. These centripetal forces cause the tidal bulge to form a roughly elliptical envelope of water around the globe.[🪶15]

Tides are caused by the movement of these large bulges as standing waves of water across the earth. The earth rotates on its axis every 24 hours and 50 minutes, and as it rotates, the tidal bulges move over its surface. If the globe were perfectly smooth and without continents, these bulges would be approximately 0.5 meters high and would travel at a speed of about 700 kilometers per hour over the earth's surface (Davis 1994). A wide variety of factors, however, complicate the magnitude and timing of tides.

Most shorelines on the earth, and all shorelines on the east coast of North America, have two tides a day, with the timing of low and high water shifting forward every day by 50 minutes. Fifty extra minutes of rotation are needed each day for the earth to catch up to the orbiting moon, which is moving in the same direction. Typically the high and low tides each day are not identical in magnitude, and are called **mixed tides**.[🪶16] The difference is caused by the relationship between the axis of the earth's rotation and the angle of the gravitational distortion of the earth's water mass. As the earth rotates around its axis, a particular location on its surface will en-

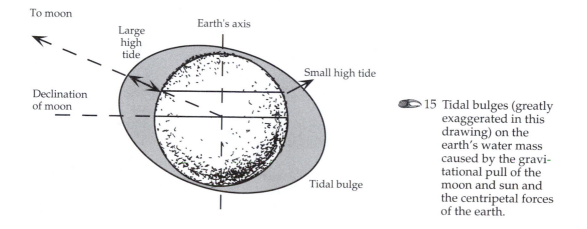

🪶15 Tidal bulges (greatly exaggerated in this drawing) on the earth's water mass caused by the gravitational pull of the moon and sun and the centripetal forces of the earth.

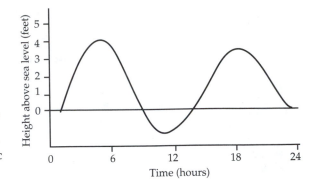

16 Mixed semi-diurnal tides, in which two high and two low tides of different magnitudes occur each day, are found on the entire Atlantic coast of North America.

counter a tidal bulge or high tide every 12.5 hours, but at most locations on the globe, these high tides will be of different magnitudes. Conversely, low tides occur when the earth rotates through the troughs between tidal bulges.

The pronounced monthly cycles in the magnitude of the tides are largely the result of interactions between the gravitational pull of the moon and the sun. When the gravitational forces of the sun and moon act together to form the tidal bulge, the gravitational distortion of the earth's oceans is at its peak, and high and low tides are their greatest. This occurs twice monthly at new moons and full moons, and the resulting extreme tides are referred to as **spring tides** (though this has nothing to do with the season "spring"). Monthly minimal tidal excursions occur when the gravitational forces of the sun and moon oppose each other. These minimal tides also occur twice monthly as the moon passes its quarter phase, and the resulting tides are referred to as **neap tides**.[👣17]

Along the east coast of North America there are always two high and two low tides each day, with each daily tidal excursion differing in magnitude due to the position of the Atlantic coast relative to the global tidal bulges and the daily rotation of the earth. This is the most common type of tidal pattern and is called a **semi-diurnal tide**. Other coastlines, due to their position on the globe and other factors, can have other types of patterns (for example, a single tide a day, or days with either one or two tides).[👣18]

In addition to this predictable monthly variation in the tides, annual variation in the intensity of tides occurs due to the position of the sun in relation to the earth. Twice a year, at the solar equinoxes (September 21 and March 21), the sun is directly over the equator and closest to the earth. At these times annual maximum tidal fluctuations occur. Conversely, at the two solar solstices (June 21 and December 21), the sun is farthest from the earth and equator, and annual minimum tidal fluctuations occur.

The exact timing and intensity of tides at a particular location is dependent on a variety of factors. Along the Atlantic coast of North America, the greatest tidal amplitudes are found in New England and the Canadian Maritime Provinces because of their global position in relation-

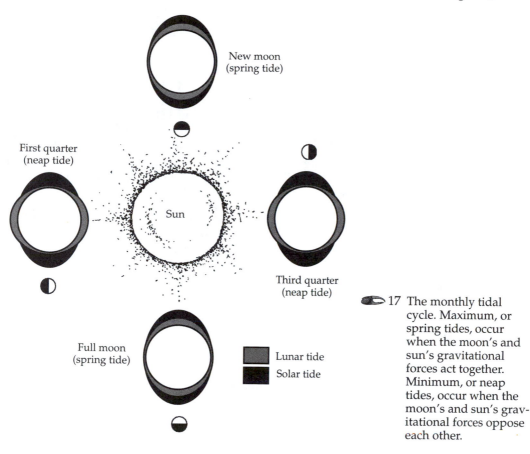

New moon
(spring tide)

First quarter
(neap tide)

Sun

Third quarter
(neap tide)

Full moon
(spring tide)

Lunar tide

Solar tide

17 The monthly tidal cycle. Maximum, or spring tides, occur when the moon's and sun's gravitational forces act together. Minimum, or neap tides, occur when the moon's and sun's gravitational forces oppose each other.

ship to the tidal bulges. On mid- and southern Atlantic shores, tidal fluctuations decrease with latitude, both because of the global position of the tidal bulges and because of the standing wave patterns of the Atlantic Ocean as a whole. South of Cape Hatteras, however, the tides in Georgia are amplified by the funnel shape of the coastline. As a consequence, the tidal range in Georgia is 3 meters, while in North Carolina and on the Atlantic coast of Florida, the tidal ranges are less than 1 and 2 meters, respectively.[🐾19] The timing of tides in particular locations is largely a product of shoreline and bottom topography and the time at which the tidal bulges reach particular locations.

Locally, the magnitude of tides can also be strongly influenced by shoreline morphology. As oceanic tidal bulges hit wide continental margins, the amplitude of tides can be magnified. Mid-oceanic islands without continental margins, for this reason, typically have very small tides. Bays and estuaries can also magnify the intensity of tides. Funnel-shaped bays, in particular, can dramatically alter tidal magnitude. The Bay of Fundy in Nova Scotia is

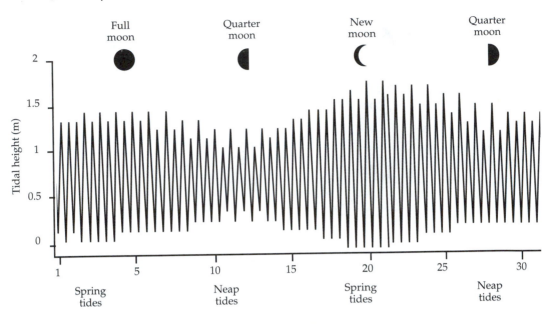

🦀 18 Lunar tidal cycle for Prudence Island, Rhode Island, January 1996.

the classic example of this effect, and has the highest tides in the world (over 15 meters). These prodigious tides are caused by the latitudinal position of the tidal bulge and the funnel shape and size of the bay.[🦀 20]

Local wind and weather patterns can also affect tides. Strong offshore winds can move water away from coastlines, exaggerating low tide exposures, whereas onshore winds, by piling up water on shorelines, can virtually eliminate low tide exposures. Particularly in small, shallow bays, wind-driven water movement can deflect predicted tidal heights by as much as 1 meter. Changes in barometric pressure can also affect local tides by as much as 0.3 meters (Little and Kitching 1996). High-pressure systems can depress sea levels, leading to clear sunny days with exceptionally low tides. Conversely, low-pressure systems associated with cloudy, rainy conditions are typically associated with tides that are higher than predicted.

WAVES AND WATER MOVEMENT

In addition to the tides that predictably submerge and expose shoreline habitats, waves and nearshore flows also play an important role in shaping both shoreline habitats and the organisms that live there. Most nontidal water movement in shoreline habitats is the result of wind-generated waves. Although the precise mechanism whereby winds pass energy to bodies of water is uncertain, the friction and pressure of wind moving over water initiates surface waves. Waves can differ greatly in magnitude, but all

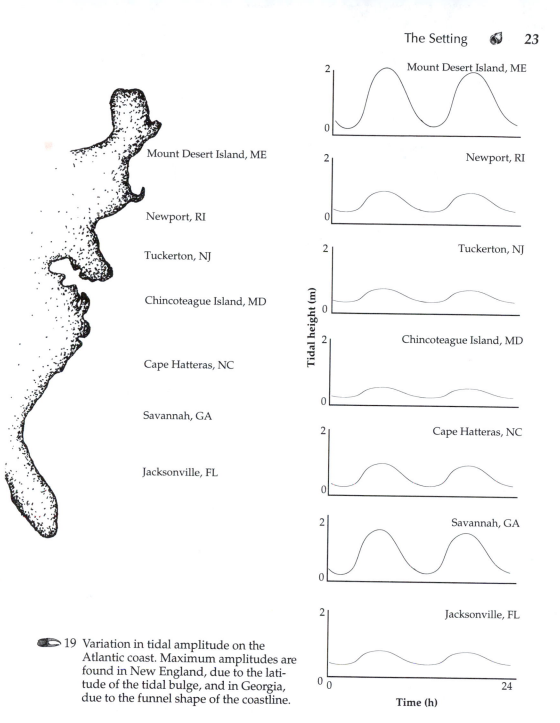

🦐 19 Variation in tidal amplitude on the
Atlantic coast. Maximum amplitudes are
found in New England, due to the latitude of the tidal bulge, and in Georgia,
due to the funnel shape of the coastline.

can be described with the same simple terminology. The high point of a
wave is its **crest**, and the low point is its **trough**. The **height** of a wave is the
vertical distance from the trough to the crest, and the length of a wave,

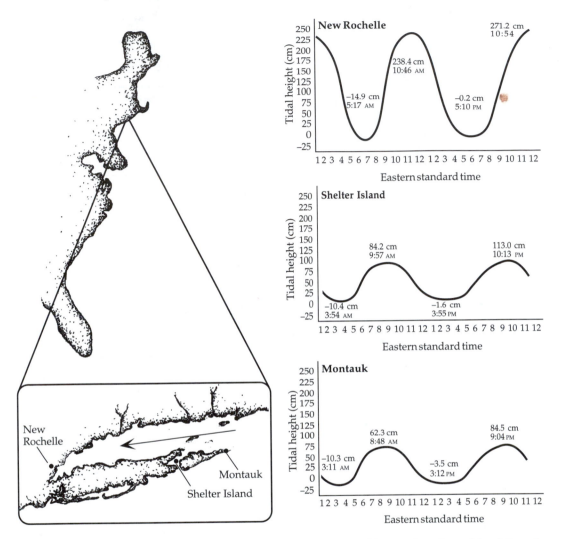

🐚 20 Amplification of tides as they enter a funnel-shaped bay such as Long Island Sound. Note the temporal staggering of the tides as they enter the Sound.

from crest to crest, is its **wavelength**.[🐚 21] The **steepness** of a wave determines its stability, and is defined as the ratio of its height to its wavelength (H/L). When a wave reaches a steepness of 7/1, it becomes unstable and collapses or breaks. The movement of a wave is referred to as its **period**, defined as the time it takes one wavelength of a wave to move past a reference point.

The magnitude of wind-generated waves depends on three factors (Bascom 1980): the speed of the winds, the **fetch,** or distance over which the

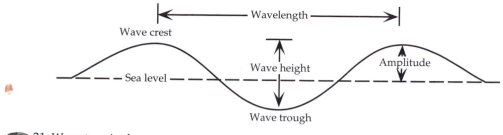

🐚 21 Wave terminology.

winds act on the water, and the length of time a wind has to generate a given wave force. The light winds that produce small ripples in calm, protected bays demonstrate the same process that generates large oceanic waves and swell. In the open ocean, strong winds of 10–25 meters per second generated by severe weather patterns, operating over hundreds or thousands of kilometers and many days, can routinely produce waves 2–8 meters in height.

When waves are associated with the winds that generated them, they are referred to as **wind waves**. In heavy winds such waves have whitecaps, since the winds often blow their crests off. Some waves, however, continue long after the winds that generated them have subsided. Oceanic waves that persist independently of their original winds are termed **swell**. Waves in the open ocean, therefore, are a mix of long-wavelength swell, dissipating and coalescing from harsh weather conditions, often far away, and smaller-scale, locally wind-generated waves.

When a wave is in deep water, its surface water does not advance forward, but rather moves in a circular orbit with the diameter of the wave height. This seemingly counterintuitive motion is obvious once you think about it. If wind waves moved horizontally over the surface of the ocean, they would sweep floating objects with them. Instead of being swept across the sea, however, floating objects bob up and down in a gentle circular motion following the water's path. Floating in the surf, you can feel yourself surge slightly forward riding up the peak of a wave, and then surge back the same distance as you slip into a wave trough.[🐚 22]

The circular motion of water molecules in waves decreases with water depth, and stops altogether at a depth equal to half the wavelength of the wave on the surface. Below the surface, water moves in circular orbits of decreasing size until that depth is reached. This means that large oceanic swells with wavelengths of up to 300 meters are not detectable at depths of more than 150 meters, and that the effects of wind-generated waves in coastal habitats with wavelengths of 2–6 meters penetrate only 1–3 meters into the water column. Water affected by all these waves, however, does not move appreciably in a net horizontal direction until the shore is reached.

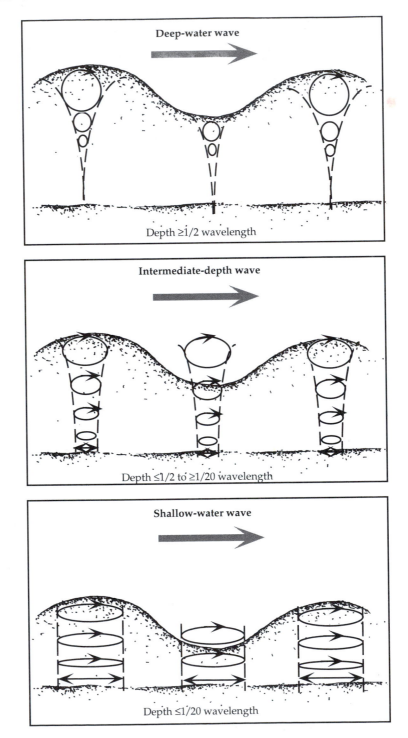

22 Movement of water molecules in waves as a function of depth.

WAVES HITTING THE SHORE

When waves are in deep water, they are not affected by the sea bottom. When waves approach the shore, however, the bottom influences their structure, and ultimately causes them to become unstable and break, a process called **shoaling**. As a wave enters shallow water, the friction of the bottom makes it travel more slowly, particularly near the bottom. This decreases the wavelength of the wave, increases its height, and squeezes the orbital paths of its water molecules into elliptical paths.[23] At the same time, bottom friction causes the wave to slow more at the bottom than at the top. As a result of all these forces, waves encountering the shoreline become unstable and break. Broken waves, however, have not entirely lost their energy, and will reform and break again in shallower water until all their energy is dissipated.

Waves break differently on shorelines depending on their wavelength as they approach the shore and the slope of the bottom (Denny 1987, 1988). Long-wavelength oceanic swell develops into the classic **plunging breakers** that surfers in Hawaii and southern California enjoy. Shorter-wavelength, wind-generated waves that hit shallow slopes result in **spilling breakers**, which are taller waves that behave much like water spilling out of a container. As the steepness of the bottom increases, both long- and short-wavelength waves take the shape of **collapsing breakers**, relatively tall waves that develop rapidly and collapse rather then break.[24]

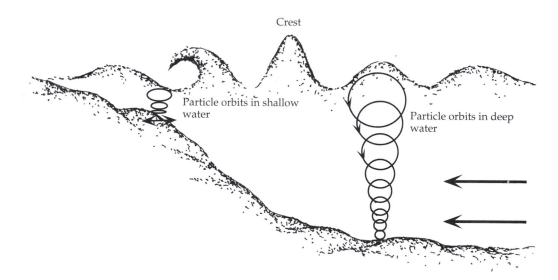

23 A wave approaching the shore encounters the sea bottom, which compresses the orbits of its water molecules and slows its movement near the bottom. This increases the height of the wave until it becomes unstable and breaks.

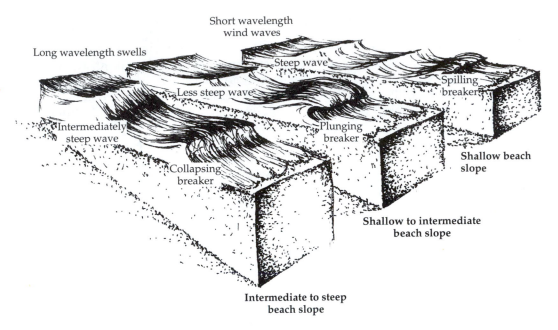

Long wavelength swells

Short wavelength
wind waves

Steep wave

Less steep wave

Spilling
breaker

Intermediately
steep wave

Plunging
breaker

Shallow beach
slope

Collapsing
breaker

Shallow to intermediate
beach slope

Intermediate to steep
beach slope

🐚 24 Types of waves hitting shorelines.

EFFECTS OF SHORELINE TOPOGRAPHY ON WAVES

Waves are also influenced by the shape of the shore and the angle at which they approach it. Waves rarely approach land with their crests parallel to the shoreline, but as they near landfall, they usually bend into the shoreline so that as they reach it, their crests are nearly parallel to the shore. This bending, or refraction, of waves is caused by the slowing of the waves by the bottom as they encounter the coast. As a wave enters shallow water at an angle, the part that first encounters shallow water is slowed, while that still in deeper water continues to move rapidly. Thus, slowing occurs at different times along a wave crest, causing it to bend into the shore.

Because of this bending of waves into shorelines, as waves meet complex shorelines, their forces are not distributed evenly across the shore. Waves converge at headlands or points, concentrating their energy at these exposed projections. This refraction of wave energy into headlands reduces the wave energy that enters adjacent embayments. As waves enter a bay, they also spread into the shape of the bay because of refraction. This further reduces the wave energy that reaches the shore of the bay.[🐚 25]

When waves encounter shorelines, they can generate impressive longshore currents, which run parallel to the shore and play an important role in transporting food, larvae, and sediments.[🐚 26] Although wave crests bend as they approach shore, they still tend to hit shorelines at an angle. When this occurs, some of the water moves back into deep water (the

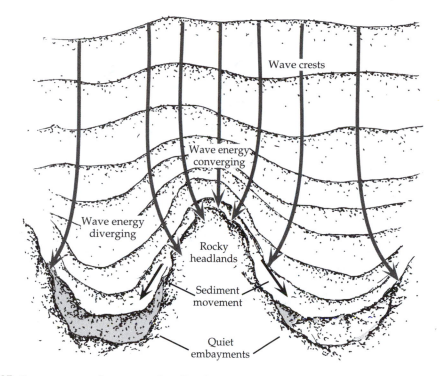

Wave crests

Wave energy
converging

Wave energy
diverging

Rocky
headlands

Sediment
movement

Quiet
embayments

🐚 25 Convergence of waves on headlands results in wave energy
being focused on headlands and dissipated in adjacent bays.

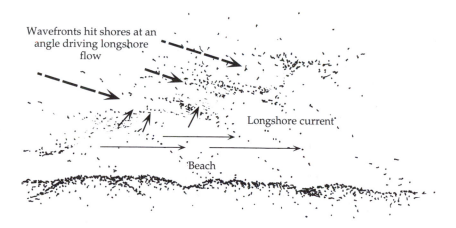

Wavefronts hit shores at an
angle driving longshore
flow

Longshore current

Beach

🐚 26 Longshore current generation by waves hitting the shore at an
angle.

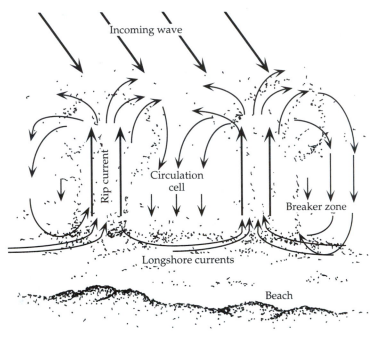

Incoming wave

Rip current

Circulation cell

Breaker zone

Longshore currents

Beach

◒ 27 As waves approach the shoreline at an oblique angle, a longshore current is produced. Rip currents form from the waves flowing back to the sea, completing a circulation cell.

undertow), and some of it moves parallel to the shore.[◒ 27] The amount of water in this longshore current typically depends on the angle at which waves hit the shore: in general, the greater the angle, the faster water moves along shore. Longshore currents are very predictable components of many shoreline habitats and have velocities ranging from 10 to over 100 centimeters per second.

WAVE EFFECTS ON SHORELINES

While shorelines strongly influence waves and nearshore water movement, waves and the currents they generate also play a major role in shaping shorelines. As waves are concentrated at headlands and dissipated in bays by refraction, headlands are eroded away while sediments are deposited in bays. Over long periods of time, typically measured in centuries, this process generally leads to exposed headlands consisting of hard rocks and soft-sediment-dominated embayments.

On both short and long time scales, the action of waves and longshore currents on soft-sediment shores leads to the development of sandbars and barrier islands. As waves pound soft-sediment shorelines, the orbital movement of water hitting the shore churns up sediments, leaving them vulnerable to erosion by longshore currents. Deposition of sediments on the edges of longshore currents, where their velocity is reduced, leads to the formation of sandbars running parallel to the shore, punctuated by channels where water (piled onshore by waves) flows back out to sea. On a larger spatial scale, this same process, coupled with rising sea level and the stabilization of sediments by vegetation, is responsible for generating the barrier islands that are a dominant feature of the southeastern coast of North America.

Seasonal variation in the intensity of wave energy often leads to seasonal patterns in the morphology of sediment beaches on wave-exposed shores. Winter conditions typically lead to high seas, high wave energy hitting the coast, and the erosion of beach sediments. This erosion often exposes bedrock as sediments are moved offshore, and leads to beaches having a shallower profile in the winter than they do in the summer. During calmer spring and summer conditions, however, beaches that were eroded away during the winter commonly experience heavy sedimentation. These seasonal shifts in wave energy can lead to dramatically different shoreline morphologies over the course of a year.[☞ 28]

WATER MOVEMENT: EFFECTS ON SHORELINE ORGANISMS

Water movement over shoreline habitats also has profound effects on intertidal organisms, populations, and communities. These effects have only recently begun to be fully appreciated. On wave-exposed shores, waves crashing into shorelines continually move and restructure intertidal habitats composed of sand and cobbles, often precluding organisms from inhabiting these highly disturbed habitats or limiting their inhabitants to rapidly colonizing species that are specifically adapted to live in unstable habitats. Surf clams, for example, migrate up and down high-energy sandy beaches with the tide, taking advantage of intertidal water movement to move and feed. Razor clams, in contrast, burrow deep into the substrate to escape shifting sediments. On wave-exposed rocky shorelines, the major adaptive challenge is typically not shifting substrate, but rather the intense force of waves hitting the shore. Strong attachments, streamlined morphology, and living in groups in which neighbors buffer one another from wave stresses are typical solutions to the problems of living on wave-exposed rocky shorelines (Denny et al. 1985). These issues will be discussed further in Chapter 5.

Not all effects of water movement are negative. Water movement over shorelines is crucial in delivering food and nutrients to intertidal organisms, and plays a major role in the dispersal of the gametes and propagules of many intertidal inhabitants. Filter-feeding organisms that rely on water movement to deliver food to them (e.g., barnacles and mussels) are particularly dependent on high rates of flow to feed and grow (Sanford et al. 1994). Algal growth is also enhanced by high rates of flow, due to increased gas exchange efficiency (Gerard 1987).

Intertidal organisms are exposed to both wave and tidally generated water movement. In open-coast habitats subject to oceanic swell and wind-driven waves, wave forces dominate the water movement patterns experienced by intertidal organisms. In protected bays and estuaries, incoming

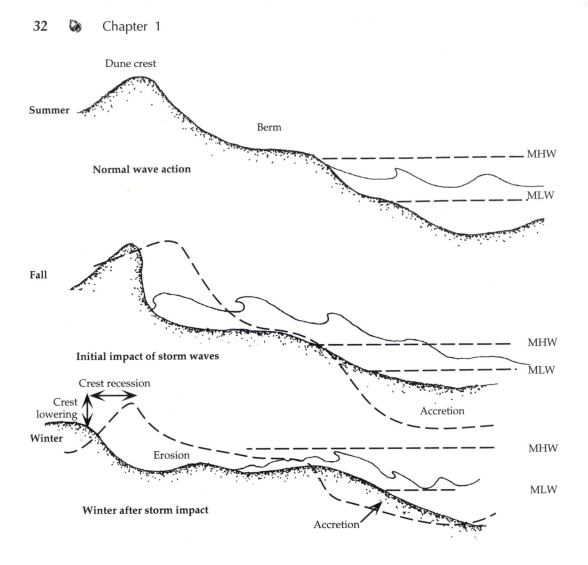

Summer

Dune crest

Berm

Normal wave action

MHW

MLW

Fall

Initial impact of storm waves

MHW

MLW

Crest recession

Crest lowering

Accretion

Winter

Erosion

MHW

MLW

Winter after storm impact

Accretion

🐚 28 Seasonal changes in a typical beach profile. Sand is accumulated in the summer when waves are small, forming a berm. The berm and beach can almost completely move offshore in the winter due to high waves and erosion. MHW = mean high water; MLW = mean low water.

and outgoing tides typically generate long, spatially and temporally predictable periods of unidirectional flow.

The hydrodynamic forces generated in exposed open-coast shoreline habitats when oceanic swell hits shorelines can be enormous. Water velocities of 10–20 meters per second and acceleration forces often exceeding 100 square meters per second are common (Denny et al. 1985). These high water velocities and accelerations result from waves breaking in shallow water, and expose intertidal organisms to drag, lift, and acceleration forces.[🐚 29]

Drag is proportional to the area of an object, results from pressure differences on the sides of an object, and acts to pull an object parallel to and in the same direction as a force. Drag can be minimized by having a low, streamlined profile. **Lift** is also proportional to the area of an object, but acts perpendicular to the force. Lift is caused by pressure differences between an object and fluid moving over the object, and is the force responsible for blowing the roofs off buildings in storms. Since lift is maximized by having a low profile, the shapes of organisms that live in wave-exposed habitats often reflect a trade-off between minimizing lift and drag forces. **Acceleration,** in contrast to lift and drag, results from the presence of an object's mass in flows, and is proportional to an object's volume, not its area. This fact has important consequences, because as organisms increase in size, acceleration forces on them increase faster than their surface area or area of attachment. Thus, if they grow larger without changing their body proportions, they will be torn from the substrate (see Denny et al. 1985 and Denny 1988 for discussion of these forces).

Tidally generated hydrodynamic forces typically differ quantitatively and qualitatively from wave-generated forces. Tidally generated forces are generally much smaller than wave-generated forces, and acceleration forces are dramatically less in tidal flows. In bays and estuaries where tidal flows dominate, the hydrodynamic forces experienced by intertidal organisms will run landward during incoming tides and seaward during outgoing tides, punctuated by relatively calm periods at low and high tide. The tidal flows experienced by bay and estuarine organisms are largely predictable, based on the geomorphology of basins. Any constriction, such as the mouth of a bay, an island that occludes tidal flow, or a shallow bottom, will predictably increase tidal flows and their consequences on intertidal organisms. This is in contrast to the bashing wave flows seen by open coast organisms.

Because of friction with the sea floor, water velocities are typically lower nearer the bottom, resulting in a layer of slower-moving water called the **boundary layer**. As a consequence, small organisms on the sea floor experi-

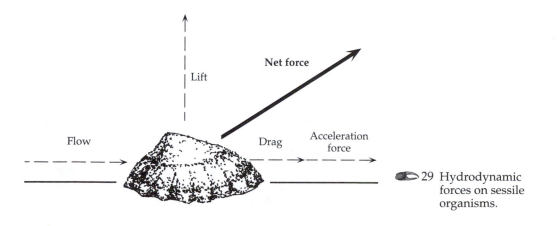

29 Hydrodynamic forces on sessile organisms.

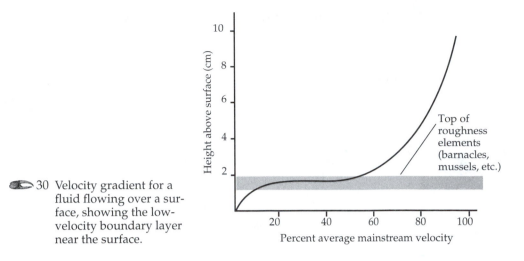

🐚 30   Velocity gradient for a fluid flowing over a surface, showing the low-velocity boundary layer near the surface.

ence a flow environment that is often very different from that higher in the water column. Very small organisms live in a boundary layer of relatively slow-moving water.[🐚 30] Boundary layers, however, are minimized by fast-moving near-bed flows and irregular surface topography. As a result, boundary layers are thinner in high wave energy or tidal flow habitats, and on rocky shores, where uneven topography generates turbulent flow conditions near the bottom. In contrast, boundary layers are more important in tidal mud and sand flats in estuaries and bays, where constant unidirectional flows over relatively flat surfaces are common.

SUMMARY

Understanding the ecology of shorelines requires an appreciation of geological history and of the intimate relationship between shorelines and oceanographic processes. The east coast of North America lies far from the edge of the North American plate, which carries the continent over the earth's surface. Sea-floor spreading has slowly pushed the plate westward for the past 150 million years. This movement has resulted in the rugged mountains characteristic of the west coast of North America, as the western edge of the plate is pushed against the Pacific plate, and an old, geologically stable eastern margin characterized by massive sediment accumulation.

On a much shorter time scale, over the past few million years, changes in sea level have also left dramatic imprints on the east coast of North America. During past glaciations, the most recent of which ended only 20,000 years ago, large ice caps covered much of North America, extending as far south as New England. These ice caps scoured sediments from Canadian and northern New England shorelines, leaving behind rugged rocky shores and large depositional basins such as Narragansett Bay and Long Island Sound. The barrier islands that characterize the east coast of North America south of New England are another conspicuous signature of the coast's history of fluctuating sea levels. When the sea level was over 40 meters lower during the most recent glaciation, waves deposited large amounts of sediment on the shore, and as the sea level rose, these shorelines were flooded, leaving the barrier islands.

The dynamic geological history of the east coast of North America has also left its mark on the organisms that live there. Glaciations over the past 3 million years, in particular, led to the extinction of rocky shoreline organisms unable to retreat south of New England, as well as many other shoreline organisms unable to cope with cold Ice Age conditions. Consequently, much of the fauna and flora of the east coast of North America has only relatively recently colonized or recolonized the coastline. In New England, most of the reinvading organisms have come from Europe and the northern Pacific, whereas in the southern Atlantic, most colonizers have come from lower latitudes. The recent history of disturbance and the relatively depauperate fauna and flora of the east coast of North America have also made it particularly vulnerable to invasion by introduced species. In the past 300 years, since colonization by Europeans, introduced species from Europe and around the globe have come to dominate shorelines on the east coast of North America. Introduced species are so pervasive that any discussion of shoreline ecology must take into account the recent and artificial nature of relationships among many shoreline organisms.

Patterns of water movement shape shoreline habitats and determine the sizes and shapes of organisms that live there. Shoreline organisms are tightly linked to these patterns, which determine their exposure to terrestrial conditions and dictate the movement of their gametes, larvae, and food. Tides are one of the most fundamental organizing forces in shoreline habitats, since they generate predictable gradients of physical stress across shorelines. Water movement also shapes shorelines by eroding high wave energy habitats and depositing sediments in low wave energy habitats.

SUGGESTIONS FOR FURTHER READING

Carlton, J. T. 1985. Transoceanic and interoceanic dispersal of coastal marine organisms: The biology of ballast water. *Oceanography and Marine Biology Annual Review* 23: 313–371.

Carlton, J. T. 1992. Blue immigrants: The marine biology of maritime history. *Mystic Seaport Museum Publication* 44: 31–36. A semipopular account of the history of marine invasions.

Davis, R. A. 1994. *The Evolving Coast*. Scientific American Books, New York. An elegant, simple overview of coastal geology.

Denny, M. W. 1988. *The Biology and Mechanics of the Wave-Swept Environment*. Princeton University Press, Princeton, NJ. A must-read for any aspiring marine intertidal ecologist.

Pethick, J. 1984. *An Introduction to Coastal Geomorphology*. Arnold Press, London. A very readable and insightful look at the dynamics of coastlines.

Vermeij, G. J. 1991. Anatomy of an invasion: The trans-arctic interchange. *Paleobiology* 17: 281–307.

Vogel, S. 1981. *Life in Moving Fluids*. Princeton University Press, Princeton, NJ. A delightfully written primer of hydrodynamics for biologists.

Chapter *2*

The Economy of the Shoreline: The Production and Consumption of Resources

*I*ntertidal communities are dynamic assemblages of organisms influenced by both resources and consumers. On the one hand, shoreline communities are tightly coupled to the delivery of resources in the water column. Nearshore oceanography determines the supply of nutrients, food, and planktonic larvae, particularly at large spatial scales, and thus plays a major role in structuring shallow-water marine communities. On the other hand, consumers also play a leading role in shaping these communities. Consumers affect the distributions of littoral organisms by limiting them to certain refuge habitats, and shape the evolution of morphological diversity by selecting for structural defenses in their prey. The structure and economy of shoreline communities is thus a dynamic balance between the supply of limiting resources and the actions of consumers.

This chapter examines the balance between production and consumption, or the economy of shoreline assemblages. I begin by introducing the primary producers of shoreline habitats and the processes that determine their productivity, then discuss the consequences of variable production for shallow-water marine communities. I then introduce the diversity of consumer strategies on shorelines. I close the chapter with a discussion of how shoreline diets are determined and the ways in which prey avoid being eaten.

☆ *Net Primary Production Estimates for Different Types of Ecosystems*	
ECOSYSTEM	NET PRIMARY PRODUCTION (DRY GRAMS/M^2/YEAR)
Marine systems	
Open ocean	125
Continental shelf	360
Upwelling zone	500
Estuary (excluding shoreline)	1500
Coral reef	2500
Kelp bed	358
Georgia salt marsh	3300
Terrestrial systems	
Tropical rainforest	2000
Temperate forest	1600
Boreal forest	800
Temperate grassland	600
Desert	90
Cultivated systems	
Wheat (world average)	344
Corn (world average)	412
Sugar (world average)	1725
Algal culture (maximum)	4530
Source: Odum 1970; Whitaker 1975.	

PRIMARY PRODUCTIVITY

Primary productivity, or the conversion of inorganic carbon to organic carbon by the harnessing of solar energy through photosynthesis, is the energetic foundation of most natural food webs. Shorelines are among the most productive of all natural habitats (Nixon 1980), and are much more productive than the open ocean. Estuarine inputs add terrestrial nutrients to shoreline habitats, and wind and tidally generated water movement enhance nutrient mixing in shallow coastal waters. As a result, shoreline habitats, particularly those near estuaries and those in upwelling areas, where nutrient-rich bottom water bathes shorelines, are highly productive. Coastal phytoplankton production is typically 3–5 times that of the open ocean, and seagrass and marsh grass primary production is comparable to that of rain forests, coral reefs, and many agricultural crops.

Primary Producers

The major primary producers in shallow coastal waters are microscopic algae, seaweeds (**macroalgae**), and rooted vascular plants. Of the single-celled microscopic producers (**phytoplankton**), diatoms and dinoflagellates are the most common.[🐚1] **Diatoms**, single-celled algae with delicate silica shells, live either individually or in long chains of genetically identical, asexually produced individuals. Diatoms are inconspicuous due to their small size, but are abundant in nearshore waters. Rocky shores are often covered with slippery diatomaceous films that can be a major food source for herbivorous snails. Diatoms are also important primary producers in soft sediments, where they are consumed by deposit-feeding crabs, molluscs, and worms. Planktonic diatoms are usually the most important food source

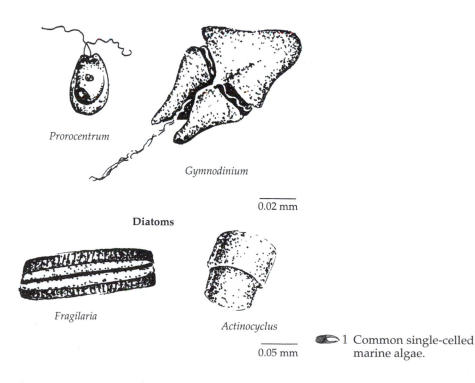

Dinoflagellates

Prorocentrum

Gymnodinium

0.02 mm

Diatoms

Fragilaria

Actinocyclus

0.05 mm

🐚1 Common single-celled marine algae.

for filter feeders such as barnacles, mussels, and clams. **Dinoflagellates**, the second most abundant microscopic algal taxon in most nearshore waters, have cellulose shells. Some dinoflagellates have elaborate horned shells that deter herbivores, while others that lack structural defenses are poisonous to potential herbivores. Outbreaks or blooms of particular species of toxic dinoflagellates are responsible for the poisonous red tides that sometimes occur in shallow, nutrient-rich waters.]

On rocky shores, seaweeds are the most conspicuous primary producers. They range from delicate green algae such as *Enteromorpha* and *Ulva*, the preferred food of many intertidal herbivores, to large, canopy-forming intertidal rockweeds such as *Fucus* and *Ascophyllum*. These large, well-defended rockweeds are readily eaten when small, but escape most consumers as adults due to their toughness and defensive chemistry. They play a critical role in shoreline communities by providing shelter for other organisms, and often enter shoreline food chains as detritus.[🐚2]

The rooted vascular plants of seagrass and marsh habitats also play a major role in providing habitats for other organisms. Seagrasses, which occur in low intertidal or shallow subtidal habitats, and marsh grasses, which occur in high intertidal habitats, are important habitat bioengineers that enhance sedimentation and stabilize sediments, protecting them from erosion. Like large seaweeds, they often enter marine food webs as detritus. [🐚3]

Factors that Limit Primary Production

Light and nutrients are the most important factors limiting marine primary production. Both light quantity and quality affect photosynthesis by marine plants and algae. Sunlight reaching the ocean surface is rapidly attenuated by the absorption and reflection of light by plankton and particulate material in the water column, so that, especially in productive nearshore waters, sufficient light for photosynthesis may not penetrate very deeply into the water column. The maximum depth at which marine plants can live is defined as the **compensation depth**, where the energy gains of photosynthesis are matched by respiratory costs.[🐚4] In clear, unproductive water, the compensation depth can be as deep as 50 meters, whereas in productive coastal habitats, high turbidity and low light transmission can lead to a compensation depth of less than 3 meters.

The spectral quality of sunlight is also strongly affected as it penetrates the water column. Longer red and infrared wavelengths are absorbed near the surface, and only shorter green and blue wavelengths penetrate deeper water. It is the dominance of blue light in deep ocean waters that gives the open ocean its characteristic blue color. Since different algal taxa have photosynthetic pigments that utilize different wavelengths, the differential penetration of light in water was long thought to strongly influence the depth distribution of algal groups (Blinks 1955; Druehl 1967; Saffo 1987). Green algae, like the common sea lettuce *Ulva*, contain chlorophylls *a* and *b*, which absorb light in the red and blue range. This was thought to restrict them to

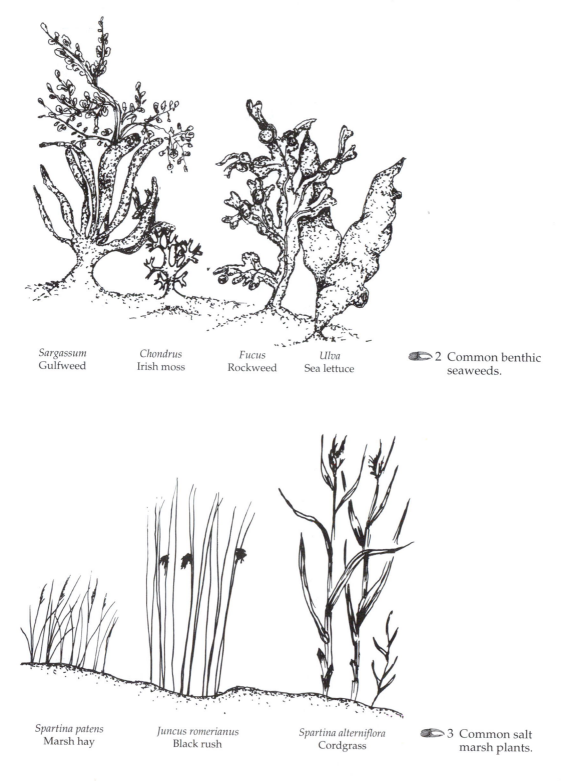

Sargassum   *Chondrus*   *Fucus*   *Ulva*   2 Common benthic
Gulfweed   Irish moss   Rockweed   Sea lettuce   seaweeds.

Spartina patens   *Juncus romerianus*   *Spartina alterniflora*   3 Common salt
Marsh hay   Black rush   Cordgrass   marsh plants.

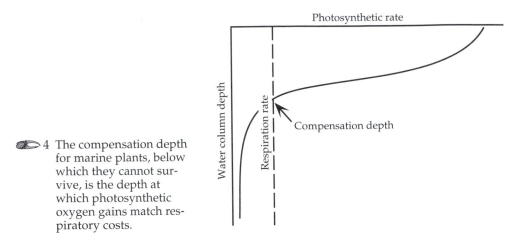

 4 The compensation depth for marine plants, below which they cannot survive, is the depth at which photosynthetic oxygen gains match respiratory costs.

living in shallow water. In contrast, brown algae, like the rockweeds *Fucus* and *Ascophyllum*, have xanthophylls in addition to chlorophyll *a*. Consequently, they were assumed to utilize deeper-penetrating green and yellow light more efficiently than green algae, making them better equipped to live in deeper water. Finally, red algae, like the Irish moss *Chondrus* and coralline crusts, were thought to be able to live deeper than either green or brown algae because their phycobilin photosynthetic pigments absorb more deeply penetrating blue and green wavelengths (Dring 1981). The simplicity of this explanation for the depth distribution of algal taxa was appealing, but has turned out to be largely incorrect. [5]

    Light quality is only one of the factors influencing seaweed depth distributions. While many have noted a superficial correlation between the pigments of algal taxa and their depth distribution (e.g., Chapman 1970), on closer examination, this correlation entirely breaks down (see Saffo 1987 for an excellent review of this issue). Some green seaweeds are typically found in deeper water than even deep-water red seaweeds. Pigment concentration, rather than type, seems to dictate the ability of seaweeds to cope with the low light levels found in deep water, and seaweeds, in general, may respond to low light conditions by increasing the concentration of photosynthetic pigments in their tissues (Ramus 1983). The biotic factors of herbivory and competition for space appear to play an even greater role in setting the vertical distribution patterns of seaweeds (Lubchenco 1978, 1980, 1983), and will be examined in Chapter 5.

    Nutrients also limit the production of marine algae and plants. In contrast to freshwater plants, which are typically phosphate-limited, marine primary producers are generally nitrogen-limited (see Valiela 1995 for a discussion). Ryther and Dunstan (1971) originally demonstrated the importance of nitrogen to coastal phytoplankton. They found that primary production and nitrogen levels in the water column both decreased from New

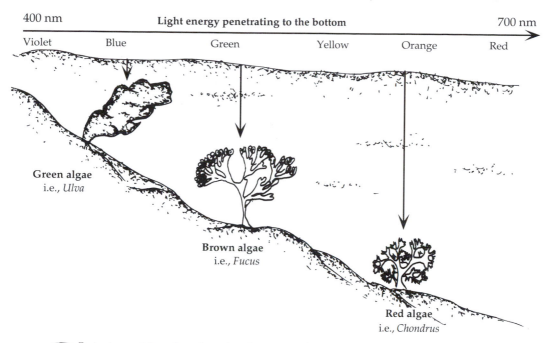

400 nm   **Light energy penetrating to the bottom**   700 nm

Violet   Blue   Green   Yellow   Orange   Red

Green algae
i.e., *Ulva*

Brown algae
i.e., *Fucus*

Red algae
i.e., *Chondrus*

🐚 5 A pigment-based explanation for seaweed zonation. Green, brown, and red algae have photosynthetic pigments that utilize different light wavelengths, a property long thought to be responsible for the depth segregation of algal taxa. Unfortunately, this attractive hypothesis has turned out to be wrong.

York Harbor to the continental shelf. Phosphate concentrations also declined, but not as precipitously. Based on this distribution of nutrients, they proposed that nitrogen, rather than phosphorus, limited offshore phytoplankton growth. They tested their hypothesis by adding nitrogen and phosphorus to plankton samples along an onshore/offshore gradient. They found that adding nitrogen, but not phosphorous, increased plankton production, demonstrating that nitrogen limitation increased with distance from the shoreline.[🐚 6]

Studies demonstrating nitrogen limitation in marine primary producers have now been done in a wide range of habitats. Experiments in New England (Valiela and Teal 1974), Delaware (Sullivan and Daiber 1974), North Carolina (Mendelssohn 1979a,b), and Georgia (Gallagher 1975) have all shown that nitrogen typically limits the primary production of salt marsh grass. Other work has shown that seaweeds (Yates and Peckol 1993; Valiela et al. 1992) and seagrasses (Williams and Ruchelshaus 1993) are also nitrogen-limited under natural conditions.[🐚 7]

What processes dictate the availability of nutrients to nearshore primary producers? Whereas the primary productivity of **benthic** (bottom-dwelling) and **planktonic** marine algae (living in the water column) and vascular plants in shallow coastal waters can deplete nutrient pools, a number of

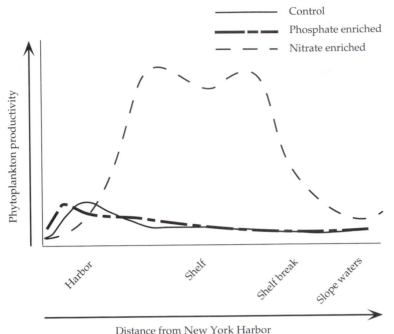

Control

Phosphate enriched

Nitrate enriched

Phytoplankton productivity

Harbor   Shelf   Shelf break   Slope waters

Distance from New York Harbor

6 Ryther and Dunstan's demonstration of nitrogen limitation of off-shore plankton primary productivity. (After Ryther and Dunstan 1971.)

mechanisms can enhance productivity by increasing nutrient supplies. Nutrient-rich runoff from watershed drainage areas is an obvious source of nutrients in estuarine habitats (Nixon 1980). Freshwater plumes from estuaries such as Narragansett Bay and Chesapeake Bay can extend hundreds of kilometers offshore and markedly affect nearshore primary productivity patterns, particularly when estuarine discharges are nitrogen-enriched by fertilizer runoff and sewage.[8]

Primary production by planktonic and benthic marine algae is confined to the **euphotic zone** where light is available. As phytoplankton and macroalgae die and decay, or are eaten and incorporated into feces, they sink to the sea floor, where they accumulate in the absence of primary producers. Three major processes commonly return these bottom nutrients to shallow water where they can be utilized in primary production (Mann 1982). [9] **Upwelling** of nutrient-rich bottom water is a common mechanism of nutrient recycling. Winds move warm surface water masses offshore, drawing cold, nutrient-rich bottom water onshore to replace the displaced surface water. Upwelling is a dominant source of nutrient enrichment in many productive shallow marine systems around the world where continental shelves are nar-

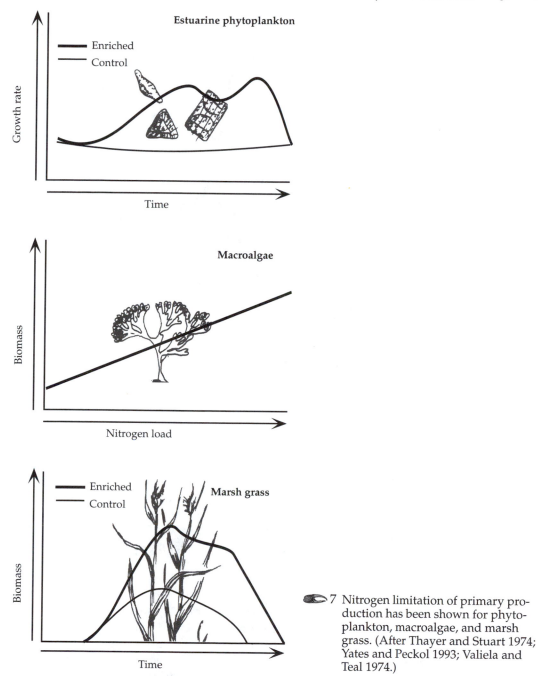

7 Nitrogen limitation of primary production has been shown for phytoplankton, macroalgae, and marsh grass. (After Thayer and Stuart 1974; Yates and Peckol 1993; Valiela and Teal 1974.)

row and ocean currents move water offshore. The coasts of Peru, California, and the Gulf of Panama are good examples of shorelines with strong upwelling. In these areas, predictable seasonal upwelling is caused by global

🌢 8 Salinity isoclines showing the extent of the plume of nutrient-rich
fresh water off Chesapeake Bay.

wind and ocean current systems. [🌢10] On the Atlantic coast of North
America, upwelling is not a dominant regional nutrient enrichment mechanism, but can be a locally important response to weather patterns that move
shoreline water masses offshore (Platt et al. 1972), particularly in Canada
(Mathieson et al. 1991).

A second mechanism that moves nutrient-rich bottom water into shoreline habitats is the **entrainment** of bottom water by freshwater runoff. When
fresh water leaves an estuary, the mixing of the lighter fresh water on the
surface with heavier salt water results in seawater being transported out of
the estuary. To replace this entrained salt water, bottom water is drawn into
shallow water, where it can replenish depleted nutrient supplies. This nutrient enrichment mechanism is important in estuaries with large freshwater
discharges and small tidal amplitudes. The Mississippi and Rio Grande
Rivers, which empty into the Gulf of Mexico, where tides are small, are good
examples of estuary systems where entrainment is an important nutrient-
enrichment mechanism.

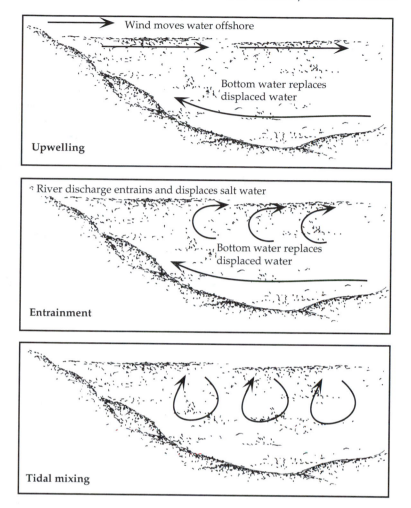

Wind moves water offshore

Bottom water replaces displaced water

Upwelling

River discharge entrains and displaces salt water

Bottom water replaces displaced water

Entrainment

Tidal mixing

9 Three mechanisms of nutrient regeneration in shallow coastal water.

Tidal mixing is a third mechanism that predictably moves bottom nutrients into shallow water. Tidal mixing is a dominant source of nutrient regeneration in systems where tidal exchange is large in comparison to freshwater input into estuaries. Under these conditions, vertical salinity gradients do not develop, and the water column is well mixed rather than stratified. This situation leads to the rapid movement of bottom nutrients into shallow water with tidal currents. Narragansett Bay is a good example of a shallow embayment where tidal mixing is the dominant nutrient regeneration process (Kremer and Nixon 1978). Other high-latitude estuaries exposed to large tidal amplitudes, such as the Saint Lawrence River in Canada and the Penobscot River in Maine, are also good examples of estuaries with strong tidal mixing.

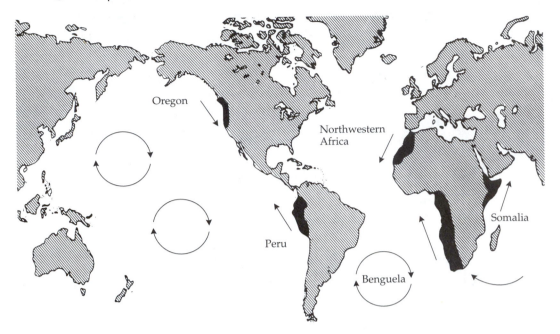

10 Major upwelling areas of the world (areas shown in black) occur where ocean circulation patterns (indicated by arrows) move nearshore water offshore, triggering upwelling.

Seasonality of Primary Production

Seasonal cycles in primary production are conspicuous in many shoreline habitats, particularly at higher latitudes where harsh winter conditions limit plant growth. In New England, for example, the vascular plants that dominate marsh habitats (e.g., *Spartina alterniflora*, *Spartina patens*, and *Distichlis spicata*) emerge from overwintering rhizomes in the spring, grow through the summer, and then in the fall reallocate their accumulated aboveground biomass into overwintering rhizomes. [11] Many of these same plants grow year-round on the coasts of North Carolina and Georgia, slowing their growth in colder months, but not "shutting down" for the winter. Since New England marsh plants grow year-round when placed in a warm greenhouse, the overwintering of northern plants is likely a simple response to harsh winter conditions, rather than a genetically programmed annual cycle.

Seasonal cycles in seaweeds can be quite different. Many seaweeds, like marsh grasses, exhibit their maximum growth rates in the summer, at the time of maximum light availability (Lobban and Harrison 1994). Other northern seaweeds, including some kelps, grow more in the winter months

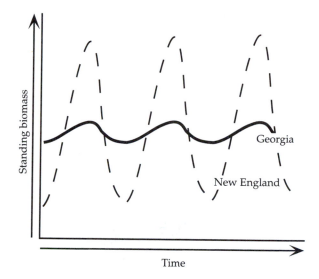

Standing biomass

Time

Georgia

New England

11 Annual variation in aboveground plant standing biomass
in New England and Georgia
marshes. In New England, high
growth rates in the summer
months are followed by a winter
dieback. In southern marshes,
aboveground growth slows in
the winter with little dieback.

than in the summer (Mann 1972). These seaweeds grow in the winter utilizing carbon resources from the previous summer, apparently to take
advantage of the availability of nutrients in the winter. Some tropical seaweeds grow at night, utilizing carbon resources acquired during the day, to
limit the exposure of new tissue to herbivores (Hay et al. 1988). How common this seaweed growth strategy is in other low-latitude species is not
known.

Phytoplankton productivity can also be highly seasonal and may be
characterized by dramatic spring blooms. At northern latitudes, phytoplankton density is low during the cooler winter months, but increases in the
early spring as temperatures and light increase. The proximate cause of this
spring bloom, however, is not simply increased temperatures or light, but
rather the development of temperature stratification in the water column
(Mann and Lazier 1991). [12] During the winter, vigorous phytoplankton growth is possible at typical low temperatures, light levels, and ambient
nutrient concentrations. Since plankton are mixed vertically throughout the
water column, however, phytoplankton can spend considerable time at
depths below their compensation depth, limiting their productivity. In the
spring, the warming of surface water stratifies the water column, limiting
the circulation of plankton to the area near the surface. This concentrates
phytoplankton, leading to bloom conditions. This high productivity, however, is typically short-lived, since the stratified water column also limits
nutrient exchange, and the phytoplankton quickly deplete their nutrient
resources. In the fall, as stratification is broken by the cooling of surface
water, increased mixing and higher nutrient levels can lead to a smaller
autumn bloom before winter conditions set in.

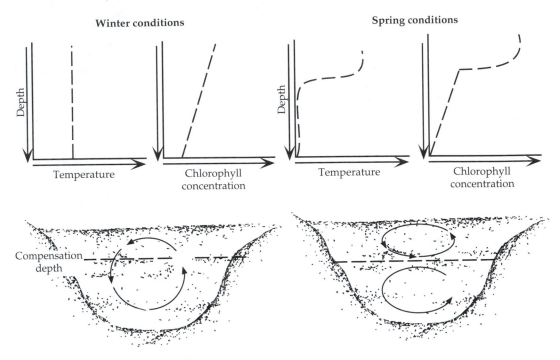

Winter conditions

Depth

Temperature

Chlorophyll concentration

Spring conditions

Depth

Temperature

Chlorophyll concentration

Compensation depth

🐚 12 Development of stratification and restricted circulation above the compensation depth in bays and estuaries.

BOTTOM-UP CONTROL OF SHORELINE COMMUNITIES

The abundances and distributions of shoreline organisms are the result of spatial variation in both resources and consumer activities. Resources set the stage by influencing patterns of growth, whereas consumers influence the abundances and distributions of organisms at lower trophic levels.

Factors that control the structure of natural communities by influencing the supply of resources (nutrients or fixed carbon) are referred to as **bottom-up forces**. Without a source of fixed carbon at their base, shoreline food webs and communities would collapse. Moreover, only about 10 percent of the energy entering a trophic level typically is passed to the next trophic level; the rest is used in growth and maintenance (Lindeman 1942). As a result, a great deal of energy is lost as energy moves through food webs. On basic thermodynamic grounds, the success of higher trophic levels is tightly coupled to lower ones.[🐚13]

Bottom-up control is evident in the response of shoreline communities to seasonal increases in production and in the responses of higher trophic levels to large-scale biogeographic variations in primary production. For example, the spring bloom of phytoplankton characteristic of the temperate zone triggers an increase in the abundances of zooplankton consumers and

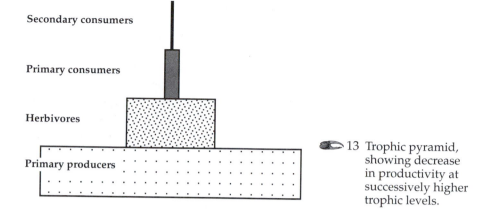

Secondary consumers

Primary consumers

Herbivores

Primary producers

13 Trophic pyramid, showing decrease in productivity at successively higher trophic levels.

of higher trophic levels, including planktivorous (plankton-eating) and piscivorous (fish-eating) fish, in turn. [🐚14] Many temperate-zone benthic organisms with larvae that feed in the plankton also appear to time their larval release to coincide with maximum water column production (Himmelman 1975).

The importance of primary productivity in regulating trophic dynamics is particularly apparent in the large-scale biogeographic variation in marine communities. The most productive shorelines in the world are found in

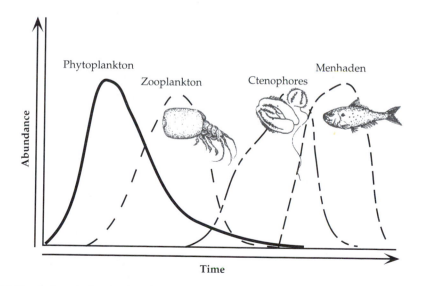

Phytoplankton

Zooplankton

Ctenophores

Menhaden

Abundance

Time

14 Typical variation in the abundances of water column organisms triggered by spring blooms. (After Gillespie 1971.)

upwelling areas, such as the coasts of California, Peru, and the Bay of Panama, and support the largest marine consumer populations in the world (Mann and Lazier 1991). On the Atlantic coast of North America, where upwelling is not prominent, the role of primary productivity in driving the abundance and distribution patterns of higher trophic levels is evident in the concentration of harvestable fish and shellfish in high-productivity areas such as Narragansett and Chesapeake Bays. Variation in primary productivity on these large spatial scales is difficult to examine experimentally, however; this difficulty has been a major obstacle in documenting linkages between primary productivity and the structure and dynamics of shoreline communities.

Because they are discrete, easily studied, and show considerable variation in primary production at relatively small spatial scales, salt marshes have been valuable in demonstrating the importance of bottom-up forces in shoreline communities. Experimental studies in which fertilizers have been added to natural marsh vegetation (Valiela and Teal 1974), and correlative studies in which marsh vegetation has been examined across known nutrient gradients (Nixon and Oviatt 1973; Oviatt et al. 1977), have both shown that increased nitrogen inputs increase marsh plant primary production. Increased marsh plant growth has also been shown to enhance the abundance and growth of marsh grass herbivores as well as higher-order consumers (Vince et al. 1981; Stiven and Kuenzler 1982) Moreover, by dictating the outcome of competitive interactions among marsh plants, nitrogen supply may control marsh plant zonation patterns (Levine et al. 1998; see Chapter 7 for a discussion of this study). Nitrogen supply, therefore, appears to act as a strong bottom-up community structuring force in marsh plant communities.

Nitrogen supply has also recently been suggested as a regulator of the structure of communities in shallow-water soft-substrate estuarine habitats. [🐚15] Shallow-water estuaries are some of the most nutrient-rich ecosystems on earth due to coastal development and the effects of urbanization on nutrient runoff. Nonetheless, the way in which this nutrient loading, or **eutrophication**, affects shoreline communities is not well understood. On Cape Cod, coastal development is positively correlated with nitrate concentrations in the groundwater, which translates into an increase in the amount of nitrogen entering local embayments (Valiela et al. 1992).[🐚16] Under natural conditions, most nitrogen entering local watersheds through rainfall and fertilizer runoff is utilized by plants. Septic systems associated with coastal development, however, are typically deeper than tree and shrub root systems, and thus nutrients that leach from septic systems are not incorporated into terrestrial biomass. Instead, they ultimately end up in the water table and drain into rivers and estuaries.

The effects of this localized eutrophication on estuarine habitats can be dramatic. Increased nitrogen supplies lead to the replacement of seagrass beds by dense, free-floating rafts of ephemeral seaweeds (*Ulva* and *Cladophora*) that may cover the bottom in a layer 50–75 centimeters thick. This accumulation of fast-growing seaweeds depletes water column nutrients and

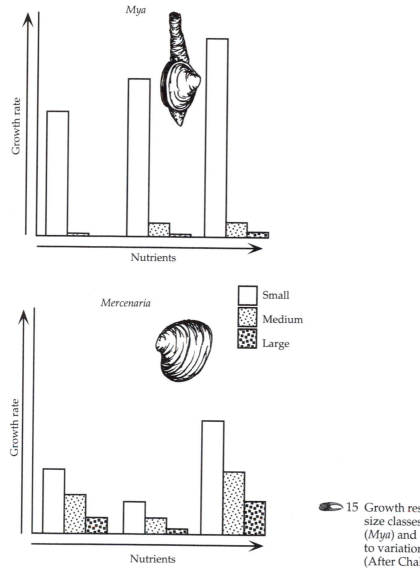

15 Growth responses of different size classes of soft-shelled clams (*Mya*) and quahogs (*Mercenaria*) to variation in nutrient supply. (After Chalfoun et al. 1994.)

dramatically alters the bottom community by limiting light penetration and water movement and by smothering organisms on the sediment surface. As a result, seagrasses, which have historically dominated New England estuaries, but have been heavily affected by disease and disturbances, have been displaced by dense algal mats. [17] Over the last half century, this process has contributed to an unprecedented decline of seagrass beds in New England. For example, 40 years ago, before the urbanization of Cape Cod, over 75 percent of Waquoit Bay was dominated by seagrass beds. Today less than 10 percent of this bay has seagrass beds, most having been displaced by

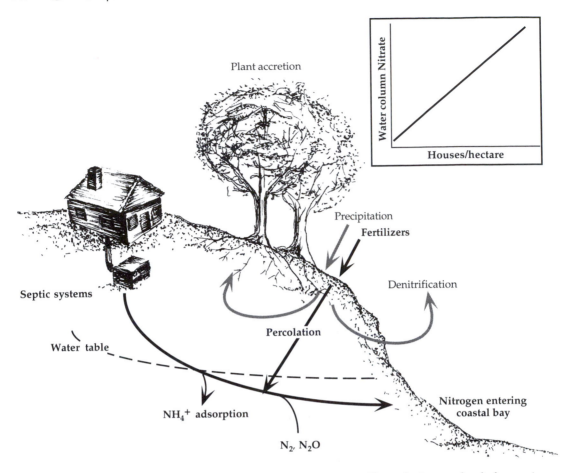

Plant accretion

Water column Nitrate

Houses/hectare

Precipitation

Fertilizers

Denitrification

Septic systems

Water table

Percolation

Nitrogen entering
coastal bay

NH_4^+ adsorption

N_2, N_2O

🐚 16 Nitrogen pathways in developed coastal estuaries. Natural nitrogen loads from rain
and runoff are typically tied up in plants or denitrified. Septic systems, however,
add nitrogen below plant root zones, leading to high levels of nitrogen in bay water.
The inset graph shows the relationship between shoreline development and water
column nitrogen concentrations in Waquoit Bay. (After Valiela et al. 1992.)

free-floating algal beds (Costa 1988).[🐚 18] Strong bottom-up linkages such
as these, in which nutrient loading affects shoreline communities directly by
differentially affecting primary producers and indirectly by influencing habitat quality, are probably common but largely unappreciated forces in many
shoreline environments.

    Bottom-up forces are probably just as important in rocky intertidal com-
munities as they are in salt marshes and shallow-water estuarine communities. Historically, most research on rocky intertidal community structure has
focused on the effects of consumers and competition for space. Recent studies, however, have revealed the importance of bottom-up forces on rocky

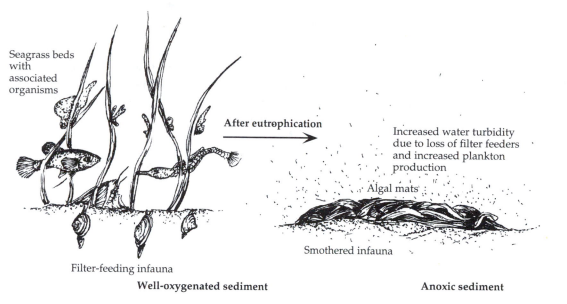

Seagrass beds
with
associated
organisms

After eutrophication

Increased water turbidity
due to loss of filter feeders
and increased plankton
production

Algal mats

Smothered infauna

Filter-feeding infauna

Well-oxygenated sediment

Anoxic sediment

17 Benthic community shift in shallow coastal waters following eutrophication. (After Valiela et al. 1992.)

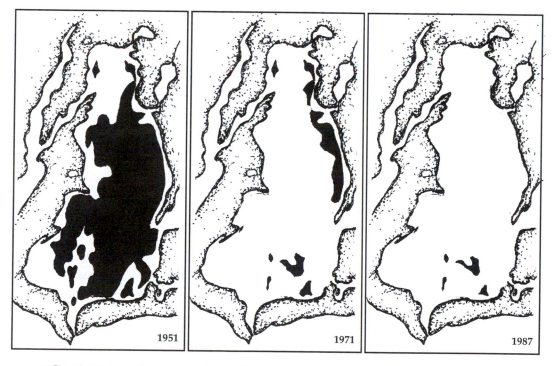

1951

1971

1987

18 Decline of eelgrass beds (in black) in Waquoit Bay, Cape Cod, over the past half century. (After Costa 1988.)

shores. Spatial variation in water column primary production appears to be directly linked to the secondary production of filter feeders, such as barnacles and mussels (Menge 1992). Large-scale spatial variation in water column productivity may play an important role in generating broad patterns in the rocky shore communities of the northern Atlantic (A. R. O. Chapman, personal communication). The highly productive shores of New England, for example, are dominated by an assemblage of filter-feeding barnacles and mussels; in contrast, the shores of the Canadian Maritimes, which are adjacent to low-productivity, **oligotrophic** waters, support only small and patchy populations of mussels and barnacles. This pattern may be at least partially the result of the scarce water column resources for filter feeders in oligotrophic habitats.

Other work on rocky shorelines, however, has suggested that variation in water flow velocities can have equally important bottom-up consequences for benthic communities. While high primary production can increase the food supply for filter feeders, high flow velocities can also increase the delivery of food to filter feeders and enhance larval supply and recruitment (Fréchette et al. 1989; Sanford et al. 1994). High flow velocities may also increase seaweed production by accelerating gas exchange rates, which can limit photosynthesis (Gerard 1987; Wheeler 1980; Gerard and Mann 1979). [🐚 19] Since seaweeds decrease their surface area in high flows to limit mechanical damage, however, the relationship between flow and seaweed productivity is not clear. Bottom-up structuring of benthic communities by flow may be substantial, but has not been well studied. Filter-feeding organisms and seaweeds, often the base of shoreline food webs, may generally have high population densities and growth rates in high-flow habitats. These flow-driven effects may increase the food supply and success of consumers at higher trophic levels, but how common this is uncertain.

TOP-DOWN CONTROL OF SHORELINE COMMUNITIES

The abundance and distribution of shoreline organisms is not simply a reflection of nutrients and resource availability, but is also affected by variations in the presence of consumers (**top-down forces**). Consumers limit the success of organisms at lower trophic levels, and thereby influence the abundance and distribution of those organisms over both ecological and evolutionary time. Over ecological time, consumers influence the distributions of their prey by restricting them to spatial and/or temporal refuges. Over evolutionary time, consumers are a potent selective force that leads to the evolution of anti-predator defenses that promote the spatial and temporal coexistence of predators and their prey.

Historically, ecologists have focused much of their attention on the role played by consumers in intertidal communities, often at the expense of ignoring bottom-up forces. Top-down consumer forces are often more amenable to experimental study, particularly when the predators are slow-moving snails and starfishes that are easily manipulated. Moreover, intertidal communities

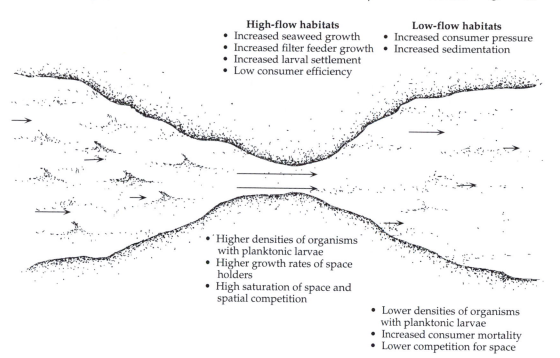

High-flow habitats
- Increased seaweed growth
- Increased filter feeder growth
- Increased larval settlement
- Low consumer efficiency

Low-flow habitats
- Increased consumer pressure
- Increased sedimentation

- Higher densities of organisms with planktonic larvae
- Higher growth rates of space holders
- High saturation of space and spatial competition

- Lower densities of organisms with planktonic larvae
- Increased consumer mortality
- Lower competition for space

19 Possible effects of accelerated flow rates at estuary constrictions on benthic communities in tidal estuaries in the Gulf of Maine. (After Leonard et al. 1998a.)

tend to be trophically simple in contrast to many terrestrial communities. In such simple communities, interactions among organisms are strong, since there is little redundancy in trophic roles, often leading to powerful top-down control (Strong 1992). In contrast, factors that influence shoreline food supply, or bottom-up forces, often operate at large spatial scales dictated by oceanographic patterns, climate, or coastline geomorphology, and are less easily manipulated (Menge 1992).

Examples of top-down control of shoreline communities are evident in most Atlantic intertidal habitats. On wave-exposed rocky shorelines in the Gulf of Maine, predaceous starfishes and snails prevent mussels and barnacles from dominating low intertidal habitats (Menge 1976), whereas on nearby rocky shores that are sheltered from waves, crab predation can limit mussel and snail abundance and distribution (Kitching et al. 1966; Leonard et al. 1998a). Similarly, in soft-substrate habitats south of Cape Cod, blue crabs and predatory fishes such as spot (*Leiostomus xanthurus*) can limit the distribution and abundance of burrowing clams and soft-bodied prey, restricting them to coarse substrates, deep sediments, or seagrass beds (Virnstein 1977; Lipcius and Hines 1986; Blundon and Kennedy 1982; Peterson 1982a; Summerson and Peterson 1984). In salt marshes, predation pressure by fish and crabs at low intertidal heights commonly limit smaller fishes (Knieb

1981, 1986), snails (Vince et al. 1976), crustaceans (Vince et al. 1976; Kneib 1982, 1985, 1997), and mussels (Stiven and Gardner 1992) to high intertidal heights and dense vegetation. How these top-down consumer effects are influenced by variation in primary production is not well understood.

The structure of food webs and the spatial and temporal abundance and distribution of shoreline organisms are clearly a complex product of both resource availability (bottom-up) and consumer (top-down) forces.[🐾 20] Increased nutrient supply in aquatic systems can lead to increased algal and benthic filter feeder production, which in turn can lead to increased consumer pressure by herbivores and carnivores (Power 1992; Menge 1992). Thus, bottom-up forces can escalate the intensity of top-down control in communities. In contrast, flow-driven, bottom-up forces that operate by influencing the delivery of nutrients, food, and propagules can negatively influence top-down forces, and can therefore decouple bottom-up and top-down effects on shorelines (Leonard et al. 1998a). High water flows with high particle fluxes can lead to high primary and secondary production, but low consumer efficiency due to decreased consumer mobility (Menge 1978) and chemoreception ability (Weissburg and Zimmer-Faust 1993). This decoupling of bottom-up and top-down forces may be a common organizing feature of estuarine communities (see Leonard et al. 1998a).

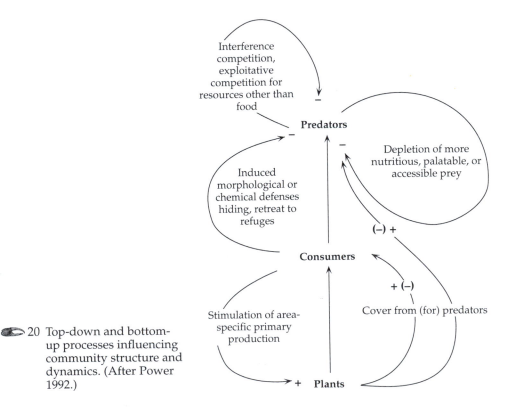

🐾 20 Top-down and bottom-up processes influencing community structure and dynamics. (After Power 1992.)

CONSUMER FEEDING STRATEGIES

Intertidal consumers come in a variety of shapes and sizes, and utilize a wide range of food resources. Filter feeders that sieve plankton from the water column, deposit feeders that sort microscopic algae and debris from sediments, herbivores that consume algal films or large seaweeds, and predators that drill or crush shelled prey are all common shoreline consumers. Below, I briefly introduce the major feeding strategies of shoreline animals before discussing consumer diets and prey defenses.

Filter Feeders

Filter or **suspension feeders**, such as clams, mussels, and barnacles, filter plankton out of the water column. They play a crucial role in the movement of nutrients and energy through shoreline communities by converting water column biomass into benthic biomass. Moreover, by reducing turbidity and increasing light penetration, and by excreting ammonium waste products, filter feeders can indirectly promote primary production.

    Passive filter feeders, such as barnacles, bryozoans, and hydroids, depend entirely on ambient water flows to deliver planktonic food particles to their filtering appendages.[21] Once food has accumulated on these appendages, it is transferred to the mouth to be processed. Barnacles, for example, slowly beat their **cirri,** or modified feet, into the flow, periodically retracting the cirri to remove captured food items. At higher flow speeds, cirral beat frequencies increase, reflecting increased food delivery, whereas in calm water cirral beating stops (Sanford et al. 1994).[22] Because passive filter feeders depend on ambient flow for the delivery of their food, the feeding success and growth of these animals is tightly coupled to flow conditions. The success of passive filter feeders, in general, probably increases

21 Typical passive filter feeders, which utilize ambient flows to deliver food to their feeding appendages, include barnacles, some tube worms, and bryozoans.

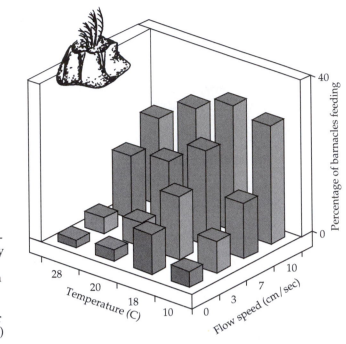

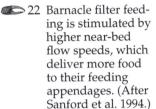 22 Barnacle filter feeding is stimulated by higher near-bed flow speeds, which deliver more food to their feeding appendages. (After Sanford et al. 1994.)

with increasing flow until flow velocities are reached at which their feeding appendages fail or their guts are saturated.

In contrast to passive filter feeders, active filter feeders, such as clams, mussels, sea squirts, and sponges, generate a current over their feeding surfaces. Clams and mussels, like most other bivalve molluscs, drive a feeding current over their gills, which sort organic material from the particles in the water. Sea squirts have a basket-shaped sieve through which they drive a feeding current. Mucus spread over the surface traps microscopic particulate food material. Sponges force feeding currents through their bodies, and food particles are absorbed and digested by cells lining the body cavity. Since these organisms are not entirely dependent on flow for food delivery, their feeding and growth is not as tightly linked to flow conditions as that of passive filter feeders. [🐚 23] High flows, however, probably still enhance the feeding success of most active filter feeders. In still water, active filter feeders can deplete local food supplies; high flows can prevent this depletion (Fréchette et al. 1989). The morphology of some active filter-feeding sponges and tunicates also might enhance passive flow over their feeding structures, decreasing their reliance on active pumping (Vogel 1974, 1977).

Deposit Feeders

Deposit feeders ingest and process sediments, removing diatoms, bacteria, and other particulate organic material. They are vital players in the consumption of detritus and recycling of nutrients in mud and sand habitats.

ﾟ 23 Active filter feeders, which move water over their feeding surfaces to collect particulate food, include most bivalves, sponges, and tunicates.

Since sediments are relatively poor sources of nutrition, deposit feeders must typically process large volumes of sediment to meet their energy requirements. The large volumes processed mean that most soft substrates are constantly reworked by deposit feeders.

Mobile deposit feeders, such as ghost and fiddler crabs, use scooplike claws to shovel sediments into their mouths, where their mouthparts comb through the sediments and extract organic material. Sedentary or slower-moving deposit feeders, such as worms and sea cucumbers, utilize a wide variety of techniques to harvest food from sediments. [ﾟ 24] Lugworms bulk-feed, like earthworms, by ingesting large amounts of sediment from which they extract organic material. Other deposit-feeding worms (e.g., *Amphitrite*) have moplike feeding appendages that lie on the sediment surface and collect organic material. Still other deposit feeders, including other segmented worms and sand dollars, live head-down in the substrate, selectively ingesting subsurface sediments and depositing processed sediments on the surface. [ﾟ 25]

ﾟ 24 Common bulk-feeding deposit feeders, which ingest sediments and remove organic content as their food, include lugworms, mud snails, and crustaceans, including fiddler crabs and mud shrimp.

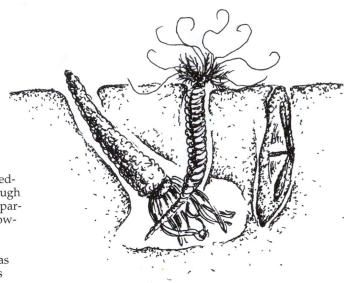

🐚 25 Selective deposit feeders, which sort through sediments for food particles, include burrowing worms such as *Pectinaria* (left) and *Amphitrite* (center) as well as sand dollars (right).

Some particularly interesting deposit-feeding worms and bivalves are facultative filter feeders. These organisms typically feed on deposits in low flows, when particle fluxes are low but the deposition of organic particles is high. However, they switch to filter feeding in high flows, when both particle fluxes and food delivery are high (Taghon et al. 1980; Olafsson 1986). These organisms illustrate a close functional relationship, and a possible evolutionary pathway, between deposit and filter feeding.[🐚 26]

Herbivores

Shoreline herbivores are very diverse, ranging from slow-moving, algal-grazing snails and urchins to the insects and geese that eat marsh grasses. Herbivorous snails use a tonguelike **radula** to rasp diatoms from hard surfaces or excavate algal cells from seaweeds. The radula is a flat proteinaceous belt with replaceable chitinous or calcium carbonate teeth. Muscles on either end of the radula pull it back and forth over a hard chitinous supporting rod (the **odontophore**). The radula of herbivorous snails is used either like a broom to sweep microalgal cells into the mouth or like a shovel to excavate cells from larger seaweeds (see Steneck and Watling 1982).[🐚 27] Periwinkles (*Littorina*) are common Atlantic shoreline snail herbivores.

Sea urchins are also important benthic herbivores. They are often responsible for the exclusion of fleshy algae from shallow subtidal habitats, and can be important seagrass consumers. Urchins scrape surfaces with a five-toothed mouth called an **Aristotle's lantern,** and are typically capable of eating all but heavily calcified or chemically defended algae.[🐚 28]

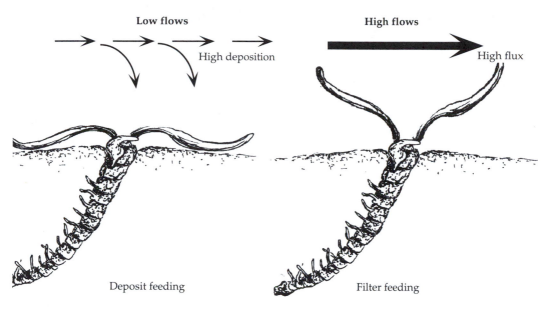

26 The feeding strategy of a number of deposit-feeding worms and protobranch (primitive) bivalves is dependent on ambient flow conditions. At low flow speeds they
deposit-feed, but at high flow speeds they filter-feed, taking advantage of the
increased delivery of potential food particles. (After Taghon et al. 1980.)

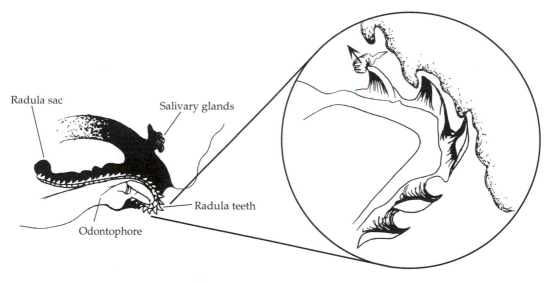

27 A simple molluscan herbivore radula, or scraping tongue. The radula is a belt of
teeth that is stretched over a supporting rod (the odontophore). The erected teeth
scrape food from the surface into mucus, which is pulled into the digestive system.

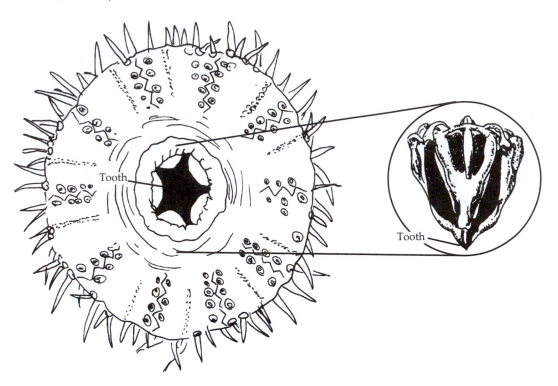

Tooth

Tooth

28 Aristotle's lantern, the feeding structure of sea urchins.

Salt marsh plants, in general, are well defended from most benthic herbivores, but have their own suite of consumers. Chewing and sucking insects are particularly abundant at higher marsh elevations. Grasshoppers and beetles are typical chewing herbivores in marshes, while aphids and leafhoppers are important sap-sucking insects inthese habitats. Other marsh herbivores include crabs and vertebrates. The Atlantic coast marsh crab, *Sesarma*, specializes in eating both above- and below-ground marsh plant tissue, particularly in southern marshes. In northern Atlantic marshes, geese can be prominent consumers of below-ground roots and rhizomes. When common, they are capable of denuding large areas of all vegetation.

Predators

Intertidal predators are also a diverse group, ranging from snails and crabs to birds and fishes. Many snails, such as dog whelks (*Nucella*) on rocky shores and moon snails (*Polinices*) on sandy shores, drill into shelled prey with a modified radula.[29] Other predatory snails common on shallow sandy beaches (such as the whelk *Busycon*) smother bivalve prey with the foot and use the radula to rasp the flesh out of their subdued victims.

Crabs are also important predators on Atlantic shorelines. In New England, the introduced European green crab, *Carcinus maenus*, and the rock

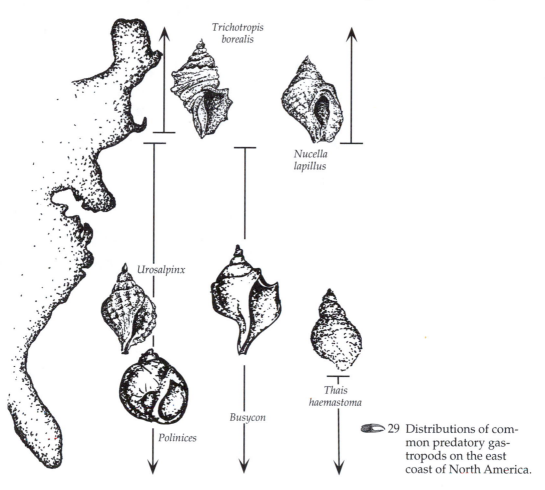

Trichotropis borealis

Nucella lapillus

Urosalpinx

Busycon

Thais haemastoma

Polinices

29 Distributions of common predatory gastropods on the east coast of North America.

crabs *Cancer irroratus* and *Cancer borealis* are common shoreline predators. A casual examination of their claws reveals their primarily predatory lifestyle. South of Cape Cod, the swimming portunid blue crab, *Callinectes sapidus*, is a voracious predator of bivalves and snails, locating buried prey by chemoreception and physical contact, digging them from the substrate, and using its strong claws to break open the shells.[30]

Birds and fish are also common predators on both rocky and sandy beaches. On rocky beaches, gulls, crows, and oystercatchers either use their beaks to hammer through prey or drop hard-shelled prey onto rocks to break them open. On sandy and muddy shores, shorebirds, as well as egrets and herons, use their long beaks to forage for mostly soft-bodied prey.[31] Fish can also be important predators in coastal communities. In salt marshes and soft-substrate habitats, killifishes and flounder are voracious predators of virtually all smaller animals, restricting many of their

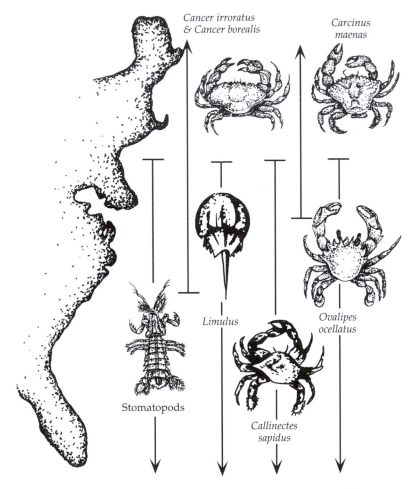

🦐 30 Distributions of some dominant predatory crustaceans along the
Atlantic coast.

prey to high intertidal refuges. Other predatory fishes, such as puffers and
tautogs, have mouthparts modified to crush shelled prey. Shell-crushing
fishes can be important mollusc consumers, particularly at lower latitudes
(Vermeij 1978).[🦐 32]

FORAGING DECISIONS: WHAT TO EAT?

The food choices of intertidal consumers play an important role in determining the distributions of prey. What determines the diets of shoreline organisms? Intertidal foragers must decide whether or not to eat any prey item
they encounter, as well as where and how to concentrate their foraging
efforts. Optimal foraging theory examines these decisions (see Hughes 1980
for a review applied to marine organisms).

31 Birds such as the snowy egret (*Egretta thula*) and the great blue heron (*Ardea herodias*) are conspicuous predators in shoreline habitats.

32 Predaceous fishes that forage in intertidal habitats include spot (*Leiostomus xanthurus*), tautog (*Tautoga onitis*), and cunner (*Tautogolabrus adspersus*).

Energy Maximization

The underlying assumption of optimal foraging theory is that consumers will maximize their net rate of energy intake per unit of foraging effort. This rate, however, is not easily calculated, given the complexities involved in searching for, handling, and ingesting prey. Searching can involve mortality risks and energetic costs, handling or subduing prey can be costly and time-consuming, and prey can differ in their caloric value and energetic costs. Nonetheless, shoreline invertebrates often appear to minimize their energetic costs and maximize the benefits gained from foraging effort.

The foraging behavior of the green crab *Carcinus* has been particularly well studied (Elner and Hughes 1978). On rocky shores in the northern Atlantic, adult green crabs have broadly omnivorous diets, but are major predators of blue mussels. *Carcinus* individuals often live associated with mussel beds, and have well-developed shell-opening behaviors for preying on the bivalves. Green crabs have unequal claws. The larger master claw has an enlarged abductor muscle that generates great force and heavy dentition that helps to crush shelled prey. The smaller claw is much thinner and is used to pick the flesh from crushed prey items. When the crab feeds on mussels, small mussels are crushed outright in the master claw, medium-sized mussels are crushed at the umbo region of the shell, and large mussels are opened either by slowly chipping the edge of the shell or by forcing the master claw between the mussel valves (Hughes and Elner 1979).

Green crabs do not randomly consume the mussels they encounter. If crabs considered only the time necessary to open and consume a mussel (its costs), they would always choose small mussels, since handling time increases with increasing mussel size. Alternatively, if crabs considered only the energetic payoff of a mussel (its benefits), they would always chose the largest mussels available, since the energetic content of mussel prey increases with mussel size. If green crabs, however, chose mussel prey so as to maximize their energy intake and minimize their foraging effort (energy gain/time spent foraging), they would preferentially consume medium-sized mussels. This is exactly what they do. Green crabs are capable of feeding on mussels of all sizes, but when all sizes of mussels are available, they eat only medium-sized mussels. Smaller or larger mussels are eaten only when medium-sized mussels are unavailable. [🦀 33]

Shoreline consumers, in general, have strong prey size preferences that maximize the energetic gain of their foraging effort. Crows and gulls commonly forage on rocky shorelines, preying on hard-bodied molluscs by dropping them onto rock surfaces to open them. Zach (1978) studied the predation techniques of crows feeding on intertidal dog whelks, and found that crows have a strong preference for large whelks, which have more food value and break more readily than smaller whelks. Crows choose prey initially by sight, selecting the largest whelks available, but when given a choice between similarly sized whelks, choose heavier ones. Once a prey item is selected, a crow flies 5–10 meters above the ground and drops it onto

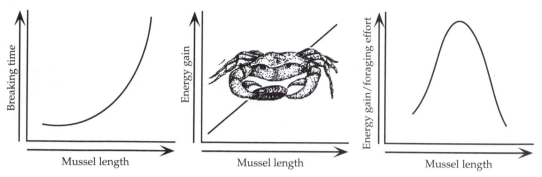

33 Green crab selection of blue mussel prey. The larger the mussel, the harder it is to open, but the bigger the energetic payoff. As a result, medium-sized mussels are the most profitable, and are preferred by the crabs. (After Elner and Hughes 1978.)

a rock surface. If it fails to break on the first try, the crow drops the prey repeatedly rather than searching for a new one, thus minimizing search time.[34]

The energy maximization principle has also been applied to filter (Lehman 1976) and deposit feeders (Taghon and Jumars 1984). With these consumers, searching and handling times are identical, so foraging decisions are reduced to the question of how rapidly to filter water or ingest sediments, and which ingested particles to eat. Both filter and deposit feeders increase their feeding rates with increasing food concentrations to the point at which the gut is fully packed, after which the feeding rate decreases due to the

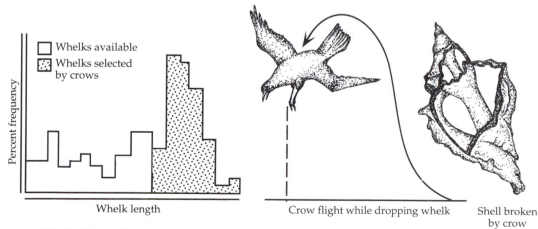

34 Size selection of whelk prey by crows. (After Zach 1978.)

inability to process higher food intake. [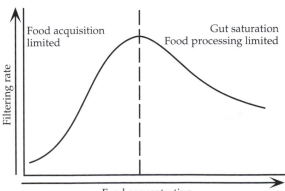35] Thus, these foragers do not expend energy ingesting more food particles than they are able to process.

Optimal Diets

Shoreline consumers are generally faced with choices among prey types as well as choices among prey sizes. The energy maximization principle can be easily extended to examine the diets of shoreline organisms faced with multiple prey choices. Given a variety of potential prey, shoreline consumers do not choose food randomly, but select diets based on prey availability, energy yield, and handling costs that maximize their energy intake. Predatory snails that utilize a radula to bore into shelled prey are valuable organisms for studying diet choices, since they are slow-moving, their victims are easily identified, and the costs and benefits of selecting different prey types are easily quantified.

Moon snails are common predators on sandy substrates. Drilling by these snails entails rasping the victim's shell with the radula and applying an acid-secreting boring organ to the drill site, which slowly dissolves the shell (Carriker 1955). Drilling proceeds slowly, at only 0.25–0.50 millimeters per hour, so that a typical prey item may take 2–3 days to drill. The moon snail *Polinices* has a distinct preference for the soft-shelled clam *Mya arenaria* over the quahog *Mercenaria mercenaria* and the blue mussel *Mytilus edulis* (Edwards and Huebner 1977). When standardized for length, *Mya* do not have the energy payoff that *Mercenaria* do, suggesting that moon snails are selecting low-quality food items. But *Mya* have thinner shells and lower drilling costs. As a result of these lower handling costs, the moon snail preference for *Mya* maximizes the energy yield per unit of foraging time. [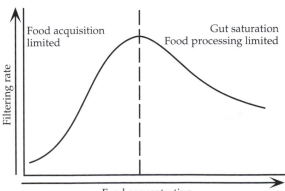36]

On rocky shores, a good example of the importance of the way in which handling costs dictate feeding preferences is found in another predaceous snail. The dog whelk *Nucella* has a strong preference for barnacles over mussels. Barnacles have fewer calories per prey item than mussels, but are preferred because they are more rapidly drilled. Whelks typically drill barnacles

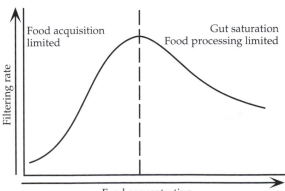

35 Maximization of foraging effort by filter and deposit feeders. Feeding rate increases with food concentration until food processing ability limits energy intake. Feeding rates are predicted to decrease at high food concentrations due to the inability to process higher food intakes.

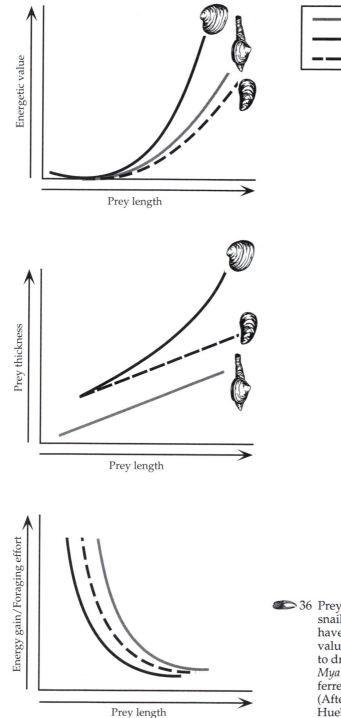

🐚 36   Prey selection by the moon snail *Polinices*. *Mercenaria* have the highest prey value, but are the hardest to drill. As a result, both *Mya* and *Mytilus* are preferred over *Mercenaria*. (After Edwards and Huebner 1977.)

at the suture lines between the plates of their tests. The sutures are thinner, weaker, and more easily drilled than the tests themselves, and make barnacle prey far more vulnerable to drilling than mussels.

Optimal foraging theory predicts that when preferred prey are not available, less preferred prey will be included in the diet in descending rank order, as long as the addition of lower-ranked prey results in a higher energy yield (E/T) than the diet without the lower-ranked item (Charnov 1976). Thus, an optimal diet will always include the preferred prey, and will include less preferred items only when preferred prey are rare. Moon snails, for example, should take *Mercenaria* only when *Mya* are rare, and dog whelks should eat mussel prey only when barnacles are rare. Thus, the inclusion of low-preference food items in predator diets is dependent on the abundance of preferred prey, not on the availability of less preferred prey.

Where to Forage

For a consumer, deciding where to eat is as important as deciding what to eat. Should a consumer spend time in only the most profitable habitats? When should a consumer leave a habitat to forage elsewhere? How well can consumers distinguish among habitats with different resource levels? Optimal foraging theory has considered these questions both theoretically and empirically. The "marginal value theorem" predicts how organisms should forage in an environment with patchy resources (Charnov 1976).

The marginal value theorem hypothesizes that when choosing foraging locations, consumers should concentrate on the most profitable resource patches, staying in a patch until the net rate of energy intake in that patch falls below the average for the habitat. Thus, as patches become depleted, more patches should be visited. The marginal value theorem assumes that foragers can distinguish among patches of different profitability and learn the spatial distribution of patches of different value. Important corollaries to the marginal value theorem are that foragers need to spend time assessing habitat resource availability, visiting even low-profitability habitats, and that greater costs of travel among patches or greater search costs within patches should increase the length of time consumers stay in patches.

The marginal value theorem has been tested with a number of shoreline organisms. Goss-Custard (1969) studied redshanks foraging on muddy shorelines in Europe, and showed that these shorebirds concentrated their foraging in areas of highest profitability, but always spent time assessing lower-quality habitats. He also found that redshank foraging shifted over time, tracking changes in patch profitability, in response to the depletion of patchy food resources and seasonal shifts in prey abundances.

Fiddler crabs have also been valuable in studying patch choice. They are easily observed on mud and sand flats, scooping sediments into their mouths, where setae separate lighter organic material from heavier inorganic particles. Weissburg (1992, 1993) found that fiddler crabs in muds forage primarily on high-profitability patches, and leave once food is depleted. Sexual dimorphism in fiddler crab claws, however, strongly affects their foraging decisions. Males have one claw that is massively enlarged for use in

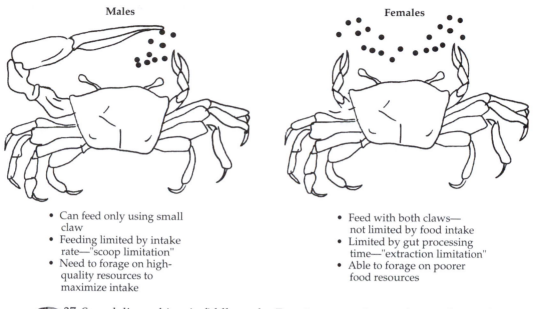

Males

- Can feed only using small claw
- Feeding limited by intake rate—"scoop limitation"
- Need to forage on high-quality resources to maximize intake

Females

- Feed with both claws—not limited by food intake
- Limited by gut processing time—"extraction limitation"
- Able to forage on poorer food resources

🦀 37 Sexual dimorphism in fiddler crabs. Foraging constraints are imposed on males by their enlarged claw, which is used for mating but is useless for feeding. (After Weissburg 1993.)

ritualized courtship and fighting (Crane 1975). The cost of this appendage is high. Not only does the enlarged claw make up 40 percent of the total body weight, but it is useless in feeding. Consequently, male crabs must feed with their small claw (Valiela et al. 1974), and females can feed twice as fast as males. This difference imposes very different foraging constraints on male and female fiddler crabs. Males are limited by the amount of food they can move into their mouths (scoop-limited), so they feed in the most profitable sediments. In contrast, female crabs are limited not by their ability to move food into their mouths, but by their ability to extract food from sediments (extraction-limited), and thus are less dependent on high-profitability sediments than males. This leads to males and females feeding in different patches (Weissburg 1992). Since females are not limited by food supply, they can forage profitably almost anywhere. Males, however, typically forage far from the safety of their burrows and vegetation cover to maximize their food intake.[🦀 37]

PREY DEFENSES: AVOIDING BEING EATEN

Consumers typically play an important role in shaping intertidal communities by restricting prey to refuges and influencing the distribution of prey populations. Consumers also exert strong selective pressure on prey defenses, leading to the evolution of chemical, structural, and behavioral defenses. Variation in consumer pressure in coastal habitats, however, is often predictably variable in space and time, leading to conspicuous patterns in prey abundance and defenses.

Patterns in Consumer Pressure

Spatial patterns in consumer pressure range from latitudinal gradients that influence biogeographic variation in prey distributions and antipredator defenses to local patterns in which vulnerable prey can be restricted by predators to living under cobbles, in crevices, or buried in sediment. The most conspicuous large-scale biogeographic pattern is an increase in consumer pressure with decreasing latitude (Vermeij 1978). In mild climates with little seasonality, consumers have low physiological maintenance costs and are able to specialize and influence their prey much more than in harsh seasonal climates (Vermeij 1987). On Atlantic shorelines south of Cape Cod, for example, the blue crab, *Callinectes*, plays a major role as a predator of infaunal organisms on muddy and sandy substrates, limiting bivalves and soft-bodied prey to refuges. North of Cape Cod, *Callinectes* is absent, and no other large, fast-moving crab predator replaces it. As a consequence, soft substrates in the Gulf of Maine are characterized by dense assemblages of bivalves and soft-bodied organisms that south of Cape Cod are restricted to refuge habitats (Woodin 1976). Southern shores are also characterized by shell-crushing (Palmer 1979; Bertness et al. 1981) and herbivorous (Randell 1967) fishes that are largely restricted to tropical and semitropical habitats. As a result of these latitudinal gradients in consumer pressure, tropical and semitropical molluscs and algae tend to be more restricted to refuge habitats, and better chemically and structurally defended, than their temperate zone counterparts (Vermeij 1978, 1987).

At smaller spatial scales, environmental variables commonly generate variation in consumer pressure. As already noted, consumers are less effective in wave-swept than in sheltered habitats, since they have reduced mobility and foraging efficiency in fast-moving water, or are simply excluded (Menge 1978; Weissburg and Zimmer-Faust 1993). Most consumers are also less effective in shallow than in deeper water because of the increased risk of aerial exposure and decreased mobility (Dayton 1971; Kneib 1997). Dense seaweed canopies, seagrass beds, and marsh grasses also limit consumer mobility (Vince et al. 1976; Peterson 1982a), but may also provide refuges for consumers (Menge 1978). As a consequence of these smaller-scale spatial patterns in consumer pressure, wave-swept open coast habitats are often refuges for algae and poorly defended shoreline invertebrates (Menge 1976). High intertidal habitats and highly structured seaweed canopy, marsh grass, and seagrass habitats may also serve as refuge areas for vulnerable prey.

Chemical Defenses

Chemical defenses are particularly common in sessile shoreline organisms that lack substantial morphological defenses or the ability to flee from consumers. Examples of sessile organisms that rely heavily on chemicals to deter consumers include seaweeds, sea squirts, sponges, and sea cucumbers.

Algal defenses against herbivory have been particularly well studied (for excellent reviews see Hay and Fenical 1988; Hay 1996). Generalist grazers such as periwinkles, urchins, and fishes are often capable of eliminating all undefended algae in shallow-water habitats (Randall 1960; Paine and Vadas 1969). Algal chemical defenses are thought to have evolved in response to these large, generalist consumers rather than to smaller snail, amphipod, and nudibranch grazers. [🐚 38] These smaller, more specialized herbivores often prefer living in association with chemically defended seaweeds, are commonly immune to seaweed chemical defenses, and utilize seaweed cover as a defense against many of the same generalist consumers that eat seaweeds (Hay et al. 1987, 1989, Hay and Duffy 1990; Duffy 1990). Defensive chemicals that inhibit fish grazing have even been shown to stimulate amphipod recruitment and feeding, apparently as an evolutionary

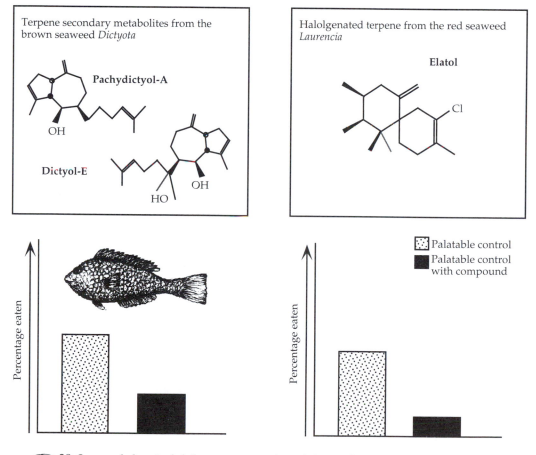

🐚 38 Seaweed chemical defense compounds and their influence on herbivore feeding. (After Hay 1996.)

response to reduced consumer pressure. Many shell-less snails (sea slugs) consume chemically defended seaweeds and utilize algal metabolites for their own defense (Pennings and Paul 1993).

The defensive chemicals of seaweeds are similar to those of terrestrial plants, except that seaweeds typically do not use nitrogen-based metabolites such as alkaloids (Hay 1996). Seaweeds are often nitrogen-limited, so that carbon-based or halogenated compounds are less costly to produce than nitrogen-based defenses (Yates and Peckol 1993). The phenolic, terpenoid, and halogenated secondary metabolites used by seaweeds for defense are typically packaged in cells near the plant surface, where they can provide a first line of defense in deterring consumers. Chemical defenses can also be concentrated in young tissue, which is structurally the least defended and the most valuable to protect (Van Alstyne 1988). Seaweed chemical defenses also vary seasonally and among populations. In New England, algal chemical defenses increase in the summer, when consumer pressure is the highest (Geiselman and McConnell 1981), and increase in response to herbivore damage (Van Alstyne 1988; Yates and Peckol 1993). Herbivore induction of defenses lets seaweeds avoid the potential costs of maintaining chemical defenses when herbivores are absent while giving them the advantages of chemical defenses when needed.

Sponges, tunicates, and sea cucumbers are among the invertebrates that rely heavily on chemical defenses. The bright colors of many of these sessile species probably advertise that they are noxious to potential consumers. Like seaweeds, warm-water sessile invertebrates are better chemically defended than their temperate relatives (Green 1977). Defensive chemicals are usually stored near surfaces where they are best positioned to deter consumers. Sea cucumbers, for example, have body wall vesicles that contain secondary metabolites or sulfuric acid that repel would-be predators. Some warm-water cucumbers have Cuvierian tubules, which can be shot out the anus at predators when the animal is disturbed. Cuvierian tubules are covered with a sticky coating of holopone, a defensive compound that sticks to fish gills and interferes with oxygen uptake (Barnes 1987).[✥39] Herbivore induction of chemical defenses is common in clonal invertebrates that can suffer partial predation losses and benefit from turning on chemical defenses in response to grazing pressure. In contrast, consumer induction of defenses is less valuable for organisms that are eaten whole, unless they can turn on chemical defenses in response to the presence of their enemies before being attacked (see Leonard et al. 1998b for an example).

Structural Defenses

Structural defenses that protect prey from their enemies include armor, defensive architecture, large body size, and weapons that sting or nip consumers. Armor that protects organisms from consumers is common in marine taxa. Tough cuticles limit the access of small grazers such as amphipods and snails to the unprotected interior cells of many large fleshy seaweeds (Gaines 1985). Heavily calcified coralline algae have the most

39 Some tropical sea cucumbers, such as species of *Holothuria*, have poisonous Cuvierian tubules, which they can eject from the anus if attacked by a predator.

extreme algal structural defenses. Encrusting coralline algae are immune to all but specialized grazers, such as some limpets, but at the cost of having extremely low growth rates (Steneck and Watling 1982).

The evolution of calcium carbonate armor in response to consumer pressure has been a pervasive feature in the evolution of many marine invertebrate groups (Vermeij 1978, 1987). Gary Vermeij has made a compelling case that the evolution of structural defenses in marine invertebrates accelerated in the Mesozoic, 65 million years ago, in response to the proliferation of predators. Primitive bryozoans, for example, had simple, interconnected calcium carbonate houses with exposed soft tissue. In response to the vulnerability of this early morphology to predators, most modern bryozoans evolved heavy calcium carbonate armor (Ryland 1970).[40] Similarly, in colonial hydroids, lineages exposed to heavy grazing, such as the fire corals,

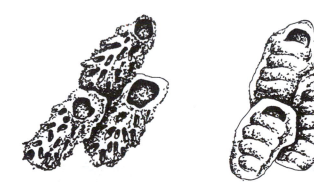

Anascan   *Ascophorans*   *Cribomorphs*

40 Primitive anascan bryozoans have fleshy prefrontal membranes, in contrast to ascophoran and cribomorph bryozoans, which have calcified or covered prefrontal membranes that are protected from predators.

evolved external calcium carbonate skeletons. Some mobile invertebrates also evolved body armors, at the cost of constrained mobility. Heavy-bodied crabs, lobsters, and molluscs are limited to relatively sedentary lifestyles in comparison to fast-moving crustaceans (such as shrimp) and molluscs (such as squid and octopus) that lack heavy body armor.

The architecture or design of calcified body armor has also clearly evolved to deter consumers. Richard Palmer (1982), for example, has argued that the reduction in number and fusion of shell plates to form a solid test in most modern barnacles has been in response to predation pressure. Drilling snails selectively attack barnacles at suture lines between plates, and in most taxa there has been a steady fusion of shell plates over time. [🐚 41] Bryozoans are not only covered with calcium carbonate, but may also be ornamented with spines that protect them from the grazing of specialized predators such as nudibranchs and isopods. In some bryozoans, such as *Membranipora*, spines are absent in individuals from habitats without predators, but develop in response to the presence of predators (Harvell 1986, 1992).[🐚 42] Inducible structural defenses have also been found in snails (Appleton and Palmer 1988; Palmer 1991; Trussell 1996) and mussels (Leonard et al. 1998b), which thicken in response to predators. Some barnacles are also known to develop predator-resistant morphologies in the presence of predators (Lively 1986a,b).

Architectural defenses have been particularly well studied in snails (Vermeij 1978, 1987). Shell-crushing predators of snails include specialized

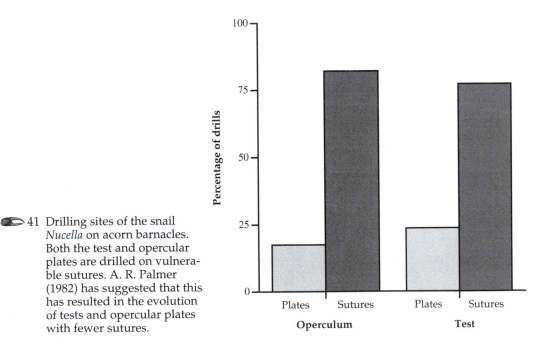

🐚 41 Drilling sites of the snail *Nucella* on acorn barnacles. Both the test and opercular plates are drilled on vulnerable sutures. A. R. Palmer (1982) has suggested that this has resulted in the evolution of tests and opercular plates with fewer sutures.

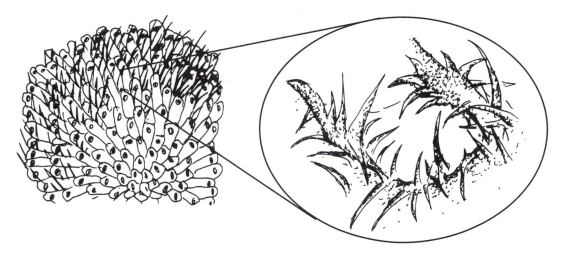

 42 The encrusting bryozoan *Membranipora* responds to nudibranch predators by producing protective spines on the colony edge. When predators are not present, spines are not produced, avoiding the cost of defense. (After Harvell 1984.)

bony fishes with heavy pharyngeal mills and large crustaceans (crabs and lobsters) that use claws or modified mouthparts to crush shells. Shell-crushing fishes, such as the spiny puffer *Diodon*, are largely restricted to warm waters, and the elaborate spines on many tropical snail shells are thought to limit the ability of these predators to get snails into their crushing mouthparts (Palmer 1979). [43]

Shell-crushing crabs usually use a modified and enlarged master claw to break shells open. The claws of crabs such as *Cancer, Callinectes*, and *Menippe* maximize mechanical advantage when applying force, and have teeth that can hold shells firmly while crushing or chipping away at the shell lip. Crab shell-crushing behavior is typically size-dependent (Vermeij 1978). Relatively small snails are crushed outright, while larger shells, too big for the crab to apply maximum force to the body whorl, are attacked either at the spire or shell lip. Tall, thin shell spires are nipped off, allowing the crab to chip away at the fractured shell and remove the soft tissue. Large shells can also be attacked by inserting a claw or claw tooth into the aperture and chipping back the shell lip. [44] The box crab *Calappa* has a master claw that is very effective at peeling back the shell lip of relatively large snails. The crab inserts the large tooth on the master claw into the shell aperture and, by repeatedly chipping the shell lip, peels back the shell aperture like a can opener.

Gastropods have evolved predator-resistant shell designs in response to these predatory tactics. Such predator-resistant traits include thick shells that limit breakage, reducing the shell spire to limit its vulnerability, occluding

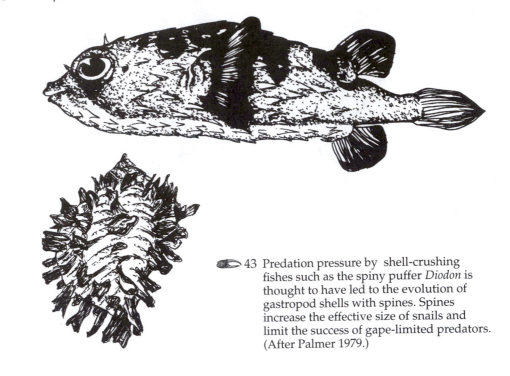

🐚 43 Predation pressure by shell-crushing fishes such as the spiny puffer *Diodon* is thought to have led to the evolution of gastropod shells with spines. Spines increase the effective size of snails and limit the success of gape-limited predators. (After Palmer 1979.)

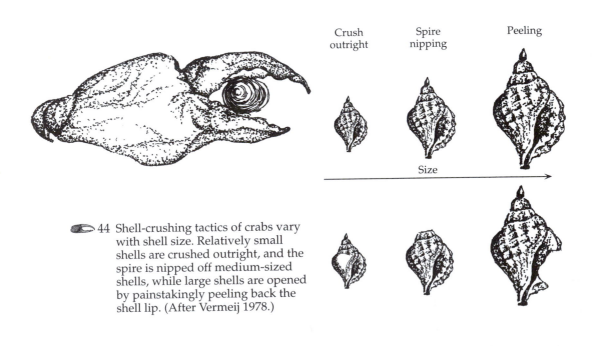

Crush outright   Spire nipping   Peeling

Size

🐚 44 Shell-crushing tactics of crabs vary with shell size. Relatively small shells are crushed outright, and the spire is nipped off medium-sized shells, while large shells are opened by painstakingly peeling back the shell lip. (After Vermeij 1978.)

the shell aperture by narrowing or adding teeth, and adding axial ribs to strengthen the shell lip and limit breakage (Vermeij 1978, 1987; Bertness and Cunningham 1981).[🐚 45]

Bivalve shell design is also influenced by shell-crushing predators. Predators of bivalves, such as the blue crab *Callinectes*, forage over soft bottoms, using tactile and chemical cues to detect their prey. Deep burial is an important spatial refuge for bivalves, since it increases their chances of not being detected and increases the costs of digging them out (Blundon and Kennedy 1982). Once a predator has located a bivalve victim, smaller clams are crushed outright, while larger clams are chipped at the edge until the soft tissue is exposed. Increased shell thickness, reinforcing ribs, inflated bodies, and tightly interlocking shell valves have all been suggested to limit crab predation on bivalves (Boulding 1984; Vermeij and Vail 1978; Vermeij 1978).

Other structural defenses are characteristic of specific invertebrate taxa. Cnidarians (anemones, hydroids, jellyfishes, and corals) have specialized stinging cells called **nematocysts** for capturing prey and antipredator defense. Nematocysts are small pressurized capsules that when tactually or chemically triggered launch a twisting harpoon, often armed with sharp

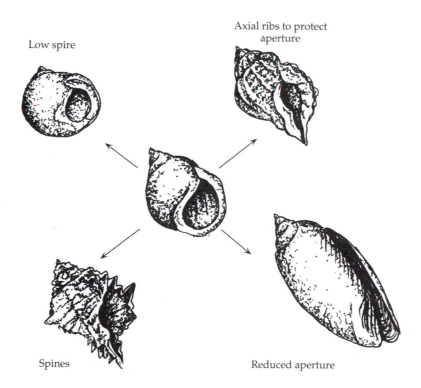

🐚 45 Morphological defenses of gastropods against shell-crushing predators.

teeth and noxious chemicals, at potential prey or enemies. While individual nematocysts are small, the tentacles of many cnidarians have very high densities of nematocysts. Nematocysts are used only once and then are rapidly replaced.[46]

Bryozoans and echinoderms also have unique structures that deter consumers. Encrusting bryozoans are vulnerable to generalized grazers such as urchins, starfishes, and snails, as well as much smaller, more specialized enemies such as nudibranchs and small crustaceans. They have specialized defensive structures called **avicularia** that are jawed and nip at potential consumers and larval settlers (Ryland 1970). Similarly, starfishes and sea urchins have pincer-jawed structures called **pedicellaria** that function in limiting larval fouling of their surfaces as well as antipredator defense. Some tropical urchins even have poison glands on their pedicellaria that, along with barbed spines, deter potential predators.

Behavioral Defenses

Behavioral defense mechanisms are also common among shoreline organisms. Many species have simple escape responses to their predators. When disturbed, filter-feeding tube worms, such as *Spirorbis*, rapidly retract into their tubes, fiddler crabs and rock crabs retreat to burrows and under rocks, and mud shrimp and killifishes rapidly move into vegetation cover or bury themselves in mud. In addition to these simple escape mechanisms, however, shoreline organisms also display a number of other behaviors that clearly minimize consumer effectiveness.

One of the most conspicuous general escape behaviors of shoreline organisms is the utilization of spatial refuges by juveniles. Typically, juve-

Cnidarian nematocyst, characteristic of anemones, hydroids, and jellyfishes

Avicularia, found in bryozoans

Pedicellaria, found in starfishes and sea urchins

 46 Antipredator defenses characteristic of specific groups of shoreline organisms.

niles are more vulnerable to consumers than adults, and live in predator-free space until they become large enough to minimize their vulnerability to being eaten. This strategy commonly leads to gradients in body size in intertidal populations, with smaller individuals occurring at high intertidal heights, where consumer pressure is less intense than at lower tidal heights. In salt marshes, juvenile fish, crabs, and shrimp heavily utilize high marsh elevations, where they are immune to large mobile predators, while their larger adult conspecifics are found at lower marsh elevations (Kneib 1982, 1985, 1997; Halpin, unpublished data). On rocky shores, juvenile periwinkles and other snails often live at higher tidal heights than larger adults, thereby avoiding their enemies (Vermeij 1972). In addition to shallow-water refuges, juveniles also use spatial refuges that are inaccessible to large mobile predators because of habitat complexity, such as marsh grass (Vince et al. 1976) and seagrass (Peterson 1982a; Thayer et al. 1979) cover in soft substrates and seaweed canopies and mussel beds on rocky shores.

Like lizards on land, many shoreline invertebrates also discard replaceable body parts when attacked by predators. This behavior, called **escape autotomy**, is found in a variety of crabs and brittle stars. The tropical rock crab *Eriphia squamata*, for example, will threaten predators with both claws extended, but if threatening does not work, will pinch the intruder and quickly flee, leaving the pinching claw attached to the smarting would-be predator. This behavior typically distracts the predator long enough for the crab (minus one limb) to escape. Fiddler and porcelain crabs have similar behaviors. Fluid loss is limited, since their limb joints have valves that seal when limbs are lost. Lost limbs are replaced, at considerable cost, when the crab molts; complete replacement typically takes two or three molts (Smith 1990). Brittle stars and many worms also lose feeding appendages or body parts to consumers and then regenerate them.

Other shoreline organisms combine cryptic behavior with coloration that blends with their backgrounds to avoid detection by predators. Some spider crabs have spines and hooks on the carapace that they use to attach seaweeds and sponges to their backs. In the presence of predators, decorator crabs will not move. These behaviors camouflage the crab from large visually hunting predators such as fish and birds.[🐚 47] Similarly, many small or juvenile crabs have color patterns and shapes that make them difficult to detect. Chip crabs, for instance, are the size, shape, and color of small sand grains, and on a sand background are almost impossible to detect.[🐚 48] Juvenile rock and fiddler crabs avoid consumers by freezing in the presence of predators. Juvenile fiddler crabs are translucent and difficult to detect, and rock crab juveniles color-match their background, as a result of either diet or pigmented chromatophores in the exoskeleton that can change color (A. Palma, personal communication). Other prominent examples of cryptic coloration and behavior in shoreline organisms include flounder and sculpins, which have camouflage coloration on sand and rock backgrounds, respectively, and the periwinkle *Littorina obtusata*, which lives on, and cryptically mimics, air bladders of the brown seaweed *Ascophyllum*.

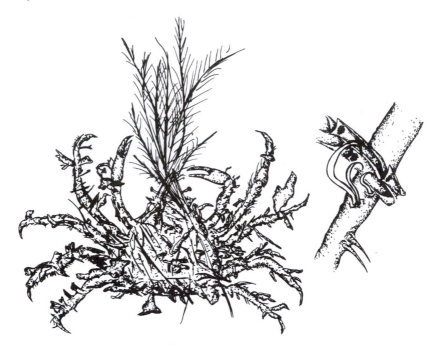

47 Decorator crabs are covered with spines and barbs, which they use to attach seaweeds and encrusting organisms to their shells for camouflage. (After Wicksten 1980.)

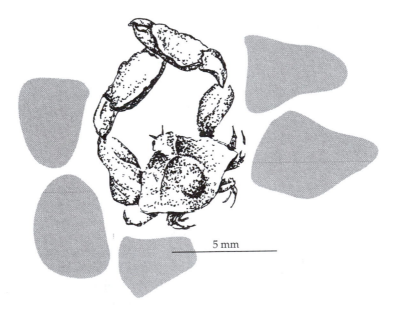

5 mm

48 Chip crabs (*Heterocrypta granulata*) are the size and shape of sand grains, making them cryptic on the sandy bottoms where they live.

Group Defenses

Group living is such a common behavioral defense against consumers that it is often overlooked. One of the more important consequences of dense aggregations or beds of mussels and oysters, for example, is the protection of juveniles from consumers. Juveniles of these species have a much wider variety of predators than adults, since they have not reached a size refuge from many of the smaller crab and fish predators to which adults are immune. By selectively settling in adult beds, juvenile mussels and oysters pay a high price, since they must compete with adults for food, but they benefit by being protected from their predators (Bertness and Grosholz 1985).

Mobile organisms that forage in groups often benefit by reducing their losses to predators (Hamilton 1971). Group members may be more responsive to the presence of enemies than solitary organisms, since group members can warn one another of danger, and predator search images of their prey can be confused by large groups of them. In coastal habitats on the east coast of North America, group foraging is particularly conspicuous in shallow-water fishes and shrimps, shorebirds, and fiddler crabs. At high tide over sandy and muddy habitats, large groups or schools of killifish, silversides, and grass shrimp typically forage together, apparently in response to predation pressure from large fish and wading birds. When approached by a predator, they move closer together, then quickly dart away in unison. This schooling behavior may also have energetic advantages, allowing individuals to reduce the costs of swimming by drafting other fish. The sand fiddler crab, *Uca pugilator*, provides probably the most dramatic example of antipredator group behavior on Atlantic shorelines. Sand fiddlers forage in groups of 25–100 individuals, moving together across sandy habitats. If approached by a potential predator, they rapidly close ranks and dart away in a dense coordinated swarm.

Many group defenses involve more than one species and are often the simple product of vulnerable organisms living in association with well-defended organisms and obtaining protection from the association. These **associational defenses** (Atsatt and O'Dowd 1976; Hay 1986) can be opportunistic or facultative, but may also be highly evolved mutualistic interactions. They may result from larval settlement patterns or habitat choice behavior, or may be the simple product of differential survival of propagules without a guiding behavioral component.

Mark Hay (Hay 1986; Hay et al. 1989; Hay and Fenical 1988) and his colleagues have examined the dynamics of associational defenses among algal species, and have shown that palatable seaweeds (such as *Ulva* and *Enteromorpha*) experience reduced grazer pressure when they live in association with chemically or structurally defended algae, as a simple consequence of grazers avoiding well-defended species. In New England, this leads to ephemeral green algae persisting epiphytically on *Ascophyllum* plants, but not on open surfaces. Similarly, in the Carolinas, palatable ephemeral seaweeds are often found exclusively on the chemically and structurally defended *Sargassum*. Some seaweeds that are poorly defended against their enemies specialize in living epiphytically on grazer-resistant plants. *Polysiphonia lanosa*, for example, is exclusively found epiphytically on *Ascophyllum* on New England rocky shores; this may be an example of an evolved defensive association.

Associational defenses are not restricted to algal associations, but are common in a wide variety of other shoreline habitats where vulnerable prey find refuge living near less preferred prey. Aaron Ellison has shown that solitary marsh plants, for example, may hide from their herbivores in dense stands of the unpalatable clonal turf plants that dominate many salt marshes (Ellison 1987b). Similarly, Skilleter (1994) has shown that dense beds of unpalatable clams can serve as a refuge for palatable soft-shelled clams, since crabs avoid digging in habitats dominated by low-preference prey items. Most recently, Sarah Woodin (personal communication) has found that in soft sediments, chemically defended worms reduce general foraging in habitats they dominate and thus reduce predation on their palatable worm and brittle star neighbors.[🦪 49]

SUMMARY

In this chapter I have examined the economy—the balance between energy production and consumption—of shoreline habitats. Shallow-water marine habitats are extremely productive due to high nutrient input to coastal habitats, which provides ideal growing conditions for plants. Two major groups of primary producers fuel littoral communities: seaweeds and vascular plants attached to the substrate and microscopic phytoplankton that live in the water column. The primary production of these groups is largely dictated by nutrient and light availability. In turn, spatial and temporal variation in the primary productivity of these plants, particularly at large spatial scales, plays a major bottom-up role in dictating spatial and temporal variation in shoreline communities.

Shoreline communities, however, are not simply a reflection of primary production patterns, but are also influenced by the actions of consumers, or top-down forces. Consumers can limit the distribution of prey organisms, as well as playing a major role in the evolution of prey defenses. Thus, the distributions and abundances of organisms, as well as the dynamics of shallow-water marine assemblages, are the product of these two opposing forces.

Understanding the economy of shoreline communities also requires an understanding of what consumers eat and why, as well as the tactics used by prey to avoid being eaten. Shoreline consumers have evolved to tap a wide range of food sources, and optimal foraging theory examines what determines their diets. In response to consumers, prey have evolved a range of behavioral, chemical, and morphological defenses.

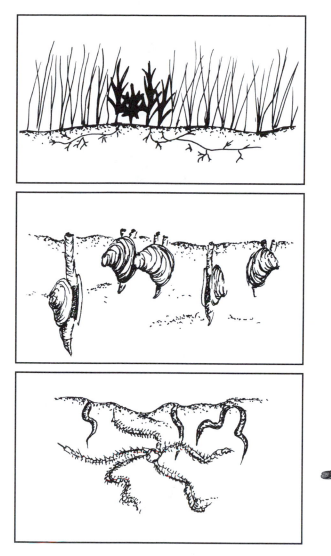

49 Associational defenses of marsh plants, burrowing bivalves, and soft-bodied worms, in which palatable prey are protected from consumers by living with undesirable prey species.

SUGGESTIONS FOR FURTHER READING

Carpenter, S. R. and J. H. Kitchell. 1993. *Trophic Cascades in Lakes*. Cambridge University Press, Cambridge.

Hay, M. and W. Fenical. 1988. Marine plant-herbivore interactions: The ecology of chemical defense. *Annual Review of Ecology and Systematics* 19: 111–145. An insightful overview of the chemical defense literature.

Hughes, R. N. 1980. Optimal foraging in a marine context. *Oceanography and Marine Biology Annual Review* 18: 423–481.

Mann, K. H. and J. R. N. Lazier. *Dynamics of Marine Ecosystems*. Blackwell Scientific, Boston. A thorough treatment of production dynamics in nearshore waters.

Power, M. 1992. Top-down and bottom-up forces on foodwebs: Do plants have primacy? *Ecology* 73: 733–746.

Strong, D. 1992. Are trophic cascades all wet? Differentiation and donor control in speciose ecosystems. *Ecology* 73: 747–756.

Valiela, I. 1995. *Marine Ecological Processes*. Springer-Verlag, New York.

Vermeij, G. J. 1978. *Biogeography and Adaptation: Patterns of Marine Life*. Harvard University Press, Cambridge, MA. A delightful examination of latitudinal variation in consumer pressure and prey defenses in marine organisms.

Chapter *3*

Reproduction and Recruitment of Shoreline Organisms

*S*uccessful reproduction has long been recognized as a potential limiting factor for the growth of marine populations (Thorson 1950). Slow-moving and permanently attached (**sessile**) marine organisms face a difficult and unusual set of problems reproducing in shoreline habitats. Fertilization often occurs after gametes are released into the water column, and the resulting dilution of gametes can reduce the sperm-egg contact rate drastically. Even if fertilization occurs, reproductive problems continue for most shoreline organisms. Over 80 percent of marine organisms have larvae that are released into the water column, where they develop for periods ranging from hours to months. Larval development in the water column, or **planktonic development**, may be a mechanism that permits juveniles to escape physically and biologically harsh benthic habitats (Strathmann 1993). Whatever its origin, planktonic development has important consequences for shoreline populations and communities.

Development in the water column subjects larvae to highly risky and uncertain conditions, and few survive. Larvae can be eaten, or starve, or be transported by currents to unsuitable habitats. Even if the larvae of shoreline organisms survive their trek in the plankton, they still face the challenge of reaching a suitable adult habitat and settling there. A critical repercussion of planktonic larval development is that local population success, measured in terms of individual growth and reproductive output, is often decoupled from **recruitment**, the successful establishment of new individuals in populations. This decoupling contrasts with the closed populations typical of terrestrial habitats, in which local population success is tightly linked to recruitment and density-dependent population processes. In the open populations of most marine organisms, gamete and larval transport processes often dictate the spatial structure of populations and become critical in understanding the dynamics of marine populations and communities.

This chapter examines the reproductive ecology and recruitment of shoreline organisms. I start by introducing problems associated with the fertilization and dispersal of marine organisms. I then examine factors that influence recruitment variation, and end by discussing some of the consequences that variable recruitment has for process and pattern in shoreline populations and communities.

FERTILIZATION

External fertilization is common in marine organisms, particularly seaweeds, sea urchins, starfishes, bivalves, and some snails.[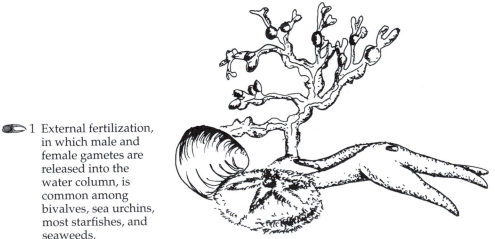1] Male and female gametes are released into the water column, where fusion and subsequent development occur. External fertilization requires a large gamete out-

1 External fertilization, in which male and female gametes are released into the water column, is common among bivalves, sea urchins, most starfishes, and seaweeds.

put. A spawning sea urchin, for example, releases up to 20,000,000 eggs or sperm during a single reproductive season. Even with this enormous gamete output, however, successful external fertilization is usually dependent on the synchronized release of gametes and the aggregation of spawning adults. In other littoral organisms, including tunicates, most sponges, and bryozoans, fertilization typically occurs internally. Sperm is released into the water column, where it is filtered out of the water by other members of the species (**conspecifics**).[🫘2] This form of fertilization with water-borne sperm is also inefficient and depends on substantial gamete production.

External fertilization is not only the most common fertilization mechanism among sessile marine organisms, but is unique to aquatic organisms and primitive plants such as mosses and ferns (Strathmann 1990). Until recently, the large gamete output of broadcast spawners, combined with mechanisms to synchronize gamete release, was believed to result in high fertilization success. In the last decade, however, it has been realized that broadcast spawning is an extremely inefficient process that may often limit successful fertilization (Pennington 1985; Levitan et al. 1992; Yund 1990). Once released, gametes remain viable for only a short time, ranging from 30 minutes to a few hours (Thorson 1950), and they are rapidly diluted.[🫘3] The dilution of gametes means that external fertilization may be effective only at distances between individuals of less than a few meters (see Levitan 1995).

The limited effectiveness of external fertilization has important consequences for sessile organisms. Local flow conditions, for example, can strongly affect fertilization success. Directional flow can impose strong spatial patterns on the fertilization success and gene flow of sessile organisms (Yund 1990), and rapid water movement can accelerate the dilution of

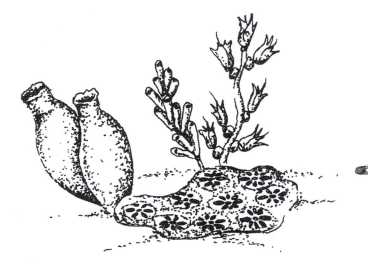

🫘2 Internal fertilization, in which sperm released into the water column is captured by females, is common among filter-feeding tunicates, sponges, and bryozoans.

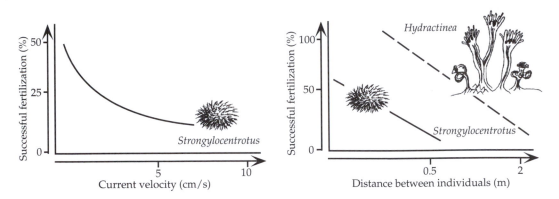

3 The efficiency of external fertilization is strongly inhibited by water movement and by distance between individuals, both of which rapidly dilute gamete concentrations and reduce fertilization opportunities. (Data from Levitan 1991; Yund 1990; Levitan et al. 1992.)

gametes and severely limit the effectiveness of external fertilization in high-flow habitats. Even relatively slow currents of 5 centimeters per second can markedly reduce fertilization success (Levitan et al. 1992). On wave-exposed shorelines, in particular, gamete dilution by wave turbulence can virtually preclude successful external fertilization (Denny and Shibata 1989).

This recent recognition of the limitations of external fertilization has called into question long-held assumptions about the population dynamics of sessile organisms with external reproduction. It was once widely believed that large body size was advantageous for sessile and slow-moving organisms (Paine 1974). Since gamete output typically increases exponentially with body size in invertebrates, large individuals were assumed to be the most successful reproductively. Realization of the limits of external fertilization, however, has suggested that low population densities, which often lead to large body sizes, may nonetheless limit the reproductive success of organisms with external fertilization. Inhibition of fertilization success at low population densities is referred to as the **Allee effect**, in recognition of W. C. Allee, who initially described the phenomenon. As a consequence of the Allee effect, the reproductive success of dense populations of small, stunted individuals with low gamete production may typically exceed that of sparse populations of large individuals with high gamete production, simply because at high population densities, the problem of gamete dilution is minimized (Levitan et al. 1992).[⬤4]

Marine organisms with external fertilization or gamete dispersal use a variety of mechanisms to enhance their reproductive success. Slow-moving external fertilizers such as sea urchins, starfishes, some segmented worms, and snails typically exhibit reproductive aggregations or swarming behavior

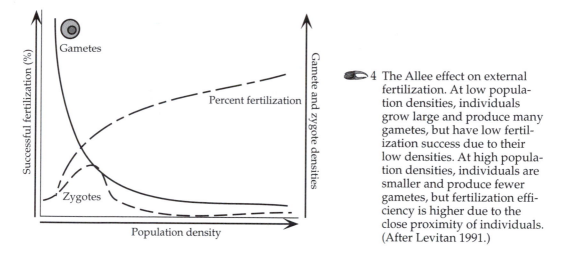

4 The Allee effect on external fertilization. At low population densities, individuals grow large and produce many gametes, but have low fertilization success due to their low densities. At high population densities, individuals are smaller and produce fewer gametes, but fertilization efficiency is higher due to the close proximity of individuals. (After Levitan 1991.)

(Thorson 1950) during the reproductive season.[5] Gamete release synchronized to tidal and lunar cues, as well as the release of gametes when other gametes are detected in the water column (Thorson 1950), should also increase the probability of successful external fertilization. The efficacy of these mechanisms, however, is not well studied, and is largely unknown.

Direct internal fertilization with copulation is certainly the best alternative to the vagaries of external fertilization. Internal fertilization clearly enhances fertilization success. It is practiced by most mobile marine organisms, including snails and all crustaceans (e.g., crabs and shrimp). Predaceous snails such as the dog whelk *Nucella lapillus* and the oyster drill *Urosalpinx cinera* form annual spring breeding aggregations, in which copulation occurs and eggs are laid. The slipper limpet, *Crepidula fornicata*, has a particularly interesting reproductive lifestyle. It is a sequential protandrous (male first) hermaphrodite (an organism that functions as both a male and a female) with direct internal fertilization and socially controlled sexual expression.[6] Juvenile *Crepidula* settle from the plankton as small, mobile males, but as they grow, they lose their mobility and join stacks of conspecifics, in which only the largest and bottom-most individual is a female. In a *Crepidula* stack, the female is inseminated by the next largest individual, which is male. When the female dies, the largest male changes sex to female, and, in turn, is inseminated by the largest remaining male.

Direct internal fertilization with copulation is uncommon among sessile organisms, as mate-finding movement is impossible. Acorn barnacles, however, are an exception. Acorn barnacles solve the problem of finding mates while sessile by gregarious settlement with conspecifics and the possession of a penis often more than five times longer than the body. Copulation in acorn barnacles reflects their descent from mobile shrimplike crustacean ancestors.

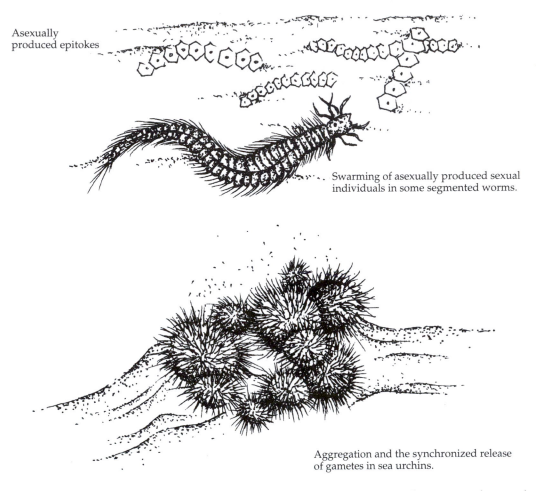

Asexually
produced epitokes

Swarming of asexually produced sexual
individuals in some segmented worms.

Aggregation and the synchronized release
of gametes in sea urchins.

5 Mechanisms used by slow-moving invertebrates to increase the success of external
fertilization.

Copulation is typically accompanied by elaborate courtship behavior
in many crustaceans. In the fiddler crab (*Uca* spp.), common in western
Atlantic salt marshes, males have one enlarged claw that is used exclusively for courtship displays and aggressive displays to male competitors
(Crane 1975). The enlarged claw of male crabs is useless for excavating
burrows or feeding, as described in the last chapter. It is assumed to be the
product of sexual selection, since except for its value in reproduction, it is
quite costly, leaving males at a disadvantage to female crabs.[7] Once
a male fiddler crab has successfully impressed a female with his displays,

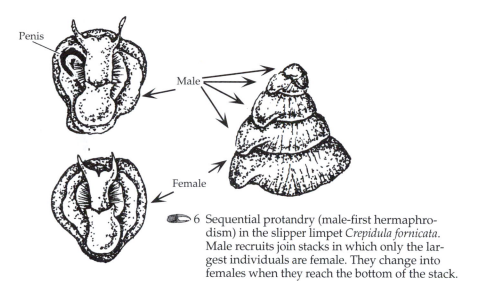

🐚 6 Sequential protandry (male-first hermaphro-
dism) in the slipper limpet *Crepidula fornicata*.
Male recruits join stacks in which only the largest individuals are female. They change into
females when they reach the bottom of the stack.

they copulate, and the female carries the embryos until they are released as
swimming larvae.

Many other crabs also have extended mating encounters preceding copulation, since typically female crabs can be impregnated only immediately
after molting, before the carapace and genital pores harden. Males may carry
or attach themselves to a female crab for days to weeks prior to her molting.
This behavior is found in many true crabs, including the northern rock crab,
Cancer borealis, and green crab, *Carcinus maenus*, in New England and the
blue crab, *Callinectes sapidus*, south of Cape Cod, as well as in the horseshoe
crab, *Limulus*.[🐚 8]

TYPES OF DEVELOPMENT

Like fertilization, the early development of marine organisms is a period of
high uncertainty and risk. As we have seen, over 80 percent of marine
organisms have larvae that develop in the plankton. Because these larvae
tend to be small in size and weak swimmers, their movement is largely controlled by prevailing oceanographic conditions, leaving them vulnerable to
both planktonic and benthic consumers.[🐚 9] Because the mortality of
planktonic larvae is typically more than 99 percent, intense selection on life
history patterns has led to a relatively small number of other reproductive
strategies (Thorson 1950; Vance 1973).

One solution to the high mortality and uncertainty of planktonic larval
development is to omit it entirely and undergo direct development in the
parental habitat. Adults can brood eggs through development into self-suffi-

Female fiddler crab

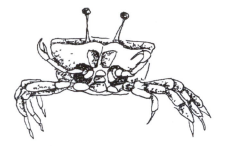

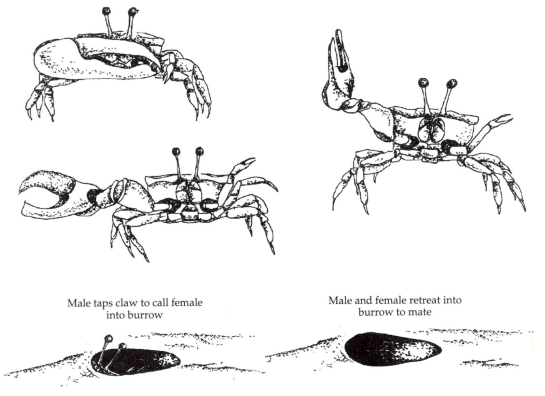

🐚 7 Generalized mating behavior of fiddler crabs.

Male fiddler crab waving claw

Male taps claw to call female
into burrow

Male and female retreat into
burrow to mate

cient, mobile juveniles that remain near their parent's habitat, as do many small crustaceans and brittle stars.[🐚 10] Some snails attach tough, chitinous egg capsules to the substrate, in which the eggs are protected as they develop into juveniles.[🐚 11] A small number of shoreline invertebrates, including the mud snail *Ilyanassa*, have mixed development, in which benthic eggs hatch into a relatively short-lived larval phase in the water column.

Copulation in the horseshoe crab, *Limulus*

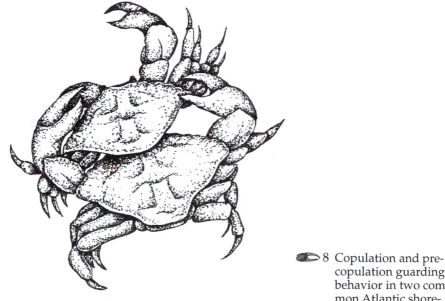

Pre-copulation guarding of mates in the rock
crab, *Cancer*

🐚 8 Copulation and pre-
copulation guarding
behavior in two com-
mon Atlantic shore-
line crustaceans.

This relatively uncommon life history strategy appears to provide the benefits of safe development in egg capsules with the dispersal advantages of
pelagic planktonic development (Pechenik 1979).

The major advantage of direct development is that it reduces mortality
in the plankton to zero and lowers benthic mortality. Egg capsules buffer
developing larvae from environmental extremes of heat, desiccation, and
osmotic stress (Pechenik 1979), and are often supplied not only with enough

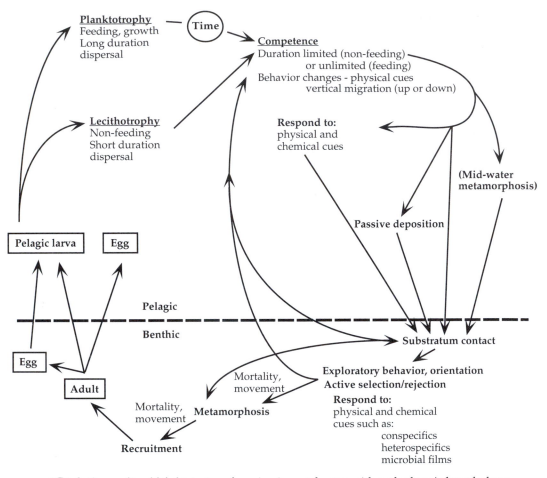

Planktotrophy
Feeding, growth
Long duration
dispersal

Time

Competence
Duration limited (non-feeding)
or unlimited (feeding)
Behavior changes - physical cues
vertical migration (up or down)

Lecithotrophy
Non-feeding
Short duration
dispersal

Respond to:
physical and
chemical cues

(Mid-water
metamorphosis)

Passive deposition

Pelagic larva   Egg

Pelagic

Benthic

Substratum contact

Egg

Adult

Exploratory behavior, orientation
Active selection/rejection
Respond to:
physical and chemical
cues such as:
        conspecifics
        heterospecifics
        microbial films

Mortality,
movement

Mortality,
movement   **Metamorphosis**

Recruitment

🦀 9 Generalized life histories of marine invertebrates with a planktonic larval phase.
(After Pawlik 1992.)

yolk to complete development, but also with unfertilized **nurse eggs** for
juveniles to consume as they emerge from their capsules (Rivest 1986). The
major disadvantage of direct development is the high parental investment
required. Mothers not only need to provide enough nourishment for the larvae to complete their development, but may also incur maintenance costs if
they carry or tend the developing eggs, or construction costs if they provide
egg capsules for the developing zygotes. As a consequence of these high
costs, shoreline organisms with direct development typically have few offspring, usually producing fewer than 100 eggs per season. In contrast,
planktonic dispersers with small eggs, such as many bivalves and segmented worms, can produce millions of eggs per season.[🦀12]
    There is also a dichotomy in the development of planktonic larvae.
Some species produce feeding, or **planktotrophic**, larvae, while others pro-

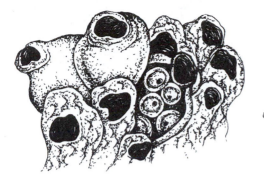

🐚 10 Brooding ovicells in the bryozoan *Drepanophora*. Ovicells are specialized individuals that store developing eggs. The ovicell in the center has had its prefrontal membrane removed to expose the fertilized eggs.

duce nonfeeding larvae that are nourished by parentally provided yolk, or **lecithotrophic** larvae. Since planktotrophic larvae utilize energy obtained while feeding in the plankton, they are cheap to produce, and since they are small, a large number can be produced by a single female. These low production costs, however, are balanced by the dependence of planktotrophic development on planktonic food supplies and exposure to planktonic mortality. Planktotrophic larval development is extremely common in marine organisms. Thorson (1950) estimated that more than 70 percent of benthic marine invertebrates have a feeding planktonic stage.

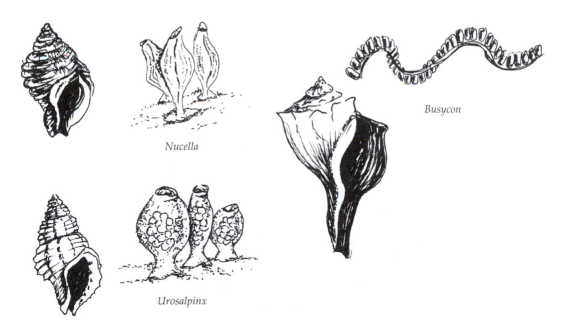

Busycon

Nucella

Urosalpinx

🐚 11 Egg capsules of common Atlantic shoreline snails.

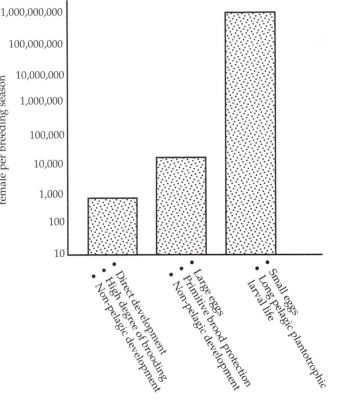

🐚 12 Variation in egg output among shoreline animals with different reproductive strategies. (After Thorson 1950.)

In contrast, larvae with lecithotrophic development do not feed in the plankton, and instead utilize yolk or parental resources to complete their development.[🐚 13] Lecithotrophic development is thus independent of planktonic food supply, but still risks exposure to mortality in the plankton. Since they must be supplied with enough energy to enable them to complete their development, lecithotrophic larvae are relatively large and costly to produce.

The prevalence of these divergent development patterns varies with latitude and organism size. Planktonic development is most common at lower latitudes, whereas direct development, including brooding and **vivipary** (bearing live young), increases toward the poles.[🐚 14] Short growing seasons and long development times in colder waters are thought to limit the effectiveness of planktonic development at high latitudes (Thorson 1950; Vance 1973). Mode of development is also strongly coupled to organism size. In a wide variety of marine taxa, including starfishes, bivalves, brittle stars, and snails, small adult body size is associated with brooding. Numerous explanations have been proposed for this pattern (see Strathmann and Strathmann 1982). Decreased brooding in larger species may simply reflect

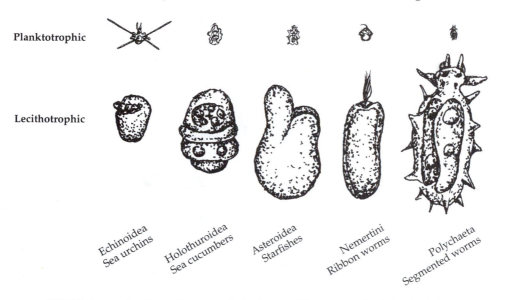

Planktotrophic

Lecithotrophic

Echinoidea
Sea urchins

Holothuroidea
Sea cucumbers

Asteroidea
Starfishes

Nemertini
Ribbon worms

Polychaeta
Segmented worms

🐚 13 Examples from five taxa of planktotrophic larvae, which feed while in the plankton, and closely related species with lecithotrophic larvae, which are nourished by maternal yolk while in the plankton. The larvae are drawn to scale to illustrate the typical size difference between the two types of larvae.

design limitations: organisms with a large egg output may simply not have sufficient space to brood their eggs. This would be true particularly if egg output increased with body volume, but brood capacity increased with some linear body dimension.[🐚 15] Alternatively, if planktonic development has a high mortality risk, small organisms may simply not become sufficiently large to produce enough larvae to make broadcast spawning a viable tactic (Menge 1975).

These larval development patterns have important ecological and evolutionary consequences for shoreline populations and communities. Because of the vagaries of life in the water column, organisms with planktonic development have long been known to have more fluctuating population densities than organisms with direct development or short-distance dispersal (Thorson 1950).[🐚 16] In addition, organisms that brood their young, have direct development, or disperse only short distances also typically have more clumped distributions than organisms with pelagic planktonic development. The surf clam *Gemma gemma*, for example, broods its young, and has a much more clumped distribution than quahogs (*Mercenaria*) or soft-shelled clams (*Mya*), which have planktonic larval stages. The most powerful ecological consequence of planktonic development, however, is that it decouples adult and juvenile success. The implications of this decoupling will be discussed at the end of the chapter.

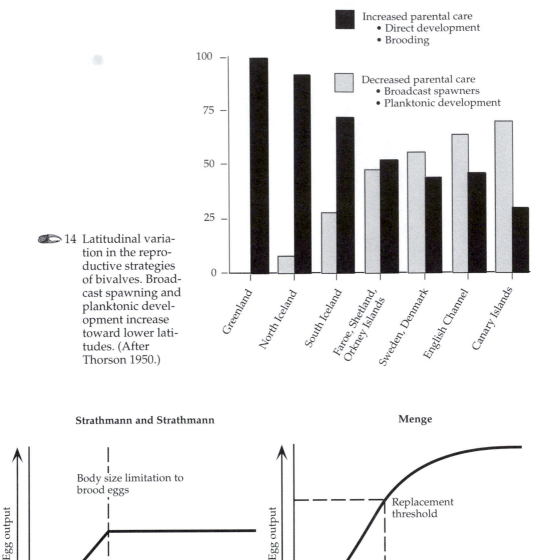

14 Latitudinal variation in the reproductive strategies of bivalves. Broadcast spawning and planktonic development increase toward lower latitudes. (After Thorson 1950.)

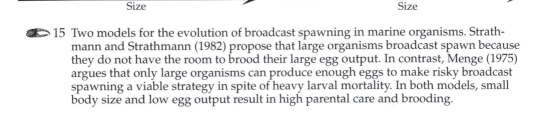

15 Two models for the evolution of broadcast spawning in marine organisms. Strathmann and Strathmann (1982) propose that large organisms broadcast spawn because they do not have the room to brood their large egg output. In contrast, Menge (1975) argues that only large organisms can produce enough eggs to make risky broadcast spawning a viable strategy in spite of heavy larval mortality. In both models, small body size and low egg output result in high parental care and brooding.

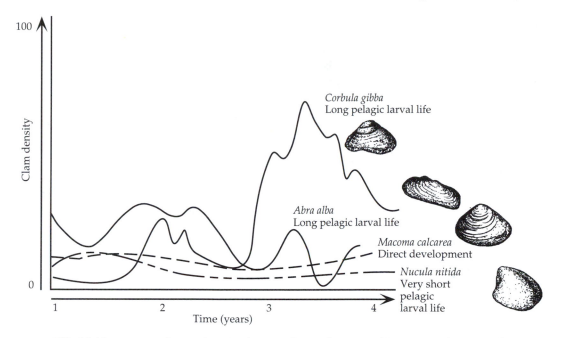

16 Year-to-year fluctuations in the population densities of European clams as a function of their reproductive strategies. Broadcast spawners have much more variable population sizes than clams with direct development or short larval life spans. (After Thorson 1950.)

Patterns of larval development can also strongly affect the genetic structure of populations and their ability to respond to local selection pressures. The dog whelk, *Nucella lapillus*, for example, a direct-developing snail with limited mobility (and thus restricted gene flow), has adapted to local thermal conditions with color morphs with different heating characteristics (Etter 1988a), and has responded historically to the introduction of a major predator, the green crab *Carcinus*, with shell thickening. In contrast, the periwinkle, *Littorina littorea*, which has planktonic development, shows little variation in shell morphology or color, and did not respond morphologically to the arrival of green crabs (Vermeij 1982a).

Mode of development can also play an important role in the **macroevolution,** or long-term evolutionary history, of marine species. The fossil record of shallow-water organisms in general shows that they are relatively immune to catastrophic extinctions, but do not experience sufficient geographic isolation to have high speciation rates (Jackson 1974). The wide environmental tolerances and low speciation rates of most shallow-water invertebrates are probably the consequence of extensive larval dispersal leading to large, resilient populations that are broadly adapted to local conditions and buffered by large numbers from local extinction (see Jackson 1974; Hansen 1978, 1980; Jablonski 1986 for discussion).

PROBLEMS FACING PLANKTONIC LARVAE

Planktonic development offers several advantages, including widespread dispersal, escape from benthic mortality while small, and the exploitation of different resources by young than by benthic adults. But planktonic larvae are also likely to suffer extremely high mortality. Gunner Thorson's seminal work, which marked the beginning of marine larval ecology (Thorson 1950), was based on trying to explain the vast "wastage" of marine invertebrate larvae. Thorson observed that marine invertebrates tend either to produce a small number of large eggs and provide high levels of parental care, or to produce an extremely large number of small eggs that undergo planktonic development. He reasoned that the advantages of planktonic development are offset by high larval mortality in the plankton, which necessitates production of large numbers of larvae. Thus, he suggested, the "wastage" of larvae is needed to ensure successful reproduction under the constraint of a risky passage through the planktonic larval stage. Pelagic larvae face three major sources of mortality: (1) finding enough to eat (food limitation), (2) unpredictable transport by currents, and (3) being eaten by planktivorous consumers (predator limitation).

Larval Food Limitation

Food supply has long been thought to be an important component of larval success, but this idea has been difficult to document or test. Planktotrophic larvae that feed in the water column are small and have high metabolic rates and energy demands, leading to speculation that food could limit their development. However, few compelling assessments of larval food limitation are available. Most studies to date suggest that marine larvae may be food-limited, but do not address the critical question of whether or not larval food limitation can affect larval mortality and recruitment (see Olson and Olson 1989 for discussion).

Low food supply can affect larval mortality directly by causing starvation and death, or indirectly by prolonging development, thus subjecting larvae to longer exposure to predators or transportation away from suitable adult habitat. Many laboratory studies of larval growth and development have suggested that natural food supply levels may limit larval development in the sea. Starved, emaciated larvae are not commonly seen in the field, however, so the extrapolation of these laboratory conclusions to field populations is questionable. Additionally, laboratory rearing studies have typically used artificial diets, so the disparity between field and laboratory studies could be due to a poor understanding of natural larval diets.

Depending on their feeding apparatus and size, marine larvae may feed on phytoplankton, detritus, bacteria, zooplankton, and/or dissolved organic material (**DOM**: see Manahan 1990). Olson and Olson (1989) have argued that different larval types rely differently on these sources of nutrition, and that the trophic status of a larva may influence the potential importance of food limitation. Larvae at higher trophic levels are thought to be more vulnerable to food limitation than those at lower trophic levels. In particular, they have suggested that omnivorous crustacean larvae (those of crabs, shrimp, and other marine organisms with jointed appendages), which feed on both phytoplankton and zooplankton, are often food-limited, and that starvation leading to low recruitment may often occur in these larvae. In contrast, molluscan larvae (those of clams, snails, and squids), which typically rely more on phytoplankton, are less likely to starve, and echinoderm larvae (those of starfishes, urchins, and brittle stars), which rely on both phytoplankton and DOM, are relatively immune to starvation because of their ability to tap into the more abundant food resources at the base of food webs. Thus, the importance of larval food limitation may vary predictably among major taxonomic groups. While crustaceans may be very sensitive to starvation, dying in only a few days without particulate food, some echinoderms are able to successfully complete their development entirely fueled by absorbing dissolved organic material from the water (see Olson and Olson 1989).[🐚 17]

Data supporting the importance of food limitation in larval development come from correlative field studies and laboratory rearing experiments. Field sampling programs have found that larval development times are shorter and that larvae grow larger in nutrient-rich nearshore waters than in nutrient-poor offshore waters (Huntley and Boyd 1984; Gaines and Bertness 1992). The results of laboratory rearing experiments, however, are often difficult to assess, since artificial diets are used, leaving in question

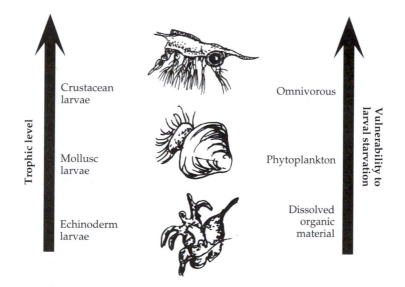

🐚 17 The Olson and Olson (1989) hypothesis of the role of larval food supply in larval mortality. Larvae at higher trophic levels are proposed to be at increased risk of starvation.

whether or not a complete diet was available. Studies using enriched or diluted natural seawater have typically found that natural food supplies can limit larval growth and lengthen development times. Adding phytoplankton to the natural diets of crustacean and mollusc larvae usually results in increased growth (Olson and Olson 1989). In one of the most compelling laboratory investigations of dietary effects on larval development, Paulay et al. (1985) reared a range of invertebrate larvae in natural seawater, filtered seawater (reduced food), and enriched seawater (natural seawater to which seawater filtrate was added). While not all larval forms responded identically, this realistic laboratory protocol showed that food concentrations in natural seawater can limit both the development times and sizes of invertebrate larvae.[⬤▷18]

Larval Transport

Transport of larvae away from parental habitats has long been thought to lead to considerable larval loss and mortality in marine organisms (Thorson 1950). Since larval swimming speeds are typically two to three orders of magnitude less than current velocities (Butman 1992), larval movement can be entirely at the mercy of currents. As a consequence, larvae of shoreline organisms can be transported by nearshore currents for hundreds of kilometers or more. An example of the role of nearshore hydrodynamics in the loss of larvae is the effect of upwelling on barnacle recruitment along the central California coast.[⬤▷19] In years with heavy upwelling, offshore winds push shoreline water and larvae away from the shore, and barnacle recruitment is low. Conversely, in years with little upwelling, the shoreline water mass is not moved offshore, and barnacle recruitment is high. The assumption that larval losses are due to offshore currents is supported by larval distributions in upwelling and non-upwelling years. During upwelling years, **competent** larvae (larvae that are ready to settle) are commonly found up to 10 kilometers from shore, where they are unlikely ever to return to the shoreline. In non-upwelling years, barnacle larvae are found mostly within 1 kilometer of the shoreline (Gaines and Roughgarden 1987). While upwelling is not a major feature of the nearshore oceanography of western Atlantic shorelines, current patterns do play an important role in the dispersal of shoreline organisms. The annual occurrence of semitropical fishes in southern New England waters in the summer demonstrates the potential role of currents in larval dispersal. These fishes are seasonally transported as larvae by the Gulf Stream well beyond their northern limit, but do not survive the winter (see Hare and Cowan 1991).[⬤▷20]

While larval loss at sea is common in littoral organisms, larval transport and losses can be reduced by larval behavior. Even strongly swimming fish and crab larvae do not swim fast enough to overcome currents, but larval behavior that regulates vertical position in the water column can regulate horizontal transport by taking advantage of predictable hydrodynamic conditions. For example, most larvae are negatively **geotactic** (move away

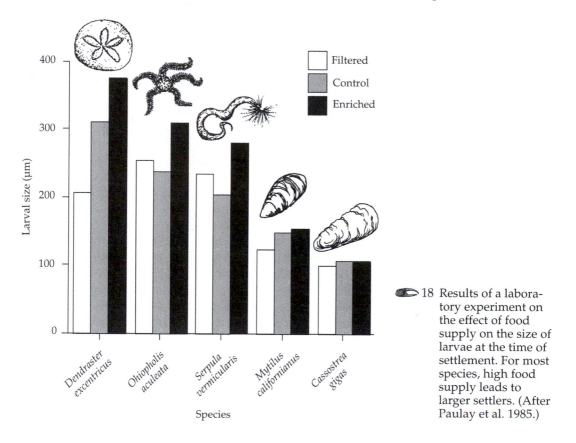

18 Results of a laboratory experiment on the effect of food supply on the size of larvae at the time of settlement. For most species, high food supply leads to larger settlers. (After Paulay et al. 1985.)

from gravity) and positively **phototactic** (move toward light) when small (Sulkin 1984). This leads newly hatched larvae to swim to the surface, where they are typically transported offshore. The position of larvae in the water column may be affected not only by swimming behavior, but also by

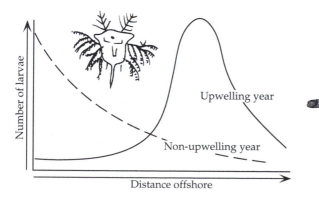

19 Offshore distributions of barnacle larvae as a function of upwelling. With upwelling, most larvae are transported offshore, where they perish. Without upwelling, larvae stay near shore, where they can settle. (After Roughgarden et al. 1988.)

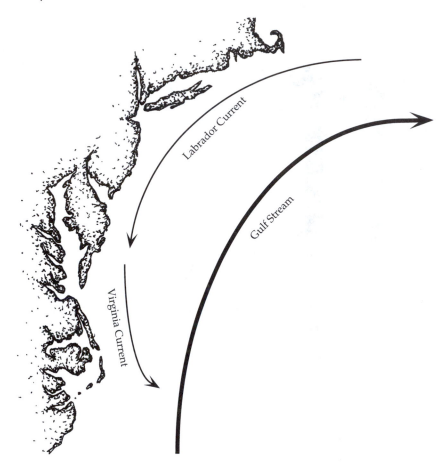

🐚 20 Major offshore currents that can influence larval dispersal on the Atlantic coast of North America.

larval size, morphology, and physiology. Small size, flattened bodies, and long spines increase larval surface area and are thought to be partially responsible for keeping larvae buoyant. The use of lipids rather than carbohydrates for energy storage may also be an adaptation to slow larval sinking (Morgan 1995).

    The transport of larvae into and out of estuaries is the best-studied case of the interaction between larval behavior and predictable horizontal water movement (for review see Morgan 1995). Estuary circulation is generally dominated by the seaward flow of a wedge of lighter, dilute river water over a denser wedge of cold, saline oceanic water, which is pushed into the estuary at high tides. Thus, estuary circulation is typically dominated by seaward-moving surface waters and shoreward-moving bottom waters. A variety of behavioral mechanisms can keep larvae in estuaries by taking advantage of estuarine current dynamics. The simplest mechanism for staying in an estuary, for example, is to sink to waters at intermediate depths and avoid surface waters flowing out of the estuary or bottom waters flowing in.[🐚 21] This picture, however, is grossly oversimplified. Physical characteristics of estuaries that affect their circulation patterns, such as depth, tidal amplitude, mouth size, and water movement patterns, probably have strong consequences for larval transport patterns, but are largely unstudied.

Considerable field evidence and laboratory larval behavioral studies have pointed to the importance of vertical positioning in regulating horizontal larval movement. In Chesapeake Bay, for example, the larvae of shallow-water species are initially found in surface waters, which transport them out of the bay, but later stages are found in bottom water, where they are transported back into the bay.[🐚 22] In contrast, larvae that go through their entire development in the bay are typically found in deeper water, where they are retained rather than exported.

Shifts in behavior that lead to the predictable export and import of larvae in estuaries over the course of an organism's development have received considerable attention. Shifts in phototactic and geotactic behavior during development can lead to the export of early larval stages but the import of later stages (Sulkin 1984). Increased size, depletion of buoyant lipid reserves, and changes in morphology during development can produce less buoyant larvae that sink to deeper water that is moving shoreward.[🐚 23]

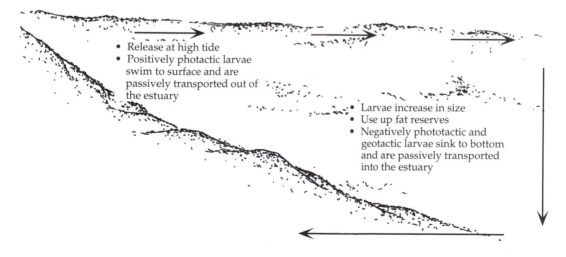

* Release at high tide
* Positively photactic larvae swim to surface and are passively transported out of the estuary

* Larvae increase in size
* Use up fat reserves
* Negatively phototactic and geotactic larvae sink to bottom and are passively transported into the estuary

🐚 21 Mechanisms proposed to transport the larvae of estuarine organisms offshore to develop and then back to shore to take up a benthic existence as adults.

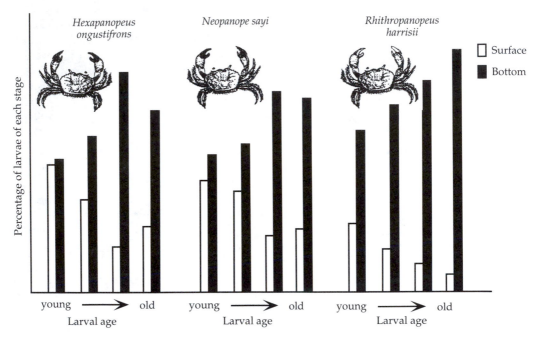

🐚 22 Distribution of early- and late-stage larvae of nearshore crabs in North Carolina. Late larval stages are concentrated near the bottom, where they are transported shoreward and have the opportunity to sample settlement habitats. (After Morgan 1995.)

Predation on Larvae

Predation is generally thought to be one of the most important sources of larval mortality (Thorson 1950), but accurate assessment of predation-related larval mortality rates is difficult (Strathmann 1990). Predators of larvae include pelagic invertebrates and fish that live in the water column, as well as benthic invertebrates that live on the bottom. Pelagic invertebrate consumers of planktonic larvae include jellyfishes and comb jellies, which are seasonally abundant along the east coast of North America.[🐚24] Jellyfishes can reduce zooplankton densities dramatically (Huntley and Hobson 1978), and in Atlantic coast estuaries, the abundant comb jelly

🐚 23 Common spawning and larval movement patterns in shoreline organisms. Larvae ▶ can stay close to shore by shifting their behavior with age (A), or by shifting their behavior to ride tidal currents (B). Larvae can develop offshore by shifting their behavior so as to be transported offshore when young and back to shore when older (C), or adults can migrate and spawn offshore, and their larvae can migrate shoreward by riding bottom currents (D). (After Morgan 1995.)

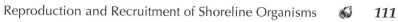

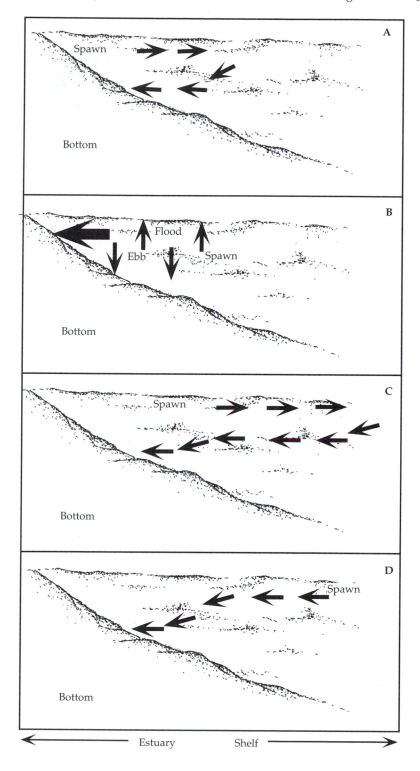

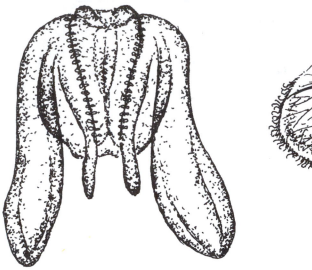

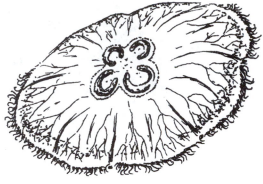

Moon jelly
Aurelia aurita

Leidy's comb jelly
Mnemiopsis leidyi

🐚 24 Some pelagic invertebrate predators
of larvae.

Mnemiopsis may be responsible for limiting larval densities (Burrell and Engel 1976).

Benthic invertebrate filter feeders may also be important predators of larvae, particularly late larval stages exploring the surface for settlement sites (but see Sebens and Koehl 1984). High densities of filter feeders, such as those found in mussel and oyster beds, are capable of filtering large volumes of water, and may filter larvae from the water as they are delivered to the shore (Woodin 1976).[🐚 25] Few studies, however, have addressed larval avoidance of filter feeders (Bingham and Walters 1989), or have estimated benthic larval losses to filter feeders. While larval consumption by filter feeders may be common, in soft-bottom communities its effect on recruitment is equivocal (Olafsson et al. 1994). Small, recently settled recruits are also vulnerable to consumers and disturbance mortality. In soft substrates, small recruits can be buried and smothered by the sediment reworking of deposit feeders (Woodin 1976) or consumed by predators and large deposit feeders. On hard substrates, recruits can be consumed or dislodged by scraping grazers such as herbivorous snails, chitons, and urchins (Menge 1976). Post-settlement mortality of small recruits is probably a very large and critical source of mortality in both hard- (Connell 1985) and soft-substrate (Olafsson et al. 1994) habitats, but this has been difficult to quantify.

Planktivorous fishes are believed to be the most predictably important predators of larvae (Morgan 1995).[🐚 26] The planktivorous juveniles of many large fishes inhabit shallow-water nursery grounds such as marshes and seagrass beds (Thayer et al. 1979). In addition, many of the permanent

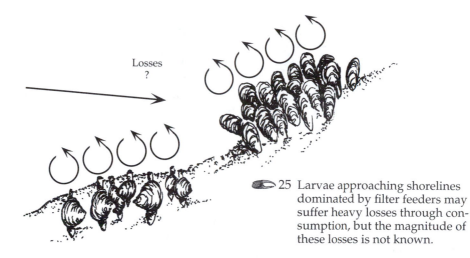

25 Larvae approaching shorelines dominated by filter feeders may suffer heavy losses through consumption, but the magnitude of these losses is not known.

residents of shallow-water habitats, such as killifish and silversides, are planktivores when small. Consequently, consumer pressure on larvae may be much higher inshore than offshore (Morgan 1989, 1992). Richard Strathmann (1993) has even argued that the long-term pelagic dispersal of many benthic invertebrate larvae is the result of selection for development in relatively safe offshore habitats.

The most compelling evidence of the importance of predation on planktonic larval forms can be seen in the prevalence of predator defenses displayed by larvae. Behaviors that limit the effects of predation include larval release during times of low predation risk (Morgan and Christy 1995), larval movement to offshore habitats with reduced predator risk (Forward 1974, 1976; Morgan 1995), daily vertical migration to limit contact with predators (Gliwicz 1986; Forward 1988) and passive sinking in response to predators.

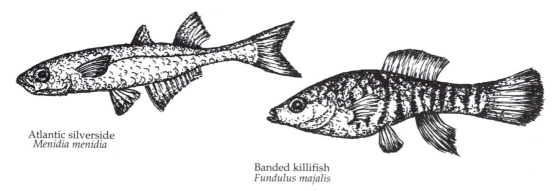

Atlantic silverside
Menidia menidia

Banded killifish
Fundulus majalis

26 Common planktivorous shallow-water fishes on the Atlantic coast of North America.

[🔊 27] Chemical defenses have been shown to limit the palatability of sea squirt and echinoderm larva (Young and Bingham 1987; McClintock 1990; Lindquist and Hay 1996).

The most conspicuous larval antipredator defenses, however, are morphological.[🔊 28] The spines and shells of many pelagic larvae increase their effective size, reducing the impact of gape-limited predators such as many fishes (Morgan 1989). Planktivorous fishes learn to avoid ingesting spiny larvae whose spines can break in their mouths (Morgan 1992). The small size, transparency, and counter-shading of the pelagic larvae of many marine invertebrates reduces their visibility to fish predators (Hobson and Chess 1976).

A particularly well documented case of the role played by planktivorous fishes in the larval ecology of benthic invertebrates is found in Steve Morgan's (1989, 1992) study of crab larvae in North Carolina. In general, crabs release pelagic larvae at flood tides, which transport the larvae rapidly away from shallow intertidal habitats characterized by intense predation. Once away from shallow-water habitats, some larvae remain near shore in bays and estuaries, presumably by behaviorally regulating their vertical position in the water column and staying near the bottom. Others, by remaining close to the surface, are transported offshore. This dichotomy in larval transport and behavior is best explained in terms of predator avoidance strategies.[🔊 29] Larvae that ride surface currents offshore are typically small, unarmored species preferred by planktivorous fish predators.

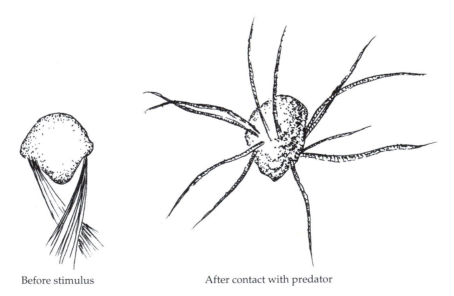

Before stimulus   After contact with predator

🔊 27 The larva of the polychaete *Sabellaria* exhibits a flaring response to predators. (After Pennington and Chia 1984.)

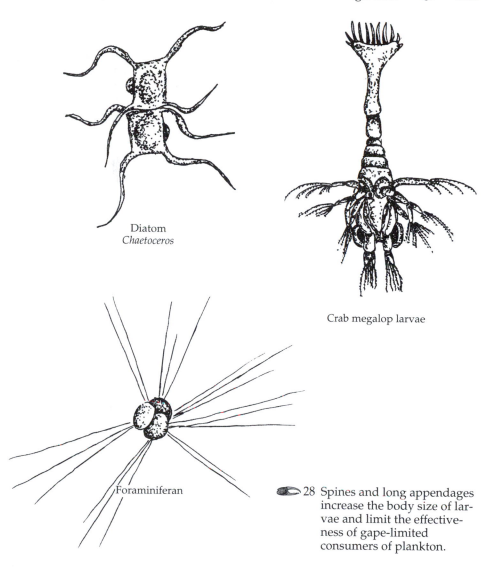

Diatom
Chaetoceros

Crab megalop larvae

Foraminiferan

🐚 28 Spines and long appendages increase the body size of larvae and limit the effectiveness of gape-limited consumers of plankton.

The behaviors of blue crab and fiddler crab larvae are good examples of this strategy. Export offshore limits exposure to predators, minimizing the need for antipredator defenses, but also probably leads to slower growth rates and development times, since offshore waters are typically nutrient-poor. In contrast, larval forms that sink to the bottom and remain in high-risk nearshore habitats are larger and have mechanisms to reduce predation losses. Larvae of the crabs *Sesarma reticulatum* and *Rhithropanopeus harrisii* and the grass shrimp *Palaemonetes pugio* remain in shallow water and are relatively large, have spines, and are avoided by fish predators. The pea crab, *Pinnotheres*, is an exception to this pattern. It has small, unarmored larvae

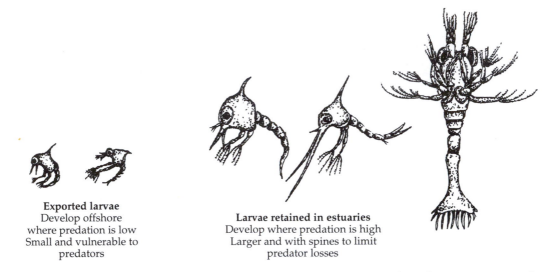

Exported larvae
Develop offshore
where predation is low
Small and vulnerable to
predators

Larvae retained in estuaries
Develop where predation is high
Larger and with spines to limit
predator losses

29 Crab larval forms that are exported offshore to develop where they are not exposed to heavy predation, and larval forms that are retained in high-productivity estuaries, where they have high growth rates but are exposed to heavy predation pressure. (After Morgan 1992.)

that remain in shallow water with high consumer pressure. When attacked by fish predators, however, pea crab larvae stop swimming and passively sink, a behavioral defense that limits losses to predators and allows the larvae to remain in highly productive nearshore waters.

Factors Regulating the Recruitment of Pelagic Larvae

Considerable recent attention has been focused on the effects of variation in recruitment on the structure of marine populations and communities. Spatial and temporal variation in recruitment is widespread in benthic invertebrates (Sutherland 1974; Gaines and Roughgarden 1985; Caffey 1985) and can strongly influence their abundance and distribution patterns. Elucidating the causes and predictability of this variation is crucial to understanding marine communities and presents a formidable challenge to shoreline ecologists.

Mechanisms responsible for generating variation in recruitment operate on a range of spatial scales.[30] At biogeographic spatial scales, measured in hundreds of kilometers, oceanographic transport of larvae is thought to be important for many types of organisms (e.g., gastropods: Scheltema 1971a,b, 1986; fishes: Parrish et al. 1981; barnacles: Roughgarden et al. 1987). Upwelling (Roughgarden et al. 1988), bay flushing (Gaines and Bertness 1992, 1993), and basin-wide circulation patterns can also determine larval concentrations by dictating larval transport and losses. At smaller spa-

Rate of arrival of larvae

Larval pool

Physical transport
processes

Disturbance and biotic
interactions: competition and
predation

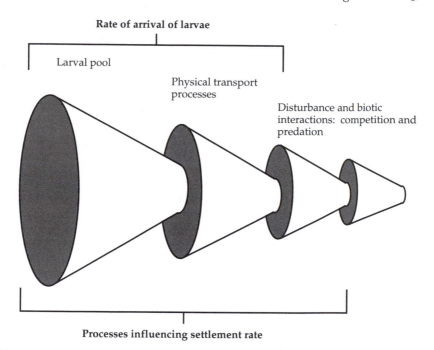

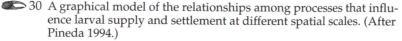

Processes influencing settlement rate

🐚 30 A graphical model of the relationships among processes that influence larval supply and settlement at different spatial scales. (After Pineda 1994.)

tial scales, in the size range of shoreline organisms and their larvae (millimeters to tens of centimeters), both hydrodynamic forces acting on larvae and larval settlement preferences are important (see Butman 1992). The importance of these small-scale processes, however, is dependent on the larval concentrations dictated by large-scale processes.

A variety of predictable large-scale oceanographic processes influence the recruitment patterns of benthic organisms with pelagic development. Oceanic circulation patterns routinely transport long-lived larval forms hundreds to thousands of kilometers, and thus play a dominant role in establishing and maintaining distribution patterns in these species as well as in mediating patterns of gene flow (Jackson 1974; Hansen 1980). Molluscs with long-lived larval forms, for example, have populations in the Atlantic Ocean with routine recruitment and gene flow between Europe and North America (Scheltema 1971b). On the west coast of North America, Point Conception in central California is a site of diverging oceanic currents and a major biogeographic boundary for shoreline organisms. Steve Gaines (personal communication) has recently shown that Point Conception is a strong biogeographic barrier to organisms with pelagic larvae, but not to organisms with direct development, suggesting that planktonic dispersal coupled with strong oceanographic circulation patterns can limit biogeographic distribution patterns. The generality of these findings is unknown. On the east coast of North America, the role of ocean currents in maintaining important biogeographic boundaries, such as Cape Cod and Cape Hatteras, is largely unstudied.

More localized oceanographic processes that influence the transport of larvae to and from adult habitats are also important. Upwelling, by moving surface water offshore, can lead to heavy losses of larvae from adult habitats, resulting in low recruitment (Roughgarden et al. 1988), but this is not a dominant oceanographic process in the western Atlantic. Embayment flushing patterns, or the rates at which water moves out of embayments into the open ocean, can also play a prominent role in dictating shoreline larval supplies, and are typically important in the estuaries of the east coast of North America (Ketchum 1954; Gaines and Bertness 1992). In Narragansett Bay, for example, annual barnacle recruitment shows year-to-year variation of over an order of magnitude that is not correlated with adult population sizes. Over 80 percent of this variation, however, is explained by interannual variation in the flushing rate of Narragansett Bay when barnacle larvae are in the water column.[🐾 31] In years with heavy rainfall, the bay flushes rapidly, transporting larvae to the open sea, with consequent low recruitment in the bay itself. In contrast, in years with low rainfall, slow flushing leads to larval retention in the bay and heavy barnacle settlement. Spatial and temporal patterns in the sizes of barnacle larvae show that this tight linkage between barnacle recruitment and bay flushing results from larval transport rather than larval mortality caused by freshwater input. Barnacle larvae that develop in the nutrient-rich bay water are much larger than those that develop in low-productivity coastal water. In years with heavy flushing, barnacle larval size distributions on the coast are bimodal, including both small (coastal) and large (bay) cyprids. In contrast, in years with low flushing, only small cyprids are found on the coast.

Wind and tidally driven currents can also be important in transporting competent larvae onshore to adult habitats. Hawkins and Hartnoll (1982) have demonstrated strong positive correlations between onshore winds, water column larval concentrations, and the daily larval settlement of acorn barnacles, suggesting that wind-driven currents can transport larvae shoreward. In Narragansett Bay, Rhode Island, when winds blow onshore, larval concentrations in the water column and shoreline settlement increase downwind, leading to wind-driven settlement signatures that reflect daily as well as annual variation in wind patterns (Bertness et al. 1996).[🐾 32]

Another mechanism that may commonly move offshore larvae shoreward is surface slicks associated with tidally driven internal waves (Shanks 1983, 1985; Shanks and Wright 1987; Pineda 1991). When tidally driven waves hit the continental shelf, they generate **internal waves** that move shoreward, pushing surface water toward the shore.[🐾 33] This can result in conspicuous **slicks,** or parcels of surface water, moving shoreward. Both barnacle and crab settlement have strong tidal cycle components driven by larvae riding these slicks to shore.

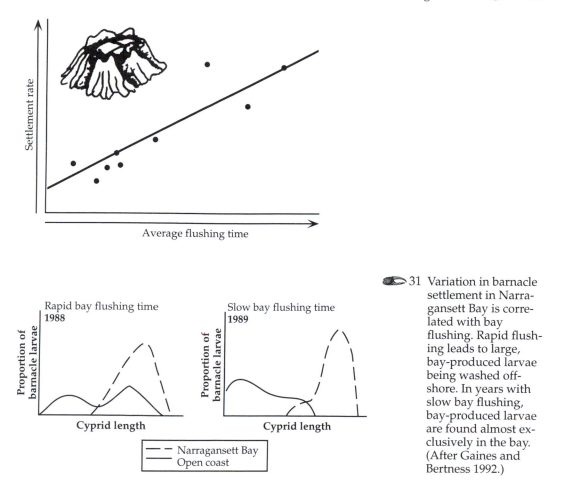

31 Variation in barnacle settlement in Narragansett Bay is correlated with bay flushing. Rapid flushing leads to large, bay-produced larvae being washed offshore. In years with slow bay flushing, bay-produced larvae are found almost exclusively in the bay. (After Gaines and Bertness 1992.)

Closer to shore, near-bed flows play an important role in dictating the delivery rate of competent larvae to available adult habitats. In laboratory flow tank studies, Joe Pawlik and his colleagues (1991) found that the settlement rate of the reef-building tube worm *Phragmatapoma* is low in still water, since larvae infrequently sample the bottom, but that settlement increases with increasing flow speed as larvae passively tumble along the bottom, frequently contacting the surface. At higher flow speeds (>15 cm/sec), larvae are eroded from the surface, so that high flows inhibit settlement.[☜ 34] Flow tank studies with barnacles have shown a similar relationship, with settlement enhanced by intermediate flow speeds (Crisp 1955; Mullineaux and Butman 1991).

Field studies have also emphasized the importance of flow in dictating larval supply patterns. Water column larval concentrations alone at a single point in time are poor indicators of shoreline larval supply, since these concentrations are often highly variable. However, when larval supply is envi-

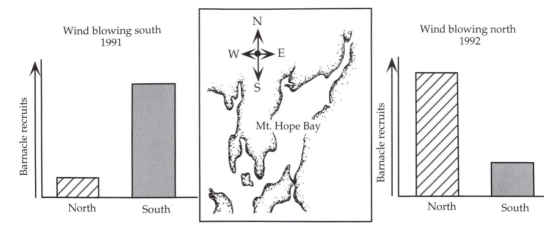

Wind blowing south
1991

Barnacle recruits

North   South

Mt. Hope Bay

Wind blowing north
1992

Barnacle recruits

North   South

🦀 32 Annual variation in barnacle recruitment in Mt. Hope Bay (an arm of Narragansett Bay, Rhode Island) is correlated with yearly wind variations. When winds blow southward, recruits accumulate on the southern side of the bay. The reverse occurs when winds blow northward. (After Bertness et al. 1996.)

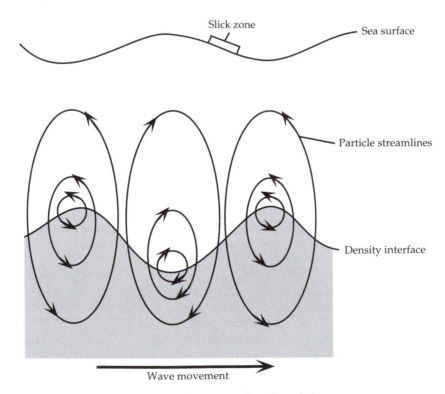

Slick zone

Sea surface

Particle streamlines

Density interface

Wave movement

🦀 33 Movement of internal waves and associated surface slicks onshore. When tidally driven waves hit the continental shelf, they generate large-amplitude internal waves of oscillatory water molecules. This slowly pushes parcels of surface water, or slicks, to shore. (After Mann and Lazier 1991.)

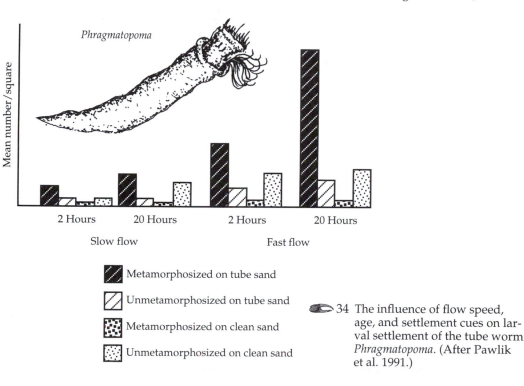

Metamorphosized on tube sand

Unmetamorphosized on tube sand

Metamorphosized on clean sand

Unmetamorphosized on clean sand

🐚 34 The influence of flow speed, age, and settlement cues on larval settlement of the tube worm *Phragmatopoma*. (After Pawlik et al. 1991.)

sioned as the flux of larvae, or the product of larval concentration and current speed, the match between larval flux and larval settlement is extremely high (Gaines and Bertness 1994).[🐚 35] Thus, while extreme flows can limit larval settlement by making it difficult for larvae to attach to the surface, and may even minimize contact with the surface, lower flows are often necessary for larval delivery.

As water masses move over suitable settlement sites and larvae settle, shoreline larval supplies can be depleted, leaving **larval settlement shadows** in habitats that experience larva-depleted water masses. Gaines and Roughgarden (1985) noted this phenomenon on intertidal rocky beaches, where the seaward edge experienced higher recruitment than shoreward habitats at identical tidal heights.[🐚 36] Larval depletion may play an important role in limiting the recruitment of shoreline organisms to high intertidal heights on both soft-substrate and rocky intertidal habitats (Underwood and Denley 1984).

Hydrodynamic forces and active larval habitat selection are clearly both important in generating settlement patterns in benthic organisms. Early in the twentieth century, marine ecologists believed that larvae randomly settled very widely, and that after differential mortality, the distribution of post-settlement survivors yielded the striking habitat segregation of most marine species. This view was challenged in the 1950s and 1960s with the discovery of active habitat selection by marine larvae (for review see

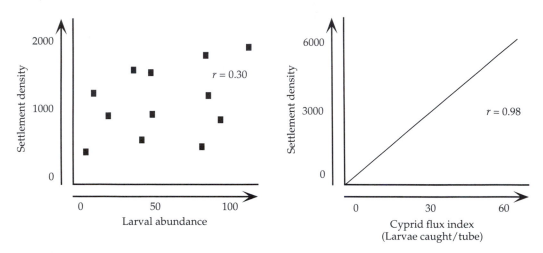

🐚 35 The relationship between water column larval concentrations and shoreline barnacle settlement. Point samples of larval abundance (upper figure) show little relation to shoreline settlement. When larval supply is examined as flux (larval concentration × flow speed; lower figure), however, a strong relationship between larval abundance and shoreline settlement emerges. (After Gaines and Bertness 1994.)

Meadows and Campbell 1971). Gregarious settlement by barnacles (Crisp and Meadows 1962) and bivalves (Bayne 1964, 1969) was shown to be the result of active larval habitat selection and larval responses to chemical cues that typically operate at short distances or require contact.[🐚 37] Settlement preferences for areas with microfloral cues indicating suitable adult habitats (Strathmann et al. 1981) and avoidance of habitats with potential competitors (Grosberg 1981) were also demonstrated. These studies led

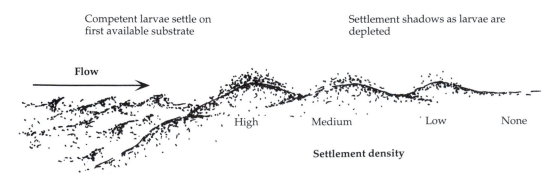

🐚 36 Larval settlement shadows occur when larvae reaching the shore settle on the first suitable substrate encountered, leaving few larvae to settle on other sites.

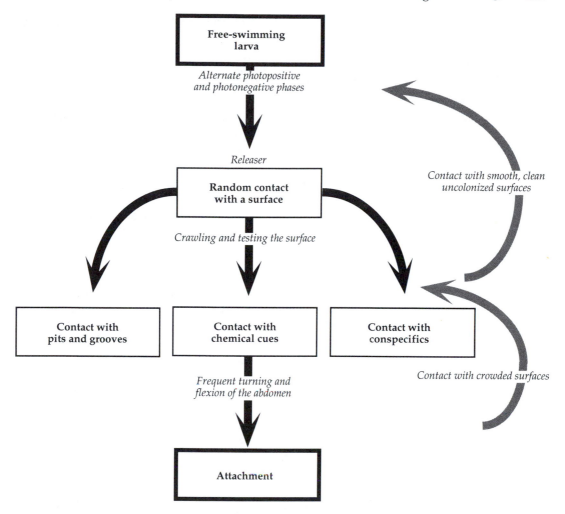

❧ 37 Larval settlement is not random, but involves active behaviors and preferences on the part of larvae, as shown in this model of the behavioral responses of larvae of the tube worm *Spirorbis* and the barnacle *Semibalanus* to benthic cues. (After Newell 1979.)

to a new recognition that larval habitat selection, rather than larval mortality, is responsible for habitat segregation by benthic marine organisms.

More recent work has emphasized the importance of passive deposition in dictating larval settlement patterns, and in particular, has shown that passive larval deposition can determine when and where active habitat selection can occur. Passive larval deposition is not equivalent to the random larval settlement envisioned by early researchers. Rather, larvae are deposited on the sea bottom as a function of their fall velocities, behavior, and bottom current flow patterns. Thus, larvae transported by currents across the bottom at velocities far in excess of their swimming speeds are deposited in areas of low flow, where their swimming and habitat selection behavior may be effective.[🐚38]

A number of studies have illustrated the role of passive deposition in dictating settlement patterns. By simply adding sewing needles to a mud flat, for instance, Eckman (1979) showed that passive deposition due to flow modification by these small structures influenced the deposition of fine sediments and larvae. The grass stalks common in marsh and seagrass habitats have also been shown to baffle near-bed flows, increasing sedimentation and larval settlement (Eckman 1983). Other studies have manipulated the characteristics of sampling devices to illustrate how passive deposition can enhance settlement (Hannan 1981; Snelgrove et al. 1993), and have revealed the importance of species-specific differences in behavior and/or fall velocities. Hannan (1981), for example, found that capitellid worm settlement in tall sampling traps was 1000 times higher than in nearby natural soft substrates, but other species settled at similar densities in the samplers and on natural substrates.[🐚39]

Interactions between flow patterns, settlement cues, and larval behaviors are probably important in dictating larval settlement patterns. Cheryl Ann Butman (1989, 1992) has argued that the weak swimming ability of most invertebrate larvae influences habitat selection at settlement by limiting their choices. Early work with the settlement responses of larvae to chemical and tactile cues (Meadows and Campbell 1971) was usually done in still water. Until recently, it was unclear whether these cues were effective under natural flow conditions, particularly since they were thought to operate on very small spatial scales, often requiring physical contact. Butman (1992) and others have suggested that, if larvae behave as passive particles, their weak swimming ability allows active habitat selection based on chemical or tactile cues only in passive deposition areas where water movement is reduced and their swimming speeds are effective, or where they are in contact with the substrate. Thus, local hydrodynamic conditions may dictate where active habitat selection can occur, and larval behaviors that influence

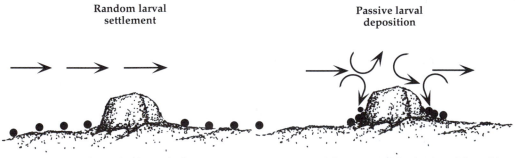

Random larval settlement

Passive larval deposition

Larvae settle randomly on the bottom

Near-bed flow patterns lead to larval deposition in areas of low flow, where they may have an active role in habitat selection

🐚38 The concept of passive larval deposition is not the same as random settlement, but determines where active settlement can occur.

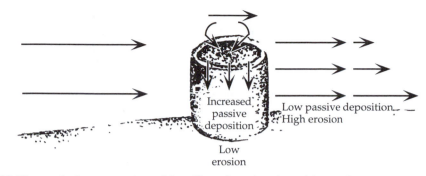

39 Hannan's demonstration of the effect of passive deposition on larval settlement patterns. Tall sampling traps accumulated 10 times as many settlers of some species as adjacent mud substrate. (After Hannan 1983.)

contact with the surface may be important determinants of where active habitat selection is possible.[40]

The results of one of the few experiments to date to examine the influence of flow on larval habitat selection stressed the importance of species-specific behavior in affecting settlement patterns (Butman et al. 1988). Competent larvae of two common soft-bottom invertebrates, the deposit-feeding worm *Capitella*, normally found on organically rich muds, and the filter-feeding quahog *Mercenaria*, normally found on coarser sand, were tested in still water and in a flume. The larvae were given a choice between mud and glass beads (representing sand) as settlement surfaces. Plastic spheres of the size and buoyancy of the larvae were used as passive controls, and were found to always have random settlement patterns. The experimental results were striking. Whereas both species chose their preferred adult habitats in still water, in a flume, *Capitella* continued to exhibit a strong pref-

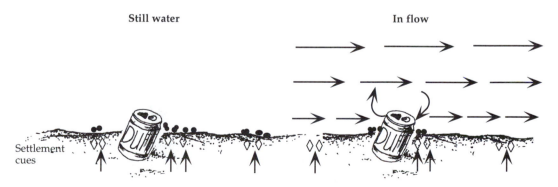

40 The influence of flow on larval settlement responses to local chemical cues. Larvae are shown as solid dots, chemical cues as open diamonds. In still water, settlement matches cues accurately. In currents, low-flow microhabitats such as the dead spaces near obstructing objects accumulate larvae and dictate settlement opportunities.

erence for mud, but the settlement of *Mercenaria* was random. The differential effect of flow on these species was a reflection of their behaviors. When competent to settle, *Capitella* larvae swim horizontally close to the surface (within 1 centimeter), frequently swimming down to test the substrate. In contrast, *Mercenaria* larvae move up and down over the bottom and do not stay as close to the surface. As a result, they are more vulnerable to flows limiting their contact with the surface.[🔹41]

POPULATION AND COMMUNITY EFFECTS OF VARIABLE LARVAL SUPPLY: SUPPLY-SIDE ECOLOGY

The pelagic development of most shoreline organisms has enormous consequences for the structure and dynamics of marine populations and communities. Until recently, however, these potential consequences were not appreciated. Prior to the 1980s, closed population models based on terrestrial systems were used to describe marine populations. These models depict local population processes as being tightly linked to successful local reproduction and recruitment. This tight linkage, however, is not found in most marine benthic organisms (see Caswell 1978; Roughgarden et al. 1988 for discussion). Most marine populations, including most common fishes and benthic invertebrates, are probably best described as open populations, in which pelagic dispersal decouples adult population dynamics from recruitment success.[🔹42]

Recognition of this decoupling of adult and recruitment success in marine populations has expanded the focus of marine population and community ecology from exclusive study of adult populations to exploration of the processes that link recruitment processes to adult populations. Early paradigms implicitly assumed that recruitment did not limit abundance or distribution patterns. As a result, many early ideas about the importance of benthic processes such as competition, disturbance, and physical stress in shoreline communities may apply only when recruitment is not limiting. This realization has focused considerable current effort on the elucidation of processes that influence larval supply, settlement, and recruitment, and has exposed many shortcomings in our current understanding of shoreline organisms.

Understanding general patterns in the relationships between adult populations and recruitment is one of the major challenges for shoreline ecologists. Whereas adult and juvenile populations are often decoupled, annual variation in recruitment is often predictably linked to environmental forces such as bay flushing (Ketchum 1954; Gaines and Bertness 1992, 1993), upwelling (Roughgarden et al. 1988), flow regimes (Sanford et al. 1994), or food supply (Olson and Olson 1992). Moreover, recruitment is often found to be predictably greater at some sites than others (Gaines and Bertness 1994). Elucidating the processes responsible for generating predictable spatial and temporal patterns in recruitment will be necessary for significant advancement of our current understanding of marine population and community regulation.

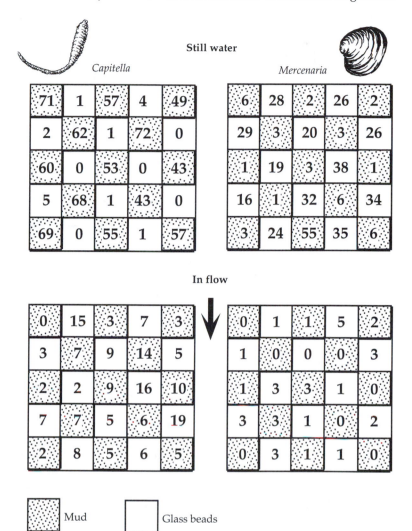

Still water

Capitella   Mercenaria

71	1	57	4	49
2	62	1	72	0
60	0	53	0	43
5	68	1	43	0
69	0	55	1	57

6	28	2	26	2
29	3	20	3	26
1	19	3	38	1
16	1	32	6	34
3	24	55	35	6

In flow

0	15	3	7	3
3	7	9	14	5
2	2	9	16	10
7	7	5	6	19
2	8	5	6	5

0	1	1	5	2
1	0	0	0	3
1	3	3	1	0
3	3	1	0	2
0	3	1	1	0

Mud   Glass beads

41 Settlement (number of larvae) of the worm *Capitella* and the quahog *Mercenaria* on a checkerboard array of mud and glass bead
substrates under still water and flume conditions. In still water,
Capitella settles in mud, while *Mercenaria* settles on glass bead
mimics of sandy substrates. In flows, *Capitella* maintains its habitat
preference, while *Mercenaria* settlement is random. (After Butman
et al. 1988.)

Larval supply can have enormous repercussions for populations of
shoreline organisms, ranging from controlling distribution and abundance
patterns to dictating the relative importance of the biological and physical
factors capable of regulating the structure of shoreline communities. Larval
supply can directly limit the distributions of populations of barnacles
(Sutherland 1974; Gaines and Roughgarden 1985), bryozoans (Yoshioka

Closed populations   Open populations

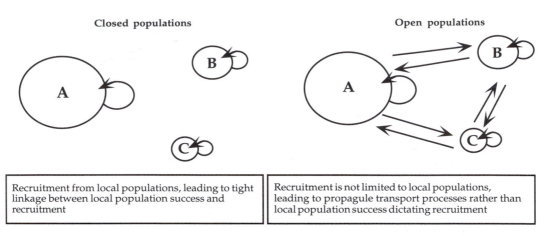

| Recruitment from local populations, leading to tight linkage between local population success and recruitment | Recruitment is not limited to local populations, leading to propagule transport processes rather than local population success dictating recruitment |

 42 Closed population models, based on terrestrial organisms, do not adequately represent the population dynamics of most marine organisms, which can be better described as having open populations.

1982; Hughes 1989), fishes (Sale 1977), and mussels (Petraitis 1991) on rocky shores. Roughgarden et al. (1985) have shown that variation in recruitment can trigger marked variation in local population processes. On the central coast of California, heavy annual recruitment of barnacles stimulates heavy starfish predation, leading to a precipitous decline in barnacle numbers and highly unstable, oscillating local populations. In contrast, with low barnacle recruitment, starfish predation is low, leading to stable, low-density populations. Thus in habitats with predictably low recruitment, stable age distributions and stable population densities occur, while in habitats with high annual recruitment, populations are highly oscillatory, unstable, and never reach a stable age distribution. A similar relationship between local population stability and recruitment intensity appears to occur in the absence of predators in Narragansett Bay (Sanford et al. 1994). At sites with low barnacle recruitment (typically low-flow, open-coast sites), low settlement rates lead to mixed age classes, relatively minor annual population oscillations, and rarely result in all available space being occupied by barnacles. In contrast, at sites that annually receive heavy barnacle recruitment (typically high-flow sites within the bay), crowding usually results in intense competition for space, heavy mortality, and dramatic annual population oscillations. Thus, in Narragansett Bay, variation in recruitment dictates variation in population structure and stability by influencing the intensity of competition, rather than predation.[✿ 43]

Variation in recruitment can also have strong effects on the processes responsible for generating the zonation of intertidal habitats. Interspecific competition for space often plays an important role in establishing and maintaining vertical zonation patterns on rocky shores. In particular, interspecific

Low recruitment

High recruitment

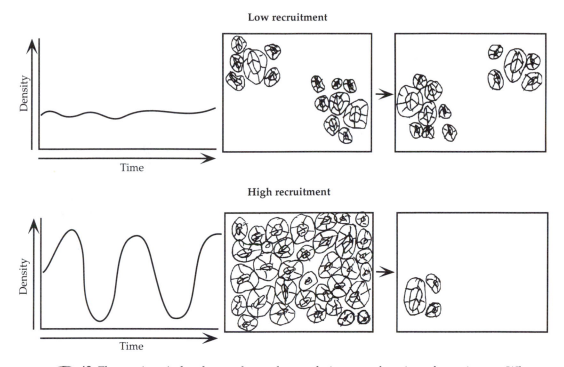

⬬ 43 Fluctuations in local acorn barnacle populations as a function of recruitment. When recruitment is low, population fluctuations are minimal and stable age distributions develop, with recruitment gains approximately matching losses. High annual recruitment, in contrast, leads to widely fluctuating local populations, since high recruitment densities result in intense competition or predation that leads to heavy mortality.

competition for space often sets the lower intertidal limits of species distributions. This is probably the case, however, only when recruitment does not limit population densities. In southern New England, for example, the high intertidal barnacle *Chthamalus* lives at high intertidal levels because it is competitively displaced by the northern acorn barnacle *Semibalanus* in the middle intertidal, but the blue mussel *Mytilus edulis* competitively displaces *Semibalanus* at low intertidal heights. This scenario, however, assumes that space is limiting. In situations in which recruits are rare, competition for space should be less important and zonation should be less distinct (see Gaines and Roughgarden 1985; Roughgarden et al. 1985 for discussion). [⬬ 44]

The roles played by disturbance and predation in intertidal communities should also be sensitive to recruitment variation. Both predators (Paine 1966, 1974) and disturbances (Connell 1978) influence the distribution and abundance of shoreline organisms by limiting the dominance of superior spatial competitors, so the role played by these factors in the structure and dynamics of these communities is clearly dependent on the assumption that

High recruitment

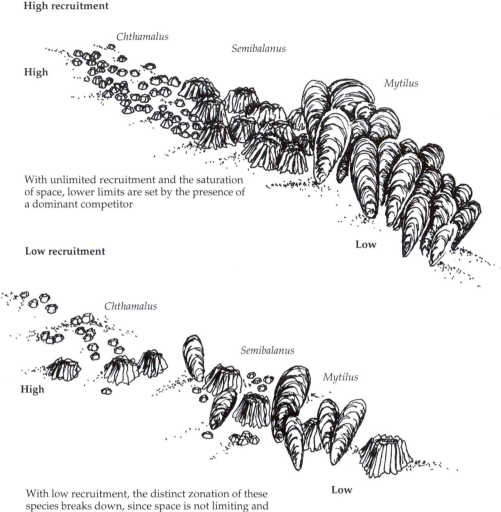

With unlimited recruitment and the saturation of space, lower limits are set by the presence of a dominant competitor

With low recruitment, the distinct zonation of these species breaks down, since space is not limiting and competition for space does not occur.

🐚 44 The influence of recruit supply on competition for space and rocky intertidal zonation.

space is a limiting resource. Neither predation nor disturbance may dramatically influence species distributions or diversity patterns if space is not limiting and low recruitment limits numbers of individual adults of each species. Recruitment-limited habitats may also not have prey populations sufficiently large to support predator populations (Sutherland 1974; Roughgarden et al. 1985, 1988). When this is the case, recruitment effects can have important negative feedbacks to higher trophic levels with potential consequences for local community structure.

Variation in recruitment may also commonly influence communities by dictating the importance of group benefits or positive feedbacks. In habitats where the presence of organisms that ameliorate harsh physical conditions (autogenic bioengineers: see Chapter 4) is necessary for the development of typical community structure, recruitment limitation of these critical bioengineers may have dramatic consequences. For example, in southern New England, the high intertidal limits of barnacles, mussels, and canopy-forming algae (*Fucus* and *Ascophyllum*) are strongly influenced by positive interactions among individuals (Bertness and Leonard 1997). In these habitats, neighbors ameliorate heat and desiccation stress at high intertidal levels, allowing groups of individuals to live closer to the upper tide mark than would be possible for solitary individuals. Without sufficiently high recruitment to realize these group benefits, these high intertidal assemblages—as well as all of the other organisms dependent on them—would collapse. [🐚 45]

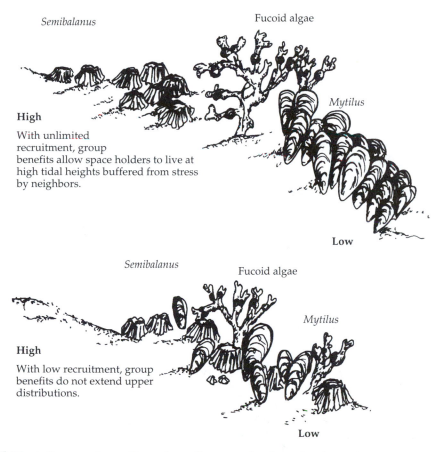

🐚 45 The influence of recruit supply on the upper borders of rocky intertidal space holders.

In marsh plant communities and soft-sediment habitats, recruitment limitation can similarly lead to a reduction in the importance of positive associations, with potentially large consequences. For plant colonizers of disturbance-generated bare space in salt marshes, for example, high recruit densities may be necessary to buffer recruits from heat and osmotic stress. Without high recruit densities, bare marsh patches may not close, or may close more slowly (Bertness and Shumway 1993; Bertness and Yeh 1994). In soft-substrate habitats, recruitment limitation of critical habitat-modifying organisms such as seagrasses, bed-forming mussels, oysters, and tube worms, which stabilize the substrate and influence near-bed flows and the deposition of sediments and larvae, clearly could lead to extensive community-wide effects. The role of recruit supply in mediating the importance of group benefits in shoreline communities has received virtually no attention in spite of its potential consequences on a range of shoreline habitats.

SUMMARY

Successful reproduction and recruitment can be a limiting factor for populations of most shoreline organisms, particularly those with external fertilization and long-lived planktonic larval stages. Recognition of the importance of reproduction and recruitment in influencing populations of littoral organisms, however, has occurred only recently. Only 15 years ago, it was widely thought that external fertilization was an efficient process, due to the high gamete output of most broadcast spawners, and that recruitment supply played little role in the patterning of intertidal assemblages. Recognition of the low efficiency of external reproduction and the pervasiveness of spatially and temporally variable recruitment in many shoreline organisms has focused considerable attention on these issues over the past decade.

External fertilization is common in marine organisms. It is inefficient because of the rapid dilution of gametes in seawater, particularly moving seawater. Organisms with external fertilization maximize their reproductive success by releasing gametes in breeding aggregations or synchronizing gamete release. The actual effectiveness of these solutions, however, is largely unknown. Direct internal fertilization is thought to be the best solution to the problems of external fertilization.

Once successful fertilization has occurred, the larvae of most shoreline organisms continue a high-risk lifestyle by developing in the plankton, where they may starve, be swept out to sea, or be eaten. While larvae have mechanisms to reduce these losses, many, if not most, larvae probably fail to survive their pelagic stage.

As a consequence of the vagaries of larval dispersal, larval settlement and juvenile recruitment is often spatially and temporally unpredictable, and is understood in very few marine organisms. Larval transport processes, in general, are important determinants of larval dispersal patterns. Benthic population success is typically decoupled from successful recruitment in most littoral organisms. Isolating the forces that generate spatial variation in recruitment, determining how predictable variation in larval recruitment is, and understanding how variation in recruitment affects the dynamics of shoreline communities are important challenges that are only now beginning to be addressed.

SUGGESTIONS FOR FURTHER READING

Butman, C. A. 1987. Larval settlement of soft-sediment invertebrates: The spatial scales of pattern explained by active habitat selection and the emerging role of hydrodynamical processes. *Oceanography and Marine Biology Annual Review* 25: 113–165. Mandatory reading for anyone interested in the history of larval settlement ideas.

Levins, L. and T. S. Bridges. 1995. Pattern and diversity in reproduction and development. In *Ecology of Marine Invertebrate Larvae*, L. McEdwards (ed.), 1–48. CRC Press, New York.

Levitan, D. R. 1995. The ecology of fertilization in free-spawning invertebrates. In *Ecology of Marine Invertebrate Larvae*, L. McEdwards (ed.), 123–156. CRC Press, New York.

Morgan, S. G. 1995. Life and death in the plankton: Larval mortality and adaptation. In *Ecology of Marine Invertebrate Larvae*, L. McEdwards (ed.), 279–322. CRC Press, New York.

Roughgarden, J., S. D. Gaines, and H. Possingham. 1988. Recruitment dynamics in complex life cycles. *Science* 241: 1460–1466.

Thorson, G. 1950. Reproductive and larval ecology of marine bottom invertebrates. *Biological Review* 25: 1–45. The classic reference for larval ecology.

Chapter 4

Process and Pattern in Shoreline Communities

*T*he processes responsible for the often striking distribution patterns of organisms in shoreline communities have long fascinated ecologists. We have already discussed how currents influence the distributions of littoral organisms by affecting fertilization success, the flux of larvae, and recruitment, as well as influencing primary production and the flow of energy through shoreline food webs. In addition to these linkages between shoreline and water column processes, interactions among established organisms can also play a major role in shaping shoreline communities. Interactions among benthic intertidal organisms in communities, in fact, are better understood than linkages between benthic and water column processes, since they are more easily studied. In this chapter I examine how competition for resources, predation, positive interactions, and recolonization after disturbance events influence the structure and dynamics of shoreline communities.

Zonation

One of the most conspicuous features of many coastlines is the pronounced **elevational zonation** of plants and animals. Especially on rocky shores, where substrate space is often entirely covered with sessile organisms, but also in salt marshes, where vascular plants cover most of the available space, organisms in shoreline habitats often occur in stereotyped elevational zonation patterns. Elevational zonation is so pronounced and repeated on temperate rocky shorelines that universal zonation schemes have been proposed for these habitats (Stephenson and Stephenson 1949, 1971), and regional zonation schemes have been proposed for temperate zone marsh plant communities (Chapman 1974).[🐚 1] The physical and biological processes that establish and maintain these distinctive distributional patterns have long been of interest to marine ecologists.

Shoreline assemblages are not unusual in having strong zonation patterns across environmental gradients. However, the steepness of gradients of physical conditions in intertidal habitats, ranging from entirely marine to entirely terrestrial over only a few meters, makes the zonation of intertidal organisms more compact and conspicuous than that along most other physical gradients. For example, plants and animals are also found in distinctive

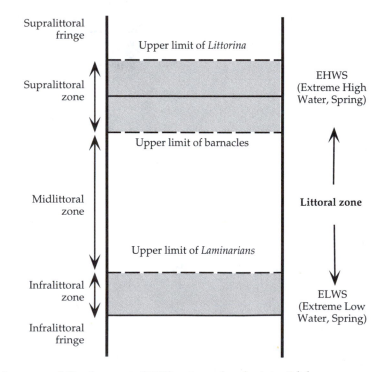

🐚 1 Stephenson and Stephenson's (1949) universal rocky intertidal zonation scheme.

elevational zones on mountainsides in response to gradients in temperature, nutrient availability, availability of water, and light intensity (Whittaker 1975). Instead of scales of tens of centimeters, however, species shifts along terrestrial elevational gradients occur on spatial scales of kilometers. The strong, compact physical gradients in intertidal habitats simply make organism zonation patterns more obvious and amenable to rigorous quantification and experimentation relative to most other habitats.

   The physical stresses that limit the distributions of intertidal organisms differ between organisms of marine and of terrestrial ancestry. For marine invertebrates and seaweeds that are vulnerable to water loss and thermal stress when out of water, physical stresses increase with increasing elevation in intertidal habitats.[🐚 2] Seawater buffers these stresses at the seaward end of the gradient, while extreme high and low temperatures and dessica-

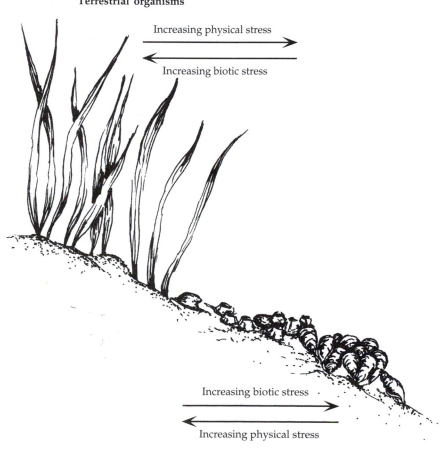

🐚 2 Stress gradients for organisms of marine and terrestrial ancestry.

tion can limit most of these intertidal organisms from living at upper tidal elevations. Conversely, physical stresses for organisms with terrestrial origins typically increase with decreasing shoreline elevation. Pulmonate snails and insects that live on marsh plants, for example, are air breathers without gills for respiration in water. Many vascular plants also have problems with respiration at low elevations, since waterlogged soils are usually low in oxygen. Consequently, these organisms do best at high intertidal elevations. Thus, for both types of organisms, the intertidal habitat represents a steep environmental gradient that typically limits their distributions. This gradient, however, runs in opposite directions for marine and terrestrial organisms.

Physical gradients that generate horizontal zonation patterns are also common, but less obvious. The physical forces that generate horizontal zonation patterns often operate on much larger spatial scales, and thus are much less conspicuous than vertical physical gradients. Physical factors that vary among habitats include thermal and osmotic stresses (which increase in estuaries in comparison to coastal habitats), wave stress (which decreases in bays and estuaries in comparison to wave-swept coastal habitats), and nutrient supplies (which are generally higher in estuarine than in coastal habitats).

Early researchers assumed that the elevational zonation of intertidal habitats was a direct consequence of the ability of littoral organisms to cope with physical conditions. Barnacles, snails, and algae were thought to be physiologically specialized for life at high intertidal elevations in rocky intertidal habitats, whereas organisms usually found at lower tidal heights were thought to be specialized for those elevations (Lewis 1964; Newell 1979). Similarly, plant zonation was initially assumed to reflect differential marsh plant success at specific tidal heights (Cooper 1982). These early interpretations of intertidal zonation as being driven exclusively by physical forces are examples of assigning causation to community patterns based on correlation only (for discussion see Paine 1994). Experimental work in intertidal plant and animal communities over the last three decades, however, has shown unambiguously that physical forces, while clearly important, are only part of the story, and that experiments can sort out the confounding of variables to show causation.

COMPETITIVE INTERACTIONS

The importance of competition for limiting resources as a determinant of the elevational zonation of intertidal organisms was first demonstrated experimentally by Joseph Connell with acorn barnacles (Connell 1961). In the middle Atlantic, as well as on many other temperate zone rocky shores, acorn barnacles of the genus *Chthamalus* are found at the terrestrial border of the intertidal habitat, while barnacles of the genus *Semibalanus* occur at lower tidal heights. By manipulating barnacle densities in the field, Connell showed that the occurrence of *Chthamalus fragilis* at higher tidal elevations in Scotland was a result of competitive displacement by the northern acorn barnacle, *Semibalanus balanoides*. *Chthamalus* larvae settle at both high and low intertidal heights, but are overgrown, crushed, or uplifted by the larger and faster-growing *Semibalanus* at lower tidal heights. Simply removing *Semibalanus* recruits from the lower zone led to the survival of *Chthamalus* there, demonstrating that *Semibalanus* competitively excludes *Chthamalus* from lower tidal elevations.[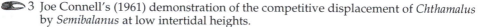3]

Connell's work led to the general rule of thumb that, in shoreline habitats where space is limited, dominant competitors displace competitive subordinates to refuge habitats. This simple rule is a robust description of spatial competition among seaweeds and sessile invertebrates on rocky shores, particularly at low intertidal heights and in subtidal habitats, and among the halophytic plants that dominate marsh plant communities (see Connell 1972, 1983; Lubchenco and Gaines 1981; Buss 1986; Keddy 1989; and Goldberg and Barton 1992 for discussion). Organisms that live in three-dimensional soft-sediment habitats are an important exception to this rule, ostensibly because other factors, such as disturbance, limit these organisms (Peterson 1979).

Any physical or biological disturbance that kills or removes parts of organisms, freeing space as a consequence, relaxes spatial constraints and potentially limits the role of competition in generating zonation in intertidal habitats (Dayton 1971). Physical stresses or factors that reduce physiological performance, such as heat and desiccation on high intertidal rocky beaches, can also limit the abundance of sessile space-holding organisms below densities at which competition for space is important (Menge and Sutherland 1976, 1987). Limited recruitment (Gaines and Roughgarden 1985; Underwood and Denley 1984) can also result in low densities at which space is not limiting.[4]

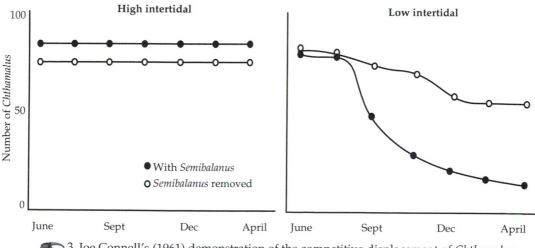

3 Joe Connell's (1961) demonstration of the competitive displacement of *Chthamalus* by *Semibalanus* at low intertidal heights.

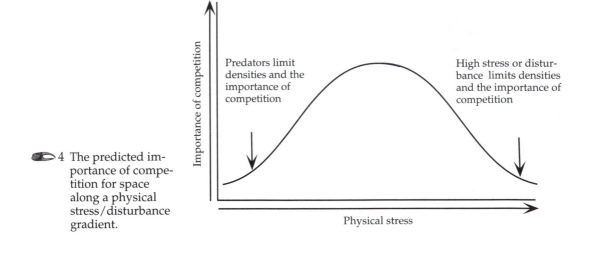

4 The predicted importance of competition for space along a physical stress/disturbance gradient.

Importance of competition (y-axis)

Predators limit densities and the importance of competition

High stress or disturbance limits densities and the importance of competition

Physical stress

Types of Competition

Competition among organisms can be of three types: exploitative, preemptive, or active interference. In **exploitative competition**, the differential ability of competitors to harvest limiting resources leads to decreased resource availability, resulting in decreased growth and reproduction for those organisms with an inferior ability to harvest resources. Common examples of exploitative competition among shoreline organisms include depletion of the planktonic food supply by filter feeders (Buss 1979a; Peterson and Black 1987); depletion of benthic diatom food resources by herbivorous snail grazers (Fenchel 1976; Levinton et al. 1985) and deposit-feeding fiddler crabs (Weissburg 1992, 1993); and hermit crabs rapidly finding and occupying the gastropod shells that they utilize for protection from predators and physical stresses (Bertness 1981; Wilber and Herrnkind 1984).

Preemptive competition occurs when a competitor recruits to and dominates a habitat, monopolizing all available space, precluding the establishment of potential competitors (Sutherland and Karlson 1977; Grace 1987). Dense beds of mussels or oysters are good examples of preemptive competition, as is the dominance of space by algal canopies on rocky shores and by marsh grass stands in salt marshes.[5]

In contrast to exploitative and preemptive competition, **interference competition** occurs when competitors physically contest access to resources. Hermit crabs fighting for the ownership of shells, neighboring barnacles crushing or uplifting one another, or mussels overgrowing and smothering barnacles are common examples of interference competition.[6]

Competition for substrate space among sessile organisms is one of the most easily studied types of competition (Jackson 1977; Buss 1986). Competition for space is not only conspicuous, resulting in superior competitors overgrowing and killing subordinates, but can also be studied by nondestructively following interactions over time. Since some competitors for space in

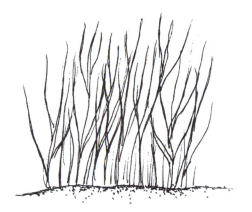

Oyster beds preventing the establishment of other space holders such as clams and mussels

Dense marsh plant cover precluding the establishment of seedlings or clonal ramets of other adults

5 Preemptive competition occurs when living space is entirely used and unavailable to potential competitors.

marine systems have calcified skeletons that often fossilize, competitive interactions among clonal invertebrates can even be studied in the fossil record.

The distinction between exploitative and interference competitive mechanisms among organisms competing for space is often fuzzy. Neighboring

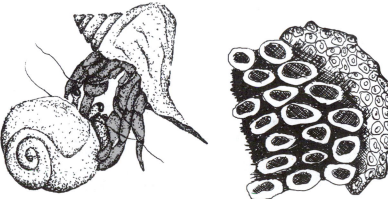

Ritualized fighting for shells in hermit crabs

Overgrowth of a bryozoan colony by another with larger individual zooids

6 Interference competition occurs when two organisms physically contest a limited resource.

Exploitative competition
Depleting food supply of neighbors

Interference competition
Crushing neighbors

7 Exploitative and interference competition in barnacles.

barnacles, bryozoans, seaweeds, and marsh grasses, for example, probably interact exploitatively before they come into physical contact, but are direct interference competitors once physical contact occurs.[7] By examining the harvesting of radioactively labeled food-sized particles by two bryozoan species, Leo Buss (1979a) found that when they were in close contact, the larger of the two bryozoans depleted the particle supply to the smaller bryozoan, leading to rapid growth for the larger bryozoan and decreased growth for the smaller one. As a consequence of this feeding imbalance, the larger species ultimately won competitive encounters and overgrew the smaller species.[8]

Competitive Dominance.

What determines the outcome of competitive encounters and thus dictates which organisms will dominate shoreline habitats where space or other resources are in limited supply? Species- or group-specific competitive mechanisms, such as the use of stinging cells and pinching devices by cnidarians (anemones, hydroids, and corals) and echinoderms (sea urchins and starfishes) and allelopathic chemicals by encrusting organisms (e.g., sponges and algae: Buss and Jackson 1979) can lead to competitive dominance. More commonly, however, large size and rapid growth rates translate into competitive dominance in shoreline organisms. Large size and rapid growth have been shown to dictate competitive dominance in intertidal algae (Lubchenco 1978, 1983), barnacles (Connell 1961; Stanley and Newman 1980) bryozoans (Buss 1979a, 1986), mussels (Harger 1971), marsh plants (Bertness and Ellison 1987), hermit crabs (Bertness 1981), fiddler crabs (Crane 1975), and many other organisms (see Buss 1986).

In addition to size and growth rate, body plans also strongly influence competitive ability. At similar body sizes, plants, sessile invertebrates, and seaweeds with **clonal** body plans are typically competitively dominant to plants and sessile invertebrates with solitary body plans. Solitary organisms must reproduce sexually to colonize new habitats or expand to overgrow a

Dominant bryozoan

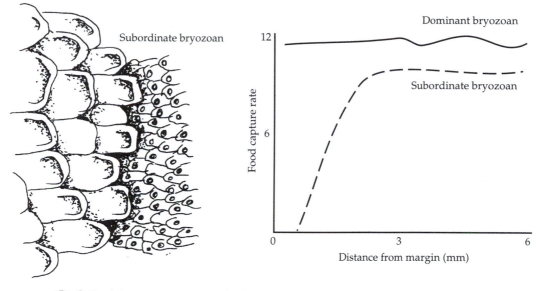

8 Exploitative competition for food between neighboring bryozoans. The colony with larger individuals monopolizes the food supply where individuals meet. (After Buss 1979a.)

potential spatial competitor. In contrast, clonal organisms, such as many tunicates and bryozoans and many plants, spread vegetatively, rapidly achieve large body sizes, and often have indeterminate growth (Sebens 1982).[🐚9] Moreover, clonal organisms are made up of repeated, genetically identical units (**ramets**) that are often physically connected and physiologically interdependent.[🐚10] As a consequence, clonal organisms can expand laterally without the necessity of sexual reproduction. This strategy avoids small, risky juvenile stages that are typically competitively subordinate and vulnerable to consumers (Buss 1979b). The physiological integration among the individuals in a clone allows clonal organisms to reach large body sizes rapidly, and when facing a potential competitor, consumer, or physical stress, allows them to share resources to overcome these difficulties. The dominance of clonal organisms in tropical reefs, subtidal rocky habitats, and salt marshes is thought to be largely the consequence of the competitive dominance of organisms with clonal designs over organisms with solitary body plans in physically stable habitats (see Jackson 1977).

Clonal architecture, or the arrangement of clonal units in space, is another important determinant of competitive ability in clonal plants and animals. Generally, sessile clonal organisms that have diffuse architecture or runner morphologies (the **guerrilla strategy**) are rapid vegetative colonizers, but poor competitors. In contrast, clonal organisms that have dense arrangements of clonal units or **turf morphologies** (the **phalanx strategy**) are typi-

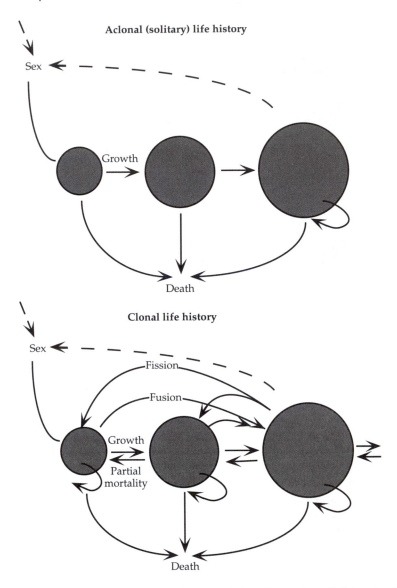

Aclonal (solitary) life history

Clonal life history

🦀 9 Schematic representation of solitary and clonal organism life histories. Solitary organisms, such as barnacles, vertebrates, and many plants, increase in numbers and size only by sexual reproduction and growth, respectively. Clonal organisms, such as tunicates, sponges, and many other plants, in contrast, can increase their numbers by fission and their size by fusion, can decrease their size by either fission or partial mortality, and can grow indefinitely. (After Jackson and Hughes 1985.)

cally slow colonizers, but dominant competitors for space and resources once established.[🦀 11]

    Since competitive encounters are often, though not always, determined by organism size, growth rate, and morphology, the outcome of competitive

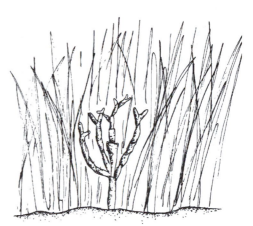

The clonal tunicate *Botryllus* overgrowing the acorn barnacle *Semibalanus*

Marsh hay *Spartina patens*, a clonal turf, overgrowing the solitary glasswort *Salicornia*

 10 Examples of the competitive dominance of clonal organisms over solitary organisms.

encounters is rarely random, but is often context-specific. Organisms can usually be ranked in competitive hierarchies based on their sizes, morphologies, and growth rates. On southern New England rocky shores, for example, the small barnacle *Chthamalus fragilis* is competitively subordinate to the larger northern acorn barnacle, *Semibalanus balanoides*. Both of these barnacles, however, are competitively subordinate to the even larger, faster-growing blue mussel, *Mytilus edulis*, which is itself competitively subordinate to the clonal tunicate *Botryllus schlosseri*. Thus, in these habitats, an unambiguous competitive hierarchy exists, with *Botryllus* competitively superior to *Mytilus*, which is competitively superior to *Semibalanus*, which is competitively superior to *Chthamalus*.[❦ 12]

Such competitive hierarchies have important consequences. First, in habitats where all competitors are capable of thriving, over time only the dominant competitor will prevail. Second, species coexistence is dependent on external forces. Disturbance, physical stress, consumers, or low recruitment is needed to prevent monopolization by the dominant competitor. In intertidal as well as most other systems, however, competitive ability and the ability to tolerate or respond to disturbance or physical stresses are often inversely correlated (see Keddy 1989 for a general discussion of the common trade-off between competitive ability and stress tolerance). Consequently, competitive hierarchies often result in organisms being sorted out in space across physical stress gradients, with dominant competitors monopolizing favorable habitats and subordinate competitors restricted to less favorable habitats. This is exactly what happens on New England rocky shores to barnacles, mussels, and tunicates. As a result of the competitive hierarchy, in the absence of disturbance and with high recruitment, *Botryllus* dominates physically benign subtidal habitats, *Mytilus* is competitively displaced to low

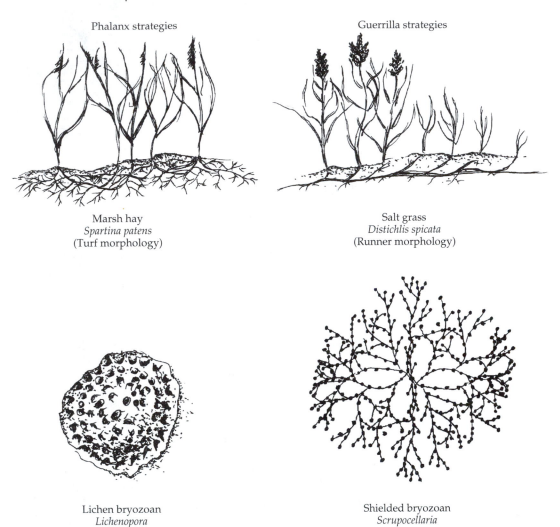

Phalanx strategies

Guerrilla strategies

Marsh hay
Spartina patens
(Turf morphology)

Salt grass
Distichlis spicata
(Runner morphology)

Lichen bryozoan
Lichenopora

Shielded bryozoan
Scrupocellaria

 11 Clonal growth strategies in marsh plants and bryozoans.

intertidal habitats, *Semibalanus* is displaced to middle intertidal heights, and *Chthamalus* is restricted to high intertidal heights. Similar scenarios, in which competitive dominants displace subordinates to physically stressful refuge habitats, leading to the strong zonation of intertidal habitats, occur in salt marsh and soft-sediment habitats.

While competitive hierarchies are a valuable heuristic tool for thinking about the dynamics of interacting species, they are typically an oversimplification of the real world. Small-scale variation in physical factors, phenotypic variation among individuals, and demographic variation in size and age all add a large element of stochasticity to the outcome of competitive encounters. Size and age structure, in particular, often reverse the outcome of competitive

PLATE 1   Shore berm of cobble and shells in Narragansett Bay.

PLATE 2   Cobble beach plant community behind a cordgrass bed in Narragansett Bay. (John Bruno)

PLATE 3   The same cobble beach plant community at high tide. (John Bruno)

PLATE 4   *Salicornia virginica* on a cobble beach. (John Bruno)

PLATE 5   Herbivorous snail removal pen in Narragansett Bay.

PLATE 6   Rumstick Cove marsh in Rhode Island.

PLATE 7 Storm swell wave hitting the Nahant Shoreline. (Jon Witman)

PLATE 8 The dense fucoid seaweed canopy at Pemequid, Maine. (Bob Steneck)

PLATE 9 Crowded hummocks of the northern acorn barnacle *Semibalanus*.

PLATE 10 A mating pair of *Cancer* crabs. (Jon Witman)

PLATE 11   The starfish *Asterias borealis* foraging on horse mussels. (Jon Witman)

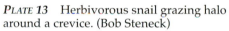

PLATE 13   Herbivorous snail grazing halo around a crevice. (Bob Steneck)

PLATE 12   The predatory snail *Nucella lapillus.*

PLATE 14   Foraging green sea urchins just below the low-water line. (Jon Witman)

PLATE 15   An aggregation of the ribbed mussel *Geukensia* attached to cord grass roots.

PLATE 16   Dead plant material (wrack) deposited in the high marsh in New England.

PLATE 17   Experimental vegetation clearings in a high New England marsh showing salt accumulation on the surface.

PLATE 18   The spikegrass *Distichlis* invading a New England marsh bare patch.

PLATE 19   Ice rafted portions of low marsh peat.

PLATE 21   A male orange-clawed fiddler crab,
Uca minax.

PLATE 20   A partially submerged 3,000 year old tree
stump in a salt marsh in
Wells, Maine is evidence
of sea level rise and marsh
growth over terrestrial
vegetation. (Michelle
Dionne)

PLATE 22   The reed *Phragmites* invading the high
marsh in southern New
England.

PLATE 23   *Borrichia* dominates this marsh on Sapelo Island, Georgia. (Steve Pennings)

PLATE 24   *Salicornia bigelovii* in a Georgia salt pan. (Steve Pennings)

PLATE 25   The common marsh killifish, *Fundulus heteroclitus*. (Ron Kneib)

PLATE 26   The marsh periwinkle *Littoraria* foraging on standing dead cordgrass. (Steve Newell)

PLATE 27   Kenan field marsh on Sapelo Island, showing drainage patterns. (Ron Kneib)

PLATE 28   Salt pan on Sapelo Island. (Steve Pennings)

PLATE 29   Oyster reefs on Sapelo Island tidal creeks. (Ron Kneib)

PLATE 30   The highly salt-tolerant plant *Batis* in a salt pan. (Steve Pennings)

PLATE 31   Worm tube reefs along the coast of Florida. (Sally Woodin)

PLATE 32   A mid-Atlantic mud flat after a ray foraging event. (Sally Woodin)

PLATE 33   Predator exclusion cages in the Bay of Fundy. (Herb Wilson)

PLATE 34   The mop-topped spaghetti worm, *Amphitrite*. (Sally Woodin)

Botryllus > *Mytilus* > *Semibalanus* > *Chthamalus*

🐚 12 Competitive hierarchy among New England rocky shore space holders.

encounters and thus add loops or reversals to competitive hierarchies. Buss (1980) has shown that while competitive encounters between bryozoan colonies are typically won by species with larger zooids (individual colony members), when individuals are similar in size, encounters between species are more stochastic because larger colonies typically win competitive encounters regardless of species identity. In marsh plant communities, plants in dense clonal turfs are usually competitively superior to the seedlings of annuals, but when the same plants compete as seedlings, annuals may be the best competitors. If competitive reversals like these are common, a system's dynamics may be very different from a strict competitive hierarchy.[🐚 13] Systems with

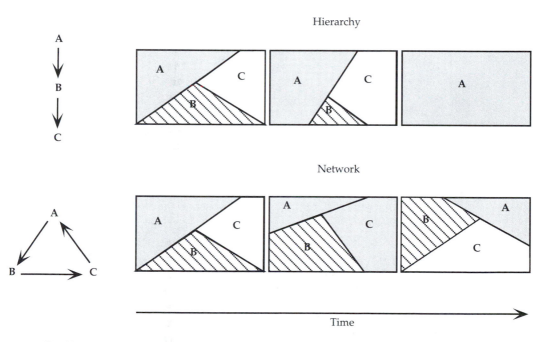

🐚 13 The influence of competitive hierarchies and networks on species coexistence. In hierarchical systems, a dominant competitor emerges, and monopolizes the habitat unless disturbances or consumers prevent it from doing so. In a network system, a dominant competitor does not necessarily emerge, and external events are not required to maintain species diversity. (After Buss and Jackson 1979.)

such competitive networks are less dependent on external events for the maintenance of species diversity than systems with competitive hierarchies. In competitive networks, a dominant competitor is not expected to ultimately displace all other competitors, and disturbances or consumers are not required to prevent habitat monopolization by a dominant competitor (Buss and Jackson 1979). While ecologists have argued over the importance of competitive reversals in natural systems (Connell 1978; Quinn 1982), natural demographic variation in the size structure of populations and spatial and temporal variation in microhabitats and the physical factors that influence competitive outcomes probably make competitive reversals and networks a common feature of most systems.

POSITIVE INTERACTIONS

Marine ecologists have focused a great deal of attention on the role of competition in influencing the distributions of plants and animals across shoreline habitats. Less attention has been given to the role played by positive interactions. Direct **positive interactions** are defined as interactions among two or more organisms that benefit one or more of the participants without negatively affecting the others. Direct positive interactions include **facilitations**, in which one organism is benefited by the presence of another that is not influenced by the association, as well as **mutualisms**, in which both participants benefit from the association. In spite of the lack of attention paid to the role of these positive interactions, they are likely to be important processes in many marine assemblages, and need to be incorporated into our understanding of these communities. Positive interactions appear to be particularly common in intertidal communities for the simple reason that the physical stresses that can limit the abundances of littoral organisms can often be ameliorated by the presence of neighbors.

On rocky shores, one of the most pervasive positive associations is the dense groups of individuals at high intertidal heights that buffer one another, both within and among species, from physical stresses. These groups allow the persistence of organisms at higher elevations than would be possible for solitary individuals without neighbors. High densities of mussels (Bertness and Leonard 1997), barnacles (Bertness 1989), and seaweeds (Hay 1981) are often necessary for persistence of these species at high intertidal heights. Mussel beds and intertidal seaweed canopies are also necessary for the persistence of many interstitial and understory organisms at high elevations, and so are also a good example of interspecific positive associations (Bertness and Leonard 1997).[🐚14] Positive interactions mediated by positive feedback between the presence of high densities of organisms and local physical conditions on rocky shores may commonly be a key interaction at high intertidal heights, even though the same organisms may be entirely competitive at lower tidal heights.[🐚15] In southern New England, for example, high barnacle and algal densities are necessary for local persistence at high intertidal heights, but at lower tidal heights similar high densities lead to intense competition for food and space.

☆ *Examples of Stress Alleviation by Neighbors in Shallow-Water Marine Habitats*

HABITAT	STRESS	MECHANISM	REFERENCE
Salt marshes	Low soil oxygen potentially limits plant growth	Dense stands of plants oxygenate soil	Howes et al. 1981; Shat 1984; Hacker and Bertness 1995b
		Fiddler crabs oxygenate soil	Montague 1982; Bertness 1985
	High soil salinity potentially limits plant growth	Plant cover shades soil, preventing soil salt accumulations	Bertness et al. 1992b; Bertness and Shumway 1993
	Low soil nutrients limit plant growth	Mussel filter feeding and feces deposition increase soil nutrients	Bertness 1984b
Mangroves	Low soil oxygen potentially limits plant growth	Mangroves oxygenate soil	McKee et al. 1988
		Fiddler crabs oxygenate soil	Smith et al. 1991
	Low nitrogen levels limit plant growth	Epiphytic sponges provide nitrogen to plants	Ellison et al. 1996
Soft bottoms	Substrate instability limits infaunal populations	Seagrasses bind substrate, limiting disturbance	Orth 1977; Irlandi and Peterson 1991
	Low nutrient levels limit seagrass colonization	Initial seagrass colonizers add organic material to substrate	Williams 1980
		Mussels attach to seagrasses; feces enhance production	Valentine and Heck 1993
	Low nutrient levels limit deposit-feeder growth	Deposit feeding enhances food supply	Levinton and Lopez 1977
Rocky shores	Heat and desiccation potentially limit sessile invertebrate and/or seaweed growth	Dense assemblages buffer group members from stress	Hay 1981; Lively and Raimondi 1987; Bertness 1989; Stephens and Bertness 1991
	Wave stress potentially dislodges sessile invertebrates and algae	Dense groups buffer members from wave stress	Holbrook et al. 1991; Denny et al. 1985

Positive feedbacks through habitat modification have been termed **positive switches** in vascular plant communities (Wilson and Agnew 1992). A typical terrestrial example of a positive switch is the invasion of forests into dry habitats. Successful recruitment of trees into a dry habitat shades soils,

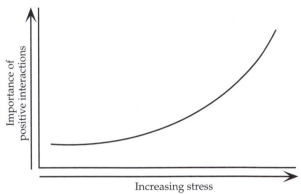

🐚 14 Positive habitat-ameliorating interactions increase in importance across shoreline physical stress gradients. Examples include heat and desiccation stresses on rocky shores, wave stress and substrate mobility in soft substrates, and soil anoxia and hypersalinity in marsh plant communities.

limits evaporative water loss, and enhances fog and humidity, leading to increased precipitation. Once trees are established, further recruitment is possible due to this habitat modification. Positive switches or feedbacks such as this are probably common but underappreciated forces in shoreline habitats. The development of intertidal algal canopies, for example, may fit a positive switch model. The upper borders of many intertidal seaweed canopies are sensitive to positive feedbacks. Without dense seaweed cover, most individuals die due to thermal and desiccation stress (Bertness and Leonard 1997). Recruitment into high intertidal habitats probably enhances

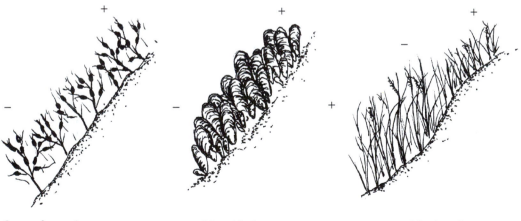

Seaweed canopies   Mussel beds   Marsh turfs

🐚 15 Variation in the nature of neighbor interactions across physical stress gradients. In intertidal seaweed and mussel beds, neighbors buffer one another from desiccation stress at high intertidal heights, but compete with one another at lower, less stressful tidal heights. In intertidal marsh grass stands, neighbors ameliorate low oxygen levels at low intertidal heights and high soil salinities at high intertidal heights.

further recruitment and growth by limiting physical stress and allowing the canopy to move to the maximum tidal height where a dense group can survive. The sharp high intertidal boundaries of dense stands of seaweeds, mussel beds, seagrasses, marsh plants, and sessile invertebrates in physically stressful habitats, in general, may be the product of the reliance of these assemblages on positive switches.[🫘16]

The dependence of habitat-ameliorating positive interactions on stressful physical conditions and high densities may lead to important, but largely unexplored, relationships between recruitment, biogeography, and the role of positive interactions in rocky intertidal assemblages. Heavy recruitment may be necessary to generate the densities necessary for organisms to survive in high intertidal environments. Thus, the supply of recruits may control positive switches. Biogeographically, the role that habitat-ameliorating group benefits play in setting the high intertidal limits of sessile space holders on rocky beaches may vary predictably with large-scale (geographical) variation in thermal stress. In particular, the importance of these interactions may increase with decreasing latitude and in bays and estuaries paralleling gradients in summer thermal and desiccation stresses.

The importance of positive interactions in intertidal systems is not limited to rocky shores. Positive interactions may be pervasive, but largely overlooked, community processes in marsh plant and soft-sediment invertebrate assemblages and wave-stressed habitats as well. In marsh plant communities, positive associations among neighbors resulting from habitat amelioration are also common, but are driven by different stresses. For the vascular plants of marshes, exposure to anoxic, waterlogged soils and the

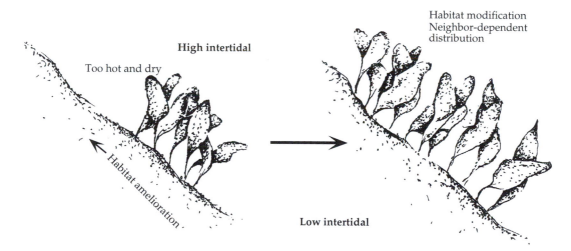

🫘 16 Habitat amelioration by intertidal seaweed canopies leads to seaweeds living at higher tidal elevations than those at which solitary individuals are able to survive. (After Bertness and Leonard 1997.)

osmotic stress caused by salty soils are often the harshest physical conditions experienced (Bertness 1992). At low intertidal heights, where soil oxygen levels are depleted, the group benefits of soil oxidization are commonly an important aspect of plant persistence (Howes et al. 1986).[🐚17] At higher marsh elevations, plant neighbors are often necessary to buffer individuals from high soil salinity, since without neighbors, evaporation of soil water can lead to salt concentrations too high for plant survival (Bertness and Hacker 1994).

In soft sediments, many organisms are buffered from heat and desiccation stresses by burial in sediment and, as such, are unlikely to benefit from neighbors in ways organisms on rocky shores do. A major physical stress in soft substrates, however, is sediment mobility, which can disrupt, move, or kill organisms. Consequently, organisms that stabilize the substrate may have strong positive effects on infaunal species. Tube-building worms (Gallagher et al. 1983; Schaffner 1990; Eckman 1979, 1983), seagrasses (Peterson 1986; Orth 1977), and marsh grasses (Eckman 1985) all play a positive role in infaunal communities by stabilizing the substrate and limiting the effects of a major physical stress.[🐚18]

Positive associations appear to be pervasive forces in intertidal assemblages for the simple reason that neighbors can often buffer physical stresses for one another, whether the stress is heat, desiccation, high soil salinity, low

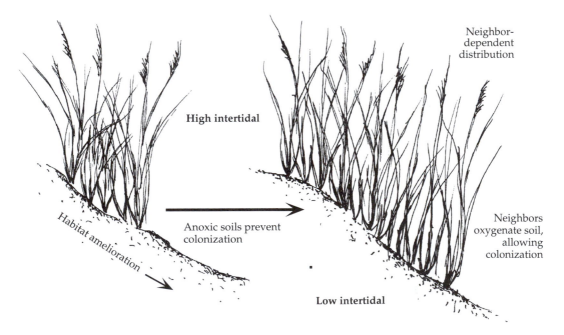

Neighbor-
dependent
distribution

High intertidal

Habitat amelioration

Anoxic soils prevent
colonization

Neighbors
oxygenate soil,
allowing
colonization

Low intertidal

🐚17 Habitat amelioration by dense stands of cordgrass. Cordgrass invades low intertidal marsh habitats, oxygenates anoxic soil, and allows colonization of habitats unsuitable for seedlings or solitary plants. (After Howes et al 1986; Bertness 1991b.)

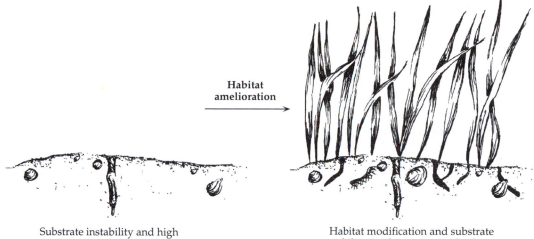

Habitat amelioration →

Substrate instability and high predation preclude a rich infaunal community

Habitat modification and substrate stabilization by seagrasses leads to a rich infaunal community

🐚 18 Habitat amelioration by seagrasses. Seagrass colonization stabilizes sediments, decreasing disturbance and limiting consumer access. This enhances further colonization and the development of a rich infaunal and epifaunal community.

soil oxygen levels, or disturbance. High-stress habitats, however, are typically characterized by low consumer pressure (Menge 1976, 1978). Conversely, at low levels of physical stress, consumer pressure is likely to be high. High consumer pressure can drive positive associations within and among species in which living in a group benefits individuals by protecting them from consumers. These **associational defenses** may entail trading off access to space, food, or light, but provide, on balance, overall benefits for participants living in habitats with high potential consumer losses.[🐚 19]

Associational defenses appear to be important community structuring forces in shoreline habitats. Mussel and oyster beds that protect juveniles and associated organisms from consumers (Bertness and Grosholz 1985; Stiven and Gardner 1992) and palatable algae and sessile invertebrates that survive in association with chemically and physically defended sessile organisms (Hay 1986; Littler et al. 1995) are conspicuous examples. A variety of fishes, shrimp, snails, clams and other small organisms live protected from their enemies by marsh grasses (Vince et al. 1976, Kneib 1997), seagrasses (Thayer et al. 1979; Peterson 1982a; Summerson and Peterson 1984), and seaweeds (Hay 1981; Duffy 1990; Carr 1994). Most of these biological refuges from consumers entail costs in the form of of decreased growth rates in comparison to conspecifics living at low densities without cover, but when consumer pressure is high, groups are necessary for survival despite the costs.[🐚 20] Based on the inverse correlation between physical stress and consumer pressure found in most intertidal habitats, positive interac-

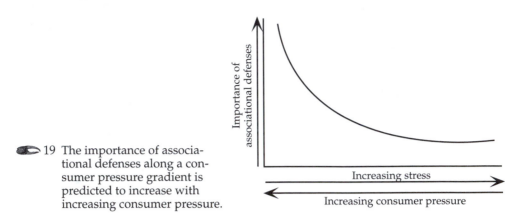

🐚 19 The importance of associational defenses along a consumer pressure gradient is predicted to increase with increasing consumer pressure.

tions should be characteristic of high and low physical stress habitats, but not common or important in habitats with intermediate stress levels.

CONSUMER EFFECTS

The general importance of consumers in affecting distributions, abundances, and species diversity in intertidal assemblages was demonstrated by Bob Paine (Paine 1966, 1974, 1976) and his students in the Pacific Northwest

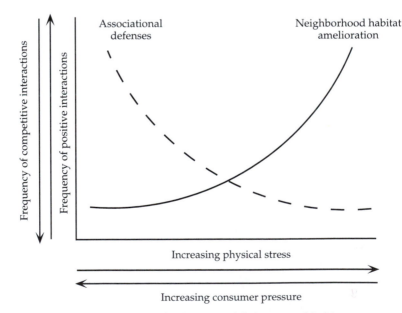

🐚 20 The predicted importance of associational defenses and habitat-ameliorating positive interactions across a physical stress gradient. (After Bertness and Callaway 1994.)

(Dayton 1971; Menge 1976; Lubchenco 1978).[🐚21] Paine's work on the role of predators in intertidal communities was the first clear demonstration that predators, by limiting the distribution of a dominant spatial competitor, could have profound top-down effects on community structure. Paine's experiment was simple. He noted that the large open-coast horse mussel, *Mytilus californianus*, was typically restricted to a band in the middle intertidal zone just above the distribution of its predator, the starfish *Pisaster ochraceous*. *Mytilus* was assumed to be excluded from high intertidal elevations due to physical stresses during low tide. It was less clear why the mussels weren't found lower in the intertidal habitat. Paine hypothesized that

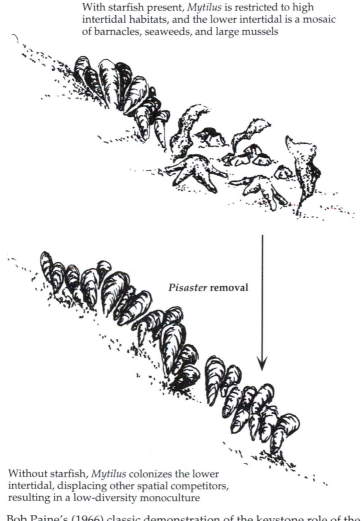

With starfish present, *Mytilus* is restricted to high intertidal habitats, and the lower intertidal is a mosaic of barnacles, seaweeds, and large mussels

Pisaster **removal**

Without starfish, *Mytilus* colonizes the lower intertidal, displacing other spatial competitors, resulting in a low-diversity monoculture

🐚 21   Bob Paine's (1966) classic demonstration of the keystone role of the starfish *Pisaster* in rocky intertidal community structure in the Pacific Northwest.

mussels were limited from living at lower elevations by *Pisaster* predation, and tested this idea by removing the starfish from large plots.

Paine's results were dramatic. Removing the starfish led to the mussels invading the low intertidal habitat and competitively excluding other sessile space holders (barnacles, other mussel species, and algae) that typically co-occurred with *Pisaster*. By limiting the success of a potentially dominant spatial competitor, starfish predation not only limited mussels to a high-elevation refuge habitat, but also maintained local species diversity by permitting the coexistence of subordinate competitors. Paine coined the term **keystone predator** to describe the major role played by *Pisaster* in these communities. The keystone predator label is now commonly used to identify consumers that have important top-down effects on the structure, dynamics, and species diversity of communities that are far greater than would be predicted on the basis of their abundance alone (see Mills et al. 1993; Power et al. 1996 for discussion of the keystone species concept). The keystone species concept has also been generalized and applied to species other than consumers that have an unusually large effect on the communities they live in (e.g., keystone positive interactions: Hacker and Gaines 1997; keystone facilitators or bioengineers: Jones et al. 1994; Bruno and Rand, unpublished data).

Direct Effects

Examples of strong top-down consumer effects on shoreline communities are found in all major habitats on the Atlantic coast of North America. On rocky shorelines in New England, starfishes and predaceous snails limit the lower vertical distributions of mussels and barnacles (Menge 1976), herbivorous snails limit the intertidal distributions of seaweeds (Lubchenco 1978, 1983; Bertness 1984a), and sea urchin grazing in shallow subtidal habitats can limit the distribution of other seaweeds (Paine and Vadas 1969; Himmelman 1983). In marsh plant communities, predation by fishes and blue crabs typically limits the distributions of small crustaceans, snails, and juvenile fishes to high marsh elevations, where they are dependent on marsh grass cover for protection from consumers (Vince et al. 1976; Van Dolah 1978; Stiven and Gardner 1992). In southern soft-sediment habitats, predation and bioturbation (sediment disturbance by feeding or mating organisms) by blue crabs, horseshoe crabs, and whelks restricts many prey species to deep sediments and shallow-water refuge habitats (Woodin 1976; Virnstein 1977; Peterson 1982; Lipcius and Hines 1986).[🐚22]

The common denominator of all these examples is that predators can restrict potentially abundant prey to refuge habitats where they escape predation, but at the cost of reduced growth. The most common refuge of intertidal organisms is living at high elevations out of the reach of marine predators. This strategy directly affects prey species by restricting them to high intertidal heights, but also indirectly affects them by forcing them to live where their growth and reproduction may be limited by physical stresses.

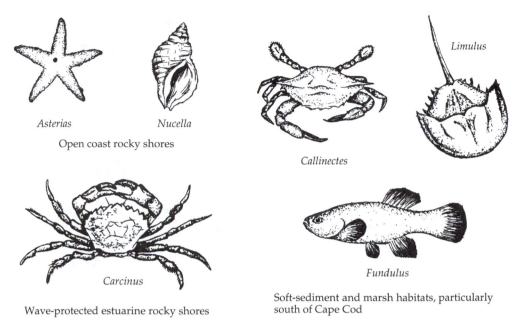

Asterias   Nucella

Open coast rocky shores

Limulus

Callinectes

Carcinus

Wave-protected estuarine rocky shores

Fundulus

Soft-sediment and marsh habitats, particularly south of Cape Cod

22 Important predators in Atlantic shoreline communities.

Indirect Effects

The indirect effects of consumers—effects other than the simple spatial restriction of prey species—can be just as important as their direct effects in the structure of intertidal communities (Menge 1995). Indirect consumer effects can result by a number of mechanisms, including restricting prey to habitats where their growth rates are low, as described above, habitat modification, trophic cascades, and positive and negative feedbacks (see Menge 1995; Wootton 1992, 1993 for detailed discussions).

Indirect consumer effects resulting from habitat modification are some of the most dramatic. Sea urchin control of kelp beds is the most widely cited example of this type of effect. Kelp beds flourish in many shallow coastal habitats, but where urchins are common, they are capable of entirely eliminating kelps from these habitats, leading to **urchin barrens**. Urchin barrens are dominated by hard substrates covered by calcified encrusting algae and little else (Himmelman 1983). When urchin populations are kept in check, however, kelp beds flourish, leading to the development of a diverse understory community (Estes et al. 1978). Kelp beds not only support large populations of mobile invertebrates and fishes dependent on kelp cover and primary production, but also alter the bottom community by slowing water movement, leading to the deposition of sediment and larvae (Eckman et al. 1989; Duggins et al. 1990). Thus, by controlling the distribution of kelp beds, sea urchins indirectly control the structure of these habitats.[23] Consumers often have strong effects on communities by controlling the abun-

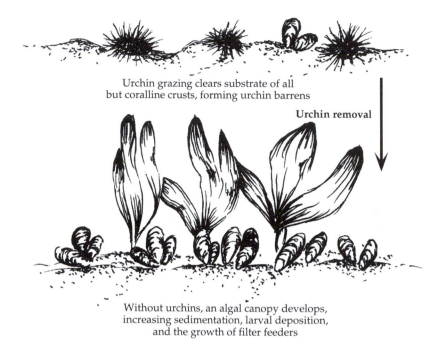

Urchin grazing clears substrate of all
but coralline crusts, forming urchin barrens

Urchin removal

Without urchins, an algal canopy develops,
increasing sedimentation, larval deposition,
and the growth of filter feeders

🐚 23 Habitat modification by sea urchins.

dance of habitat-modifying organisms. Periwinkle snails that control sedimentation on rocky beaches (Bertness 1984a, see Chapter 5); starfishes and snails that control the distribution of mussel beds, which stabilize soft sediments and provide habitat for a large assemblage of associated organisms (Suchanek 1986); and rays and crabs that dig up sediment-stabilizing seagrass beds, leading to large shifts in bottom assemblages (Orth 1975) all provide good examples of consumer-mediated habitat modification of Atlantic shorelines.

A much more subtle, but probably more typical, indirect effect of predators on shoreline communities involves trophic cascades. **Trophic cascades** occur when consumers limit the abundance of their prey, releasing lower trophic levels from consumers and thus increasing population densities at lower trophic levels. Trophic cascades have been shown to maintain populations of small anemones (*Nematosella*) that are common in marsh habitats (Kneib 1988). These anemones are the preferred prey of grass shrimp, which are themselves eaten by adult killifish. By limiting grass shrimp populations, killifish predation leads to the persistence of anemones in association with killifish.[🐚 24] Similarly, on rocky shores, bird predation on sea urchins can have strong indirect effects on algal abundance by releasing seaweeds from their consumers (Lubchenco 1978; Dumas and Witman 1993). Trophic cascades may be particularly common in marine habitats with simple food

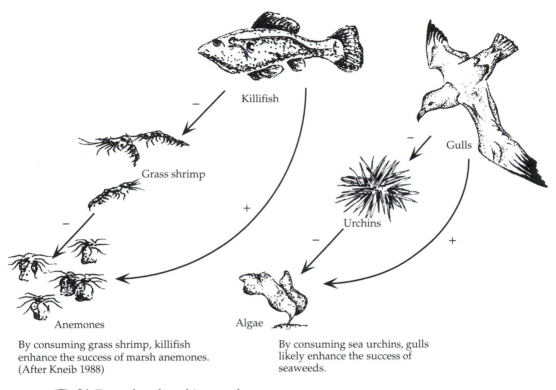

Killifish

Grass shrimp

–

+

–

Anemones

By consuming grass shrimp, killifish enhance the success of marsh anemones. (After Kneib 1988)

Gulls

–

Urchins

–

+

Algae

By consuming sea urchins, gulls likely enhance the success of seaweeds.

🌀 24 Examples of trophic cascades.

webs, since responses to consumers are relatively direct and rapid (Strong 1992). In more diverse systems, trophic cascades may be dampened by the presence of interchangeable (overlapping) members at each trophic level.

Other indirect effects are the consequence of organisms sharing resources and/or consumers, which can result in either positive or negative effects. The coexistence of barnacles, algal crusts, and grazing snails on many rocky shores is a good example of a system strongly influenced by indirect effects (Dungan 1986).[🌀 25] In this system, algal crusts keep densities of barnacles low by limiting barnacle larval settlement, since barnacle larvae avoid algal crusts. Grazing on algal crusts by herbivorous snails, however, increases barnacle cover by providing settlement space for barnaclelarvae, which in turn reduce the density of grazing snails by preempting foraging space. As a consequence of this chain of indirect effects, the fates of barnacles, algal crusts, and snails are tightly linked, and factors that influence the success of any of these species can have strong indirect effects on the other components.

Indirect effects leading to commensalisms or mutualisms may be common in marine intertidal communities, but require painstaking experimental work to uncover (for discussion see Wootton 1992, 1993). One of the best-

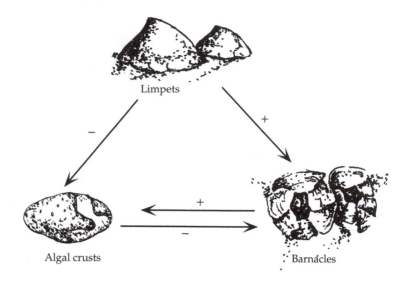

Limpets

Algal crusts

Barnacles

Barnacles positively affect their algal crust spatial competitors by limiting limpet foraging efficiency. Limpet grazing on algal crusts provides settlement space for barnacles. (After Dungan 1987.)

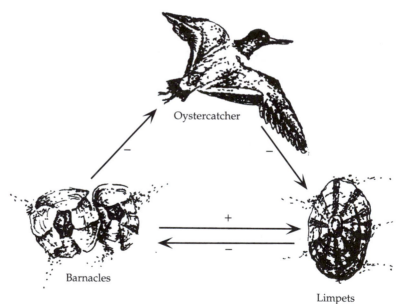

Oystercatcher

Barnacles

Limpets

25 Indirect consumer effects in shoreline habitats.

Limpets and barnacles compete for space, but barnacles camouflage limpets from their bird predators. (After Wootton 1993.)

studied indirect interactions of this type is the relationship between chitons and limpets in the San Juan Islands of Washington State.[◐ 26] Both chitons and limpets are herbivorous grazers that utilize a scraping radula to feed on algae. Chitons and limpets had long been thought to be competitors.

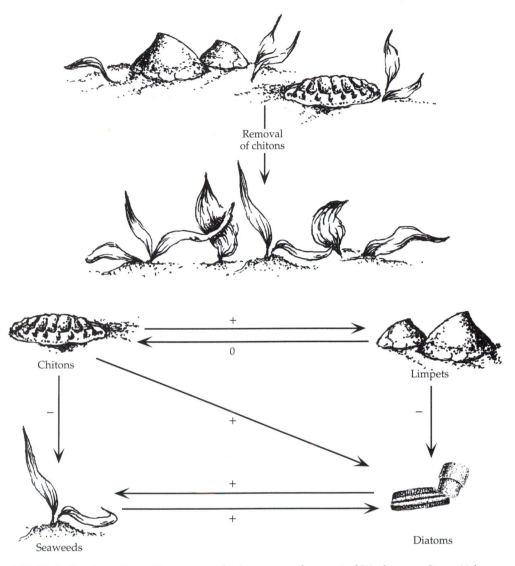

26 Indirect positive effects among herbivores on the coast of Washington State. (After Dethier and Duggins 1984.)

Experimental removal of chitons, however, led to a rapid decline and local extinction of limpets, showing that limpets are dependent on chitons (Dethier and Duggins 1984). This surprising result was due to the food preferences of these herbivores and the competitive relationship between their algal food sources. While chitons prefer macroalgae, limpets feed primarily on diatoms, which are competitively subordinate to seaweeds. As a result, limpets depend on chitons to remove seaweeds to increase their diatom food supply.

Many other indirect effects are negative. **Apparent competition**, for example, occurs when two organisms do poorly when found together due to a shared enemy rather than competition for a shared resource. In apparent competition, consumer efficiency is enhanced by the presence of alternative prey, and thus prey populations are adversely affected by the presence of alternative prey (Holt 1977). This interaction is particularly difficult to identify because the joint depression of populations when species co-occur can be easily falsely interpreted as evidence for competition for limited resources. Consequently, it is not known how common apparent competition is. It has been demonstrated in subtidal rocky habitats in southern California. In these habitats, spiny lobsters and octopus prefer eating jewel box clams on rocky beaches, but also eat herbivorous snails in adjacent cobble areas (Schmidt 1987). Manipulation of clam and snail densities, however, revealed that both prey types were subject to more predation in the presence of the other than when alternative prey were not available. Such reciprocally negative effects among populations that are driven indirectly by predators rather than by competition for resources may be common (also see Grosholz 1992).

INTERACTIONS AMONG THE FORCES THAT SHAPE SHORELINE COMMUNITIES

Although competition, positive interactions, and predation all influence the structure of shallow-water communities, these forces do not act independently to generate patterns. To develop a working model of the forces responsible for the widespread spatial patterns found in marine communities, we need to understand how those forces interact with one another. In particular, we need to know how these biological factors interact across the steep physical gradients found in most intertidal habitats and are affected by variation in recruitment.

Bruce Menge and John Sutherland (1976, 1987) have developed a conceptual framework for examining how the community structuring processes we have just discussed interact to influence patterns. Menge and Sutherland's heuristic model predicts how the relative importance of biotic processes in structuring communities will change across gradients of physical stress and recruitment intensity.[🐚 27] With a high supply of recruits, their model predicts that competition will play little role in shaping communities at either high or low levels of physical stress. Under high-stress conditions, populations will be limited by physical stress, precluding competition from limiting population densities. Conversely, at low levels of physical stress, competitive effects will be minimal because of reduction in densities by consumers. Thus, the model predicts that competition for space and other limiting resources will be a dominant community structuring force only at intermediate levels of physical stress, where consumer pressures are not severe. A key component of the model is an inverse relationship between physical stress and consumer pressure. This relationship has been empirically demonstrated in a variety of shoreline assemblages (Menge 1976, 1978; Paine 1974; Dayton 1971) and appears to be the simple product of predators having difficulty foraging in physically stressful habitats and failing to meet their metabolic demands.

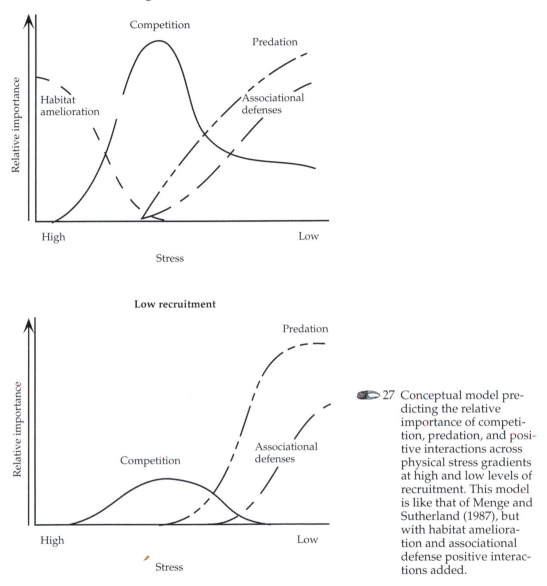

27 Conceptual model predicting the relative importance of competition, predation, and positive interactions across physical stress gradients at high and low levels of recruitment. This model is like that of Menge and Sutherland (1987), but with habitat amelioration and associational defense positive interactions added.

With low recruitment, the model predicts that space will not be limiting, and thus that competition will play a reduced role.

Positive interactions can be easily added to Menge and Sutherland's model (Bertness and Callaway 1994). With high recruitment and high levels of physical stress (where competitive effects are small), habitat amelioration leading to positive interactions among neighbors is predicted to be important. In contrast, at low levels of physical stress (where consumer pressure is usually high), associational defenses in which neighbors benefit one another to reduce predation losses are predictably common (see Hay 1986). At intermediate levels of physical stress and consumer pressure, positive interactions are predicted to be weak. Thus on rocky shores with high recruitment, the revised Menge and Sutherland model predicts that (1) habitat-ameliorating positive interactions will be important at stressful high intertidal heights, (2) competition for space will be critical at intermediate tidal heights, and (3) both predation and associational defenses will be important at low intertidal heights characterized by minimal physical stress.

These predictions change markedly when recruits are in low supply. Neither habitat-ameliorating nor associational defenses are predicted to be important, since low recruitment leads to densities insufficient to yield these benefits. Similarly, competition for space and resources plays a much less important role in shaping low-recruitment communities, since population densities may not be high enough for resources to be limiting. Predation, however, can still play a dominant role in recruitment-limited communities unless the predators themselves are recruitment-limited. Applying these model modifications to rocky intertidal communities with low recruitment generates predictions that are very different from those for communities with high recruitment.

The relationship between physical stress, biotic interactions, and species diversity, or the number of species present in a community, has also been examined with a simple didactic model (Menge and Sutherland 1976, 1987; Connell 1978; Hacker and Gaines 1997).[28] In this model, predicted species diversity is examined across a gradient of physical stress, assuming the same negative relationship between physical stress and consumer pressure described above. With unlimited recruitment, low species diversity is predicted at high levels of physical stress due to stress limitation of population densities, typical of many high intertidal habitats. Low species diversity is also predicted at low levels of physical stress due to consumer limitation of species numbers. This occurs at low intertidal heights on rocky shores. At intermediate levels of physical stress and consumer pressure, maximum species richness is found, since neither stress nor consumers entirely limit species numbers. Positive interactions influence this model at high recruitment densities by increasing species numbers at both high physical stress and high consumer pressure levels by habitat amelioration and associational defenses, respectively (Hacker and Gaines 1997).

Like the Menge and Sutherland model that predicts the relative importance of different biotic processes across a physical stress gradient, the species diversity model also makes very different predictions at low recruitment densities. With low recruitment, neither habitat-ameliorating nor associational defense-positive effects are expected to occur, and competition for space and other resources is minimal. At low recruit densities, the supply of recruits alone may commonly dictate species numbers.

NATURAL DISTURBANCE PROCESSES

While it is appealing to think of communities as stable assemblages of organisms in which levels of physical stress, consumer pressure, positive interac-

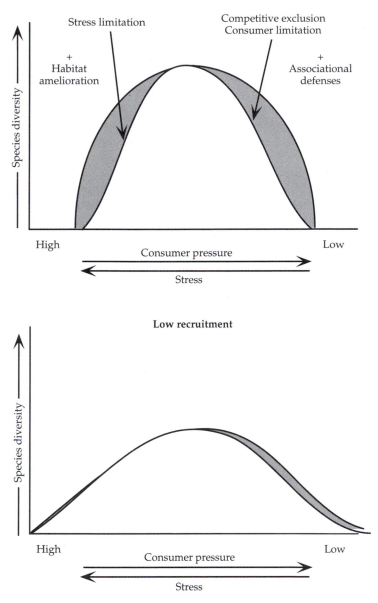

28 The influence of physical stress, consumer pressure, and positive interactions on community species diversity at high and low levels of recruitment. The shaded area shows the contribution of positive interactions. (After Hacker and Gaines 1997.)

tions, and recruitment dictate community process and composition, this simple picture is rarely accurate. In reality, most natural communities are disturbed frequently enough so that they are best envisioned as nonequilibrium assemblages, or mosaics of habitat patches in different stages of successional recovery from natural disturbances. Physical or biological disturbances that kill organisms can affect patterns of abundance and distribution. In terrestrial plant communities, fires, hurricanes, and herbivory play a major role in shaping communities (Pickett and White 1985). A variety of disturbances are commonly important in coastal habitats as well. Currents and waves can dislodge organisms (Denny et al. 1985), ice scour can destroy organism on all types of shorelines, water-borne objects can crush sessile organisms (Dayton 1971), and on rocky beaches, cobbles can be overturned, killing attached organisms (Osman 1977; Sousa 1979a,b). Soft-sediment habitats are particularly vulnerable to disturbance, since the sediments themselves are easily transported. Water movement (Grant 1983), ice scour (Whitlatch 1981), and foraging crabs (Woodin 1978) are all common disturbance agents in soft-substrate habitats. In marsh and seagrass habitats, floating plant debris (**wrack**) can cover and smother dense plant stands (Reidenbaugh and Banta 1982).

Disturbance can strongly affect the types and numbers of organisms inhabiting a shoreline. Heavily disturbed habitats are often dominated by organisms that can either resist the disturbance or rapidly colonize after a disturbance. Wave-swept shores, for example, are usually dominated by seaweeds and small sessile organisms that can withstand strong wave forces, along with ephemeral algae that can rapidly invade denuded rock surfaces.

The **intermediate disturbance hypothesis**, originally proposed to describe species richness patterns in tropical rain forests and coral reefs (Connell 1978), attempts to explain species abundance and distribution patterns in a wide range of habitats (Petraitis et al. 1989).[⬤▷ 29] In both terrestrial and marine habitats, maximum local species diversity is commonly found in habitats with an intermediate frequency and intensity of disturbance. At low levels of disturbance, competitive dominants often monopolize habitats, displacing competitive subordinates. At high levels of disturbance, few species are able to persist or invade. At intermediate levels of disturbance, however, competitive dominants coexist with competitive subordinates and fugitives, since disturbance is too frequent to allow competitive exclusion but too infrequent to totally eliminate subordinates.

Particularly good examples of the role played by disturbance in maintaining species numbers in shallow-water marine communities are found in cobble beach and marsh plant habitats. On cobble beaches, small cobbles are vulnerable to overturning and usually harbor few species, while larger cobbles typically are covered with competitive dominants. In contrast, medium-sized cobbles often harbor a mixture of ephemeral fugitives and dominants (Osman 1977; Sousa 1979a,b).[⬤▷ 30] Similarly, in northern salt marsh habitats, floating plant debris is often a severe, but spatially predictable, disturbance. Marsh areas that are rarely disturbed are usually monopolized by competitive dominants, whereas marsh areas that are heavily disturbed every year are inhabited by resistant species that can survive or rapidly recover from disturbance. Marsh areas that suffer an intermediate amount of disturbance usually support mixtures of both competitive dominants and fugitives (Bertness and Ellison 1987).

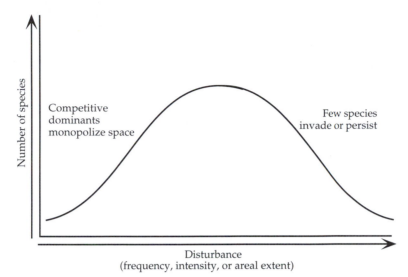

Number of species

Competitive
dominants
monopolize space

Few species
invade or persist

Disturbance
(frequency, intensity, or areal extent)

❧ 29 The intermediate disturbance hypothesis. (See Connell 1978.)

In habitats occupied by competitive dominants, fugitive species are often dependent on disturbance for local persistence. One of the best examples of this sort of disturbance dependence is the sea palm, *Postelsia palmaeformis*, which lives on wave-exposed shores off the west coast of North America. *Postelsia* lives in bare patches within beds of the large open-coast horse mussel, *Mytilus californianus*. *Mytilus* is the competitive dominant in these habitats, and when disturbance-generated holes are made in mussel beds, mussels rapidly recruit, either sexually or by lateral movement to fill

Algal turfs and high barnacle densities completely displace other organisms, resulting in low species diversity

A mixture of rapidly colonizing ephemerals and competitive dominants coexist at intermediate disturbance levels

Only rapidly colonizing ephemeral algae are present, resulting in low species diversity

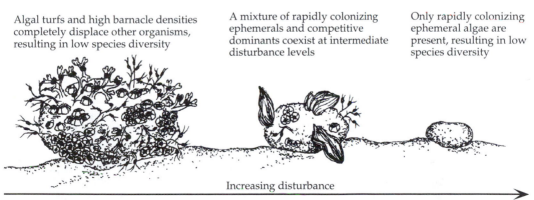

Increasing disturbance

Large cobble that rarely moves

Small mobile cobble

❧ 30 The intermediate disturbance hypothesis illustrated on mobile substrate on cobble beaches. (After Osman 1977; Sousa 1979a.)

the patch. *Postelsia* is entirely dependent on frequent disturbances for local persistence. It is displaced in low-disturbance habitats by mussels, but can tolerate extreme wave stress conditions (Dayton 1973; Paine 1979).[🐚31] In New England salt marshes, the fugitive annual plant *Salicornia europaea* is another good example of a disturbance-dependent fugitive. *Salicornia* recruits to disturbance-generated bare patches, but is rapidly outcompeted by the invasion of clonal turf plants. *Salicornia* persistence in these marshes is entirely dependent on its ability to colonize new disturbance-generated bare space (Ellison 1987a).

Intertidal habitats, in general, are best characterized as mosaics of disturbed habitat patches in varying stages of recovery. Examples include rocky intertidal habitats (Dayton 1971; Paine and Levin 1981; Sousa 1979a, Osman 1977), marsh plant communities (Bertness and Ellison 1987), and soft-substrate habitats (Woodin 1978, 1981; Gallagher et al. 1983; Zajac and Whitlatch 1982, 1985). Paine and Levin's (1981) study of wave-swept rocky shores of the Pacific coast of Washington State is one of the classic examples illustrating the importance of patch dynamics in intertidal communities. Paine and Levin monitored the formation and closure of wave-generated bare patches in *Mytilus californianus* mussel beds for over a decade. They found that most bare patches are generated when winter storms rip mussels from beds, leaving gaps of up to 35 square meters, which require two to four years for recolonization. The recovery of these bare patches is very predictable.[🐚32] Following mussel removal, bare substrate is initially colonized by diatoms, which are followed over the next year by a variety of fugitive space holders. These fugitives include seaweeds, acorn barnacles, and the smaller, rapidly colonizing blue mussel, *Mytilus edulis*. After three years, patches are dominated by blue mussels, coralline algae, and gooseneck barnacles. After four years, most patches are again dominated by horse mussels, the competitive dominant in the system.

🐚31 The sea palm, *Postelsia*, lives in mussel bed gaps on the coast of Washington State.

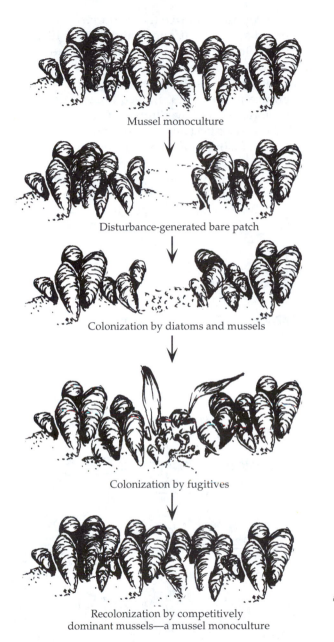

Mussel monoculture

Disturbance-generated bare patch

Colonization by diatoms and mussels

Colonization by fugitives

Recolonization by competitively
dominant mussels—a mussel monoculture

32 Recovery from disturbance to mussel beds. (After Paine and Levin 1981.)

MECHANISMS OF SECONDARY SUCCESSION

A disturbance event may be followed by new individuals invading the disturbed habitat. The recolonization of disturbed habitats often triggers interactions among the colonizers that lead to predictable species replacements and the ultimate dominance of specific species. This process is referred to as **secondary succession**.

Three mechanisms have been proposed to drive the secondary succession of natural communities after a disturbance (Connell and Slatyer 1977). In the **inhibition** model, the initial colonizers, by monopolizing resources, inhibit the recruitment of other potential colonizers by preemptive competition. In this model, therefore, secondary succession is basically a first-come, first-served process. In the **tolerance** model, the initial colonizers do not entirely preclude other potential colonizers, and over time, the most tolerant colonizers persist. In this model, the species best adapted to local conditions ultimately dominate the habitat. In the **facilitation** model, the initial colonizers modify the habitat, improving conditions for new colonists. Initial colonization is necessary for succession in this model, and the ultimate dominants are not able to invade bare patches on their own without some form of habitat amelioration by the initial colonizers.[🫘 33]

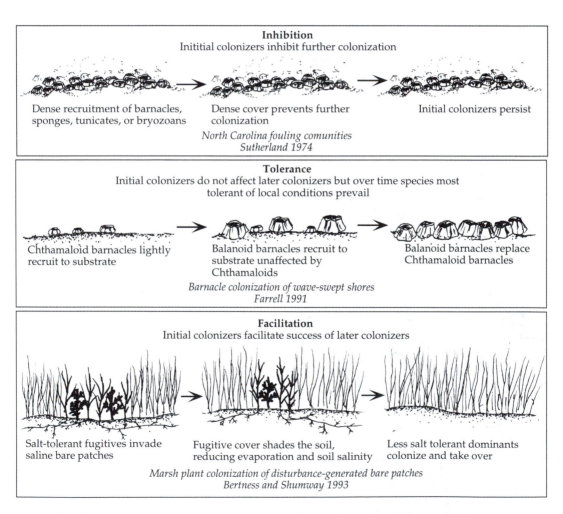

Inhibition
Inititial colonizers inhibit further colonization

Dense recruitment of barnacles, sponges, tunicates, or bryozoans

Dense cover prevents further colonization

Initial colonizers persist

North Carolina fouling comunities
Sutherland 1974

Tolerance
Initial colonizers do not affect later colonizers but over time species most tolerant of local conditions prevail

Chthamaloid barnacles lightly recruit to substrate

Balanoid barnacles recruit to substrate unaffected by Chthamaloids

Balanoid barnacles replace Chthamaloid barnacles

Barnacle colonization of wave-swept shores
Farrell 1991

Facilitation
Initial colonizers facilitate success of later colonizers

Salt-tolerant fugitives invade saline bare patches

Fugitive cover shades the soil, reducing evaporation and soil salinity

Less salt tolerant dominants colonize and take over

Marsh plant colonization of disturbance-generated bare patches
Bertness and Shumway 1993

🫘 33 Mechanisms of secondary succession. (After Connell and Slatyer 1977.)

Evidence for the relative importance of these different successional models in shallow-water marine communities is mixed. In many space-limited hard-substrate habitats with unlimited larval recruitment, early recruits often saturate the substrate and inhibit further settlement (Sutherland 1974; Sutherland and Karlson 1977). Recruitment to soft substrates of bioturbating deposit feeders, which can kill later recruits, has also been suggested to lead to inhibition succession (Rhoads and Young 1970, 1971). Less evidence supports the tolerance model of succession in intertidal habitats, but this may reflect a bias in the locations where studies have been done rather than the true importance of this mechanism in nature. The classic example of the tolerance model is in forests, where seedlings recruit and persist as stunted individuals in the understory until canopy trees die, allowing the tolerant understory trees to ultimately dominate. Examples of the tolerance model in intertidal habitats are rare, since initial recruits are usually found to have at least some negative effects on further recruitment. Terry Farrell (1991) has demonstrated that the tolerance model can describe rocky shore succession where initial recruitment is not high enough to saturate substrate space and limit further colonization. With high levels of recruitment, however, most examples of succession on rocky shores are probably best described by the inhibition model.

The facilitation model of succession was once widely believed to be the most common succession mechanism in nature (Clements 1916; Odum 1969), even though little empirical evidence actually supported this view (Connell and Slatyer 1977). Recent evidence from a wide range of marine communities, however, has suggested that facilitated succession, while not the rule in benthic marine communities, is not uncommon. On rocky shores, barnacles facilitate the recruitment of seaweeds and mussels by providing suitable settlement sites and limiting herbivore pressure (Menge 1978; Lubchenco 1980, 1982, 1983; Farrell 1991). In soft substrates, tube-building colonizers stabilize the substrate, limit mortality, and enhance the passive deposition of larvae, leading to facilitated succession (Eckman 1979, 1983; Gallagher et al. 1983). Finally, in marsh plant and seagrass communities, early colonizers often appear to facilitate the success of later colonizers. By stabilizing sediments and enhancing sediment nutrient content, initial seagrass colonizers facilitate further colonization (Williams 1980). In marsh plant communities, early colonizers commonly facilitate further colonization by shading soils, limiting the evaporative buildup of potentially limiting soil salts, and oxidizing reduced soils (Bertness and Shumway 1993; Callaway 1994).

On balance, evidence to date suggests that no single successional model dominates shoreline systems, but that each of the models may be predictably common under certain conditions.[◐ 34] The inhibition model appears to be common in physically benign habitats that experience high recruitment and densities of primary space holders that can limit further colonization (Sutherland 1974; Sousa 1979a,b; Breitburg 1984; Dean and Hurd 1980; Sutherland and Karlson 1977). The tolerance model appears to be common where recruits are in low supply, space is not limiting, and initial recruits have little effect on later arrivals (Farrell 1991). Facilitated succession

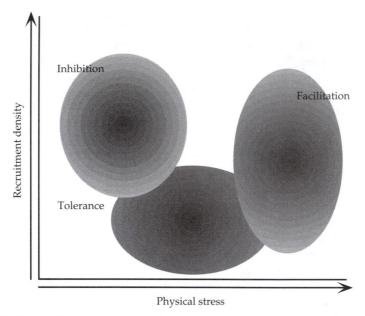

🦞 34 The predicted occurrence of mechanisms of secondary succession as a function of physical stress and recruitment density.

is common under harsh physical conditions where early recruits can ameliorate stress (Connell and Slatyer 1977; Farrell 1991; Bertness and Callaway 1994), and may be relatively insensitive to the supply of recruits. Further work is needed to determine whether these different models of succession occur predictably under specific physical and biotic conditions across different habitat types and recruitment regimes.

Bioengineering

One of the most conspicuous deficiencies in our current understanding of how natural communities are organized is a lack of appreciation for the role played by habitat-modifying organisms. The activities of these organisms provide habitats for other organisms, and often lead to entire communities being dependent on them. Habitat-modifying organisms play leading roles in most shoreline communities, but have generally not been appreciated as critical structuring forces. Current models of shoreline community organization focus on interactions among competititors, consumers, physical stress, and disturbance, and treat habitat modification effects as idiosyncratic features of systems. The effects of habitat-modifying organisms, however, cannot be ignored if we are to adequately understand the forces that generate patterns in shallow-water marine communities.

Clive Jones, John Lawton, and Moshe Shachak (1994) have recently suggested that organisms that affect communities by providing habitats or converting resources into usable forms be termed **bioengineers**. They argue that bioengineers play a major role in the structure and function of most natural communities and need to be incorporated into the conceptual framework of ecology. They further suggest that bioengineers that provide living space for others be termed **autogenic bioengineers,** and that those that convert resources from one state to another be termed **allogenic bioengineers**. Both autogenic and allogenic engineering are important in many marine communities.

☆ *Prominent Atlantic Coast Bioengineers*			
EXAMPLE	**HABITAT**	**EFFECT**	**REFERENCES**
Autogenic engineers			
Seagrasses	Shallow, subtidal soft sediments	Stabilize soft substrates, enhance settlement, predator protection	Orth 1977; Peterson 1982a; Irlandi and Peterson 1991
Marsh grasses	Intertidal soft sediments	Stabilize soft substrates, enhance settlement, predator protection	Vince et al. 1976; Kneib 1982, 1991, 1997; Eckman 1979, 1985
Seaweed canopies	Intertidal hard substrates	Enhance settlement, predator protection, ameliorate physical stresses	Menge 1976; Eckman et al. 1989; Duggins et al. 1990
Mussel beds	Intertidal and subtidal hard and soft substrates	Stabilize and bind substrates, ameliorate physical stresses, predator protection	Suchanek 1986; Bertness and Grosholz 1985; Seed 1969; Dame 1996
Allogenic engineers			
Mussel and oyster beds	Intertidal and subtidal hard and soft substrates	Convert plankton into nutrients available to benthos	Kuenzler 1961; Jordan and Valiela 1982; Bertness 1984b; Dame 1976
Infaunal bivalves	Intertidal and subtidal soft substrates	Convert plankton into nutrients available to benthos	Doering et al. 1986, 1987

Many strong autogenic engineering effects on intertidal communities have already been discussed as positive interactions. Autogenic engineers, however, can have either positive or negative effects on other species, and in general, any given autogenic engineer will positively affect some species while having negative effects on others. Seagrass beds, marsh grasses, algal canopies, and shellfish beds are all good examples of autogenic engineers. Dense assemblages of all of these organisms modify habitats, creating habitats for some species while simultaneously destroying habitats for others. A dense cover of marsh plants or seagrasses, for example, binds substrates, increases sedimentation, and provides refuge from some predators. These services enhance the populations of many small mobile organisms and filter feeders. These same dense stands of vascular plants, however, exclude many burrowers and deposit feeders unable to cope with root mats and inhibit organisms that are sensitive to sedimentation. Whether these effects are positive or negative, it is clear that these autogenic engineers play key roles in dictating process and pattern in shoreline communities.

Allogenic engineers are also common in marine habitats and can also play major roles in structuring nearshore assemblages. Bioturbating deposit feeders provide particularly good examples of allogenic engineers. By continually turning over sediments, deposit feeders oxygenate sediments and increase drainage. This enhances the success of some community members, such as rooted plants, but can virtually preclude most larval settlement. Filter-feeding organisms such as oysters, mussels, and clams are also important allogenic engineers because they transform plankton and nutrients from the water column into feces, making them available to other benthic organisms.[🐚 35]

The task of incorporating bioengineering into our understanding of shallow-water marine communities has yet to be formally undertaken, but holds great potential for elucidating process and pattern. Bioengineering

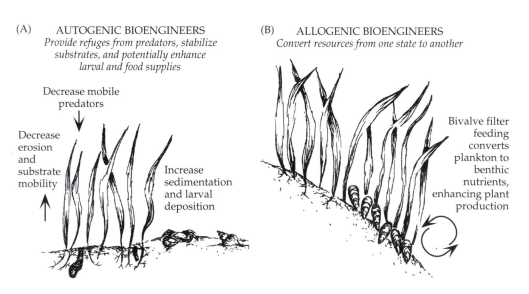

(A)   AUTOGENIC BIOENGINEERS
Provide refuges from predators, stabilize substrates, and potentially enhance larval and food supplies

Decrease mobile predators

Decrease erosion and substrate mobility

Increase sedimentation and larval deposition

(B)   ALLOGENIC BIOENGINEERS
Convert resources from one state to another

Bivalve filter feeding converts plankton to benthic nutrients, enhancing plant production

🐚 35 Bioengineer types. (A) Vascular plants in marine soft sediments are autogenic bioengineers. (B) The filler-feeding bivalves associated with the vascular plants are an example of allogenic bioengineers. (After Jones et al. 1994.)

may be a particularly important process in soft sediments, in which bioengineers can stabilize or destabilize sediments, leading to dramatic switches in communities.

SUMMARY

This chapter has examined benthic processes that influence the distribution and abundance of organisms across shoreline habitats. In contrast to water column processes such as larval dispersal, benthic processes have historically received a great deal of attention, since they are often conspicuous and easily studied. Without sufficient recruitment, however, benthic processes cannot dictate community patterns, and the relative importance of benthic processes in influencing shoreline communities is strongly linked to successful recruitment.

Most coastal habitats are characterized by striking vertical or elevational zonation of organisms across intertidal stress gradients. Benthic processes are often important in generating these zonation patterns. At high intertidal elevations. only organisms capable of living out of water for extended periods persist. Competition for space is particularly common in physically benign habitats such as the low intertidal, where physical stresses do not limit organism abundances. When recruitment does not limit population densities, attachment space may do so, leading to competition for space. Large body size and clonal organization are often associated with superior competitive ability, and large clonal organisms dominate many low intertidal, rocky intertidal, and marsh plant habitats.

Consumers often alleviate spatial limitation and reduce the role played by competition for space in benthic marine assemblages. Many slow-moving predators such as snails and starfishes, like their prey, do best in physically benign habitats, so that these consumers limit prey densities in these habitats. Consumers, under these conditions, can restrict prey to physically stressful high intertidal habitats. Consumers also have a wide variety of indirect effects on communities that can be as important, but less conspicuous, than their direct effects.

Positive interactions among organisms are also important structuring forces in littoral habitats. In physically benign habitats, associational defenses, or positive associations in which neighbors benefit one another by limiting consumer success, are common. In contrast, in many physically stressful habitats, organisms benefit from habitat amelioration due to group alleviation of thermal, desiccation, or disturbance stresses. In contrast to the attention given to the negative roles of competition and consumers, the role played by these positive associations in shoreline communities has been largely overlooked. To begin to incorporate positive interactions into our understanding of shoreline communities, I present a revised version of the didactic models of Bruce Menge and John Sutherland incorporating positive interactions.

I end this chapter by discussing the roles played by physical disturbance and habitat modification or bioengineering by organisms in coastal habitats. Physical disturbances that kill organisms and trigger recolonization and sec-

ondary succession are common in all shoreline habitats and play a conspicuous role in the patterning of most intertidal landscapes. The secondary succession mechanisms that follow disturbances, however, vary widely, and the
factors likely to determine colonization dynamics are discussed. Bioengineers, which influence shoreline communities by affecting physical habitat
characteristics or processes of resource regeneration, play a large, but largely
overlooked, role in all shoreline habitats. Examples of important bioengineers are given for all major shallow-water marine habitats of the Atlantic
coast of North America.

SUGGESTIONS FOR FURTHER READING

Bertness, M. D. and G. Leonard. 1997. The role of positive interactions in communities: Lessons from the intertidal. *Ecology* 78: 1976–1989.

Buss, L. W. 1986. Competition and community organization on hard substrates in the sea. In *Community Ecology*, J. Diamond (ed.), 517–536. Harper and Row, New York.

Farrell, T. M. 1991. Models and mechanisms of succession: An example from a rocky intertidal community. *Ecological Monographs* 61: 95–113.

Jones, C. G., J. H. Lawton and M. Shachak. 1994. Organisms as ecosystem engineers. *Oikos* 69: 373–386.

Little, C. and J. A. Kitching. 1996. *The Ecology of Rocky Shores*. Oxford University Press, New York.

Lubchenco, J. and S. D. Gaines. 1981. A unified approach to marine plant-herbivore interactions: I. Population and communities. *Annual Review of Ecology and Systematics* 12: 405–437.

Menge, B. A. and J. Sutherland. 1987. Community regulation, variation in disturbance, competition and predation in relation to environmental stress and recruitment. *American Naturalist* 130: 730–757.

Paine, R. T. 1994. *Marine Rocky Shore and Community Ecology: An Experimentalist's Perspective*. Ecology Institute, Oldendorf/Lohe, Germany.

Petraitis, P. S., R. E. Latham and R. A. Niesenbaum. 1989. The maintenance of species diversity by disturbance. *Quarterly Review of Biology* 64: 393–418.

Raffaelli, D. and S. Hawkins. 1996. *Intertidal Ecology*. Chapman and Hall, New York.

Underwood, A. J. and E. J. Denley. 1984. Paradigms, explanations and generalizations in models for the structure of intertidal communities on rocky shores. In *Ecological Communities*, D. Strong, D. Simberloff, L. G. Abele and A. Thistle (eds.), 151–180. Princeton University Press, Princeton, NJ.

Wootton, T. 1993. Indirect effects and habitat use in an intertidal community: Interaction chains and interaction modification. *American Naturalist* 141: 71–89.

Chapter 5

Rocky Shores

R ocky intertidal habitats are home to many sessile and slow-moving invertebrates, as well as algae capable of dealing with widely fluctuating physical conditions. In contrast to soft-sediment shorelines, where organisms are often buried in the substrate and thus buffered from exposure, sessile and slow-moving organisms on rocky shores routinely face extreme temperature and desiccation stresses during low tides. Rocky intertidal organisms can also be subjected to severe wave stresses that can dislodge them from the substrate, limit their mobility, and greatly influence their size and morphology.

Numerous advantages counter these disadvantages of rocky intertidal habitats, however, and they typically harbor rich assemblages of algae and sessile invertebrates. Rocky shores are bathed in plankton-rich water, and water movement over littoral habitats typically enhances the delivery of particulate food to filter-feeding invertebrates. Moreover, conditions for algal growth are excellent, and rocky intertidal habitats usually have high algal productivity, supporting large herbivore and carnivore populations.

Rocky intertidal habitats have been particularly valuable study systems for ecologists because their strong zonation patterns and their relatively simple and easily observed communities make them especially suitable for manipulative experiments. A number of classic experiments performed in these habitats have greatly influenced modern ecology. Work on rocky shores has elucidated the roles of consumers (Paine 1966; Menge 1976; Lubchenco 1978), competition (Connell 1961), disturbance (Dayton 1971; Sousa 1979a), and recruitment (Gaines and Roughgarden 1985) in communities. As a consequence of their accessibility, ecologists have paid a great deal of attention to these habitats, and so we understand them better than most other communities.

On the east coast of North America, rocky intertidal habitats are primarily found north of Cape Cod, where glaciers from the most recent Ice Age have scoured the shore clean of sediments. In comparison to the west coasts of North America and Europe, rocky shores in New England and Canada have a relatively depauperate fauna and flora. The low species diversity of these rocky shorelines is a result of geologically recent large-scale disturbances. Pleistocene glaciers covered the east coast of North America to southern New England as recently as 10,000 to 20,000 years ago. Their scouring of sediments from the shore led to the extinction of marine hard substrate-dependent organisms, since these species could not retreat to southern shorelines, which lacked hard substrates (Vermeij 1978, 1987). Consequently, New England shorelines have been recolonized and recovering from these local extinctions only for the past 10,000 to 20,000 years, and are geologically very young. Severe winter ice scouring has continued to occur frequently enough in Canadian and New England intertidal habitats over the last few centuries to prevent the reestablishment and persistence of most organisms other than the fastest-growing, earliest-reproducing species (Wethey 1985).

In this chapter I examine the ecology of western Atlantic rocky shores. I begin by introducing the organisms that occur on these shores, paying particular attention to the biology of the numerically dominant organisms. I then discuss the unusually extreme physical stresses that influence the distributions of organisms in this habitat, and describe the striking zonation of organisms on rocky shores that has long attracted the interest of marine ecologists. I end by touching on some of the mounting conservation concerns pertaining to rocky shore communities.

ROCKY INTERTIDAL ALGAE

Two sources of primary production form the base of rocky intertidal food webs: planktonic diatoms and benthic algae. Planktonic diatoms are microscopic and inconspicuous, but serve as the primary food source for the sessile filter-feeding organisms, such as barnacles and mussels, that characterize most rocky shorelines. In contrast, the benthic algae common on rocky shores range in size from microscopic diatoms to large seaweeds and kelps. They are grazed by a variety of herbivores, including snails, urchins, and fishes.[🐚 1]

Seaweeds are extremely productive along most shoreline habitats. The high levels of nutrients, light, and water movement that characterize many

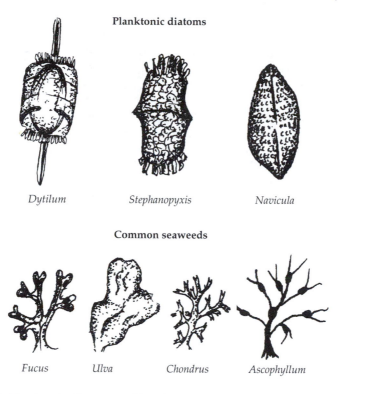

Planktonic diatoms

Dytilum   *Stephanopyxis*   *Navicula*

Common seaweeds

Fucus   *Ulva*   *Chondrus*   *Ascophyllum*

🐚 1 Planktonic diatoms and benthic algae, consumed by filter feeders and benthic grazers, respectively, are at the base of rocky shore food webs.

intertidal habitats promote primary productivity. High nutrient supplies in New England nearshore waters are the result of terrestrial runoff and the nearshore oceanographic processes that bring nutrient-rich bottom water to nearshore habitats (Mathieson et al. 1991; see Chapter 2). In contrast, the nutrient-poor, oligotrophic conditions in the nearshore waters off New-foundland and Labrador are the result of low nutrient inputs (Chapman and Craigie 1977). The high levels of light and water movement found in rocky intertidal habitats also enhance seaweed productivity (Gerard and Mann 1979; Gerard 1987; Leigh et al. 1987). The influence of light on seaweed distribution and production is discussed in Chapter 2. Water movement enhances gas exchange (CO_2 absorption and O_2 release), and by keeping algal fronds in motion, ensures that light reaches more algal surfaces than if the algae were stationary and constantly shading their neighbors. Seaweed morphological responses to water movement, however, may also act to decrease the surface area available for photosynthesis, and thus negate potential productivity advantages for seaweeds in high flows.

The high productivity of seaweeds has made them a valuable resource for human populations, which have harvested them for a variety of purposes over many centuries (see Chapman 1970). In the seventeenth and eighteenth centuries in Europe and North America, large brown seaweeds or "kelps" were used for fertilizer and livestock fodder, or dried and burned for potash (potassium salts), or used in pottery glazes and the manufacturing of glass and soap. The word *kelp*, in fact, originally referred to brown seaweeds that were ashed in large rock-lined pits near shorelines for their potash, rather than live seaweeds. In the early nineteenth century, the discovery of other potash sources and the high iodine content of some seaweeds shifted the focus of seaweed harvesting to iodine extraction. In Europe, harvesting of intertidal seaweeds was so intense that government regulation of harvesting and processing became necessary by the middle of the seventeenth century.

The use of seaweeds for potash and iodine ended early in the twentieth century when mining mineral deposits for these resources became more cost-effective than processing seaweeds. Currently, seaweeds are harvested globally for gel and emulsifier products. The common red seaweed *Chondrus crispus* (Irish moss) is heavily harvested in New England and the Canadian Maritime provinces. Irish moss is valued as a source of carrageenan, which is widely used as a stabilizer and emulsifying agent in lotions, toothpastes, ice creams, pie fillings, jellies, and more. Natural harvesting of Irish moss has declined over the last few decades as mariculture techniques have been developed to grow seaweeds under artificial conditions. Many seaweeds, however, continue to be harvested and cultured for food, particularly in Asia. In the Canadian Maritime provinces the abundant shallow-water brown seaweed *Ascophyllum nodosum* (knotted wrack) continues to be harvested for its alginate, a thickener used in everything from salad dressing to oil-drilling lubricants.

In addition to serving as food sources for people and intertidal consumers, some seaweeds play a vital role in intertidal communities by providing refuges from the physical stresses of heat and desiccation (Dayton 1975; Menge 1978; Denny et al. 1985). In the Gulf of Maine, the knotted wrack (*Ascophyllum*), in particular, often forms dense intertidal canopies on rocky shores that are not exposed to direct wave shock. These canopies of long-lived, often slowly recruiting algae buffer understory organisms from heat, water loss, and wave forces, and they can support a rich assemblage that often would not exist in these habitats without the modifying effects of the canopy.

An important distinction to make among seaweeds, which is not always taxonomic, is between ephemeral and leathery seaweeds. **Ephemeral seaweeds**, such as *Ulva* and *Enteromorpha*, are usually short-lived, rapidly recruiting species that are easily eaten. In contrast, the **leathery seaweeds**, such as *Ascophyllum* and *Fucus*, are heavier, less easily eaten, slower-recruiting, and often serve as refuges for other intertidal organisms.

Major Seaweed Taxa

Seaweeds are primitive photosynthetic organisms that lack the internal water transport systems of higher plants. They range in complexity from undifferentiated sheets of cells to highly differentiated macroalgae with many of the structural characteristics of higher plants. Seaweeds are phylogenetically divided into three major taxa based on their photosynthetic pigments. The **green algae**, or **chlorophytes**, utilize primarily green chlorophyll pigments. Common intertidal green seaweeds include the sea lettuces (*Ulva*), the filamentous *Enteromorpha*, and dead man's fingers (*Codium*). The **brown algae**, or **phaeophytes**, utilize special xanthophyll pigments such as fucoxanthin in addition to chlorophylls. Common brown algae include the rockweeds (*Fucus*), knotted wrack (*Ascophyllum*), and the kelps (*Laminaria*). The **red algae**, or **rhodophytes**, utilize phycobilin and carotenoid pigments in addition to chlorophylls, but not fucoxanthin. Common red algae include Irish moss (*Chondrus*) as well as fleshy and calcareous algal crusts such as *Hildenbrandia* and *Lithothamnium*, respectively.

Seaweed Life Histories

Seaweed life histories are highly variable, but often include an alternation of haploid and diploid generations (in which there are one and two copies of the genetic instructions, respectively). Typically, a diploid sporophyte produces haploid spores through a reduction division (meiosis). These spores develop directly into haploid gametophytes, which produce haploid gametes. These gametes fuse to form a juvenile diploid sporophyte, which grows into a mature adult form. Modifications of this basic scheme, however, are extensive.[⬤▷2]

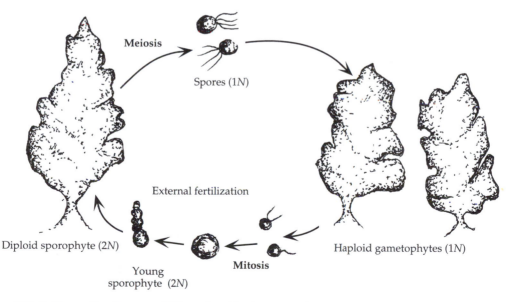

Meiosis

Spores (1*N*)

External fertilization

Diploid sporophyte (2*N*)

Young
sporophyte (2*N*)

Mitosis

Haploid gametophytes (1*N*)

⬤▷2 Generalized seaweed life cycle with alternation between similar (isomorphic) diploid (2*N*) and haploid (1*N*) generations. Common examples of seaweeds with this type of life cycle are *Ulva* and *Enteromorpha*.

Some common seaweeds, such as the sea lettuce *Ulva* and the filamen-tous green alga *Enteromorpha* have sporophyte and gametophyte stages that are indistinguishable (isomorphic). Others have heteromorphic gametophyte and sporophyte stages that are structurally and functionally different. Seaweeds from a number of taxonomic groups alternate between fleshy erect forms (typically the sporophyte), which have high growth rates but high vulnerability to herbivory, and encrusting forms, which are well defended against herbivores, but have low growth rates.[⟡ 3] Examples of Atlantic coast algae with this type of heteromorphic life cycle include *Scytosiphon* and *Petalonia*.

An interesting variation on this theme is found in the sea laver, *Porphyra*, a common seaweed of New England that is a culinary delicacy used in soups, salads, and sushi in Japan. It alternates between a fast-growing, sexu-ally reproducing, fleshy erect blade form, found in the high intertidal zone during the winter, and a subtidal conchocelis (shell-boring) stage that bores into mollusc shells during the rest of the year. As was the case for many sea-weeds with heteromorphic life histories, the conchocelis stage of *Porphyra* was long thought to be a separate species.[⟡ 4]

Encrusting or boring heteromorphic stages of seaweeds are thought to be adaptations for avoiding herbivores and/or physical stresses (Lubchenco and Cubit 1980; Dethier 1981). The fast-growing fleshy stages of these algae have a high reproductive output, but are vulnerable to grazing and physical

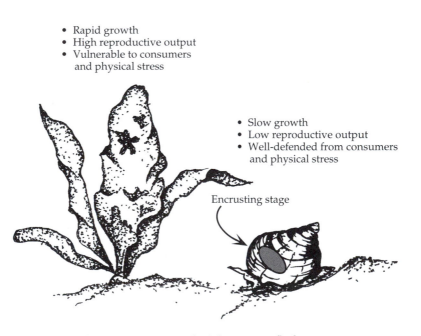

- Rapid growth
- High reproductive output
- Vulnerable to consumers and physical stress

- Slow growth
- Low reproductive output
- Well-defended from consumers and physical stress

Encrusting stage

⟡ 3 Alternation of generations in *Petalonia* between a fleshy erect stage and an encrusting stage on a snail shell.

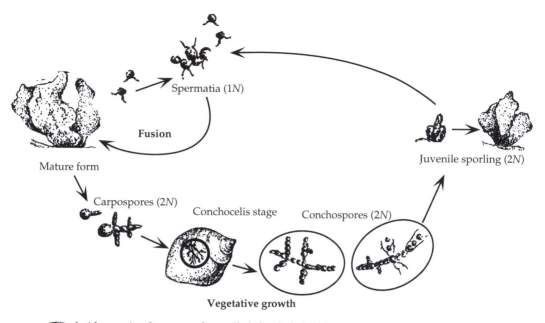

Spermatia (1N)

Fusion

Mature form

Carpospores (2N)

Conchocelis stage   Conchospores (2N)

Juvenile sporling (2N)

Vegetative growth

4 Alternation between the well-defended shell-boring conchocelis and rapidly grow-ing, upright fleshy forms of *Porphyra*.

stresses, whereas the encrusting or boring stages are slow-growing, but resis-tant to grazing. On the coast of Oregon, where the fast-growing erect forms of these algae are typically found only in the winter, experimental removal of herbivores leads to erect fleshy forms persisting throughout the year (Lubchenco and Cubit 1980). In New England, however, fast-growing fleshy forms of many heteromorphic seaweeds can usually be found year-round.

Not all seaweeds have alternation of generations. *Fucus* and *Codium*, two of the most common Atlantic coast seaweeds, have a more animal-like life history. In these species, diploid adults give rise to gametes by meiosis, and fertilized gametes develop directly into diploid adults.[⬧5]

Settlement and recruitment are the least understood aspects of seaweed life histories. Seaweeds do not have any resistant resting stages such as hard-shelled seeds or spores for getting them through tough times (dehydra-tion, freezing, grazing), so they depend on alternating forms to escape these problems. Algal propagules appear to be passive particles with little control over their movement. Propagule dispersal distances, however, vary widely (for examples see Dayton 1973; Burrows and Lodge 1950), suggesting that dispersal mechanisms, propagule buoyancy, and the length of time that propagules remain viable all influence algal dispersal. It is likely that algal settlement is strongly influenced by local hydrodynamic conditions, with propagules accumulating in depositional microhabitats, such as pits and grooves, but we know surprisingly little about actual algal settlement and recruitment patterns under field conditions (see Chapman 1986 and Vadas et al. 1992 for reviews of what is known).

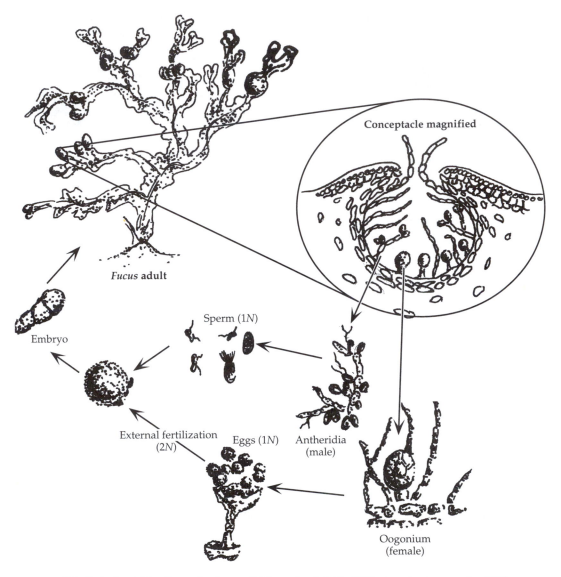

Conceptacle magnified

Fucus adult

Embryo

Sperm (1N)

External fertilization
(2N)

Eggs (1N)

Antheridia
(male)

Oogonium
(female)

5 Animal-like *Fucus* (rockweed) life cycle. Adults (2N) give rise to gametes (1N) through meiosis. External fertilization results in young diploid plants that grow to the adult form. The life cycle does not include an alternation of generations as in the green algae.

Seaweed Functional Groups

The sizes and morphologies of shallow-water seaweeds are influenced by a variety of physical and biological factors. Coping with wave forces, minimizing heat and desiccation stresses, and maximizing nutrient and gas exchange are some of the physical challenges that affect algal design.

Resisting consumers and competitors is also an important biological consideration. Since dealing with any one of these problems can adversely affect the ability to cope with others, compromises and trade-offs are apparent in algal design (Littler and Littler 1980). The highest nutrient and gas exchange rates, for example, are found in single-celled algae (diatoms), which have high surface-to-volume ratios and little structural support, and in some undifferentiated sheets of cells like those of the sea lettuce *Ulva*, which is two cells thick. Maximizing gas exchange, however, leaves these seaweeds extremely vulnerable to herbivory and physical disturbance. In contrast, seaweeds that invest heavily in structural support to resist disturbance, such as kelps with long, flexible, structurally complex stipes (or trunks) and blades, or seaweeds that invest in antiherbivore defenses, such as algal crusts, may do so at the cost of reduced gas exchange and slower growth. The ruffled shape of some kelp blades is thought to increase gas and nutrient exchange (Gerard and Mann 1979).

A functional grouping of algae that reflects common morphological solutions to these problems of design has been proposed by Steneck and Dethier (1994). At the ephemeral end of the spectrum are the structurally simple, small, and fast-growing algal groups (microalgae such as diatoms; filamentous and foliose algae such as *Ulva* and *Enteromorpha*). These are rapidly colonizing, fast-growing algae that are typically found in recently disturbed, early successional habitats. Other seaweeds invest more in structural features that increase their persistence, competitive ability, and resistance to disturbance and herbivory; the trade-off for these species is reduced growth and reproductive output. The structurally complex, leathery macroalgae (Irish moss, *Chondrus crispus*; kelps, *Laminaria*; rockweeds, *Fucus*) provide examples of this strategy. The calcified erect algae (*Corallina*) and coralline and fleshy algal crusts (*Lithothamnium* and *Ralfsia*) represent extremes of investment in structural defenses, resulting in extremely low growth rates.[🐚6]

Steneck and Dethier (1994) suggest that the occurrence of these functional groups can be predicted by two environmental features: the disturbance potential and the productivity potential of the habitat (see Grime 1977 for a similar model for terrestrial vascular plants). Disturbance potential is defined as the rate of herbivore-induced or physically induced tissue loss or mortality. Productivity potential is defined as the potential for primary productivity in a habitat, given specific nutrient and abiotic conditions. This model thus predicts the spatial and temporal distributions of seaweeds. In habitats with low disturbance potential and high productivity potential, long-lived, persistent, competitively superior algal forms are dominant. Shallow, subtidal, wave-exposed kelp bed habitats are a good example of this competitive strategy. If disturbance potential (herbivory) is increased, but productivity potential remains high, algal crusts capable of persisting under intense herbivory dominate (the disturbance-tolerant strategy). Intensively grazed shallow subtidal habitats are a good example of this situation. When disturbance potential is high and productivity potential is low,

Functional group	Examples	Morphology	Anatomy (cross-section)	Grazing difficulty
Microalgae	Diatoms Blue-green algae			Increasing toughness
Filamentous algae	*Cladophora* *Ectocarpus*			
Foliose algae	*Ulva* *Porphyra*			
Macrophytes	*Chondrus* *Mastocarpus*			
Leathery macrophytes	*Laminaria* *Fucus* Non-calcareous crusts			
Articulated calcareous algae	*Corallina*			
Crustose coralline algae	*Lithothamnium* *Clathomorphum*			

6 Algal functional groups. (After Littler and Littler 1980; Steneck and Dethier 1994.)

such as in constantly disturbed cobble fields on exposed coasts with low nutrient levels, no viable strategy is possible, and seaweeds do not occur. When both productivity potential and disturbance potential are low, stress-tolerant strategists, including algal crusts and microalgae, prevail. Between these extremes, weedy seaweeds with intermediate solutions—typically involving rapid reproduction—are found.[7]

Seaweed Defenses

Seaweeds commonly have both structural and chemical defenses. The only seaweeds that lack any real structural defenses are the rapidly growing ephemeral green algae such as *Ulva* and *Enteromorpha*, which are the preferred food of most intertidal herbivores. The tough, leathery cuticles of corticated seaweeds (e.g., *Chondrus* and *Fucus*) act as a structural barrier to limit herbivory by small crustacean grazers such as amphipods (Gaines 1985).

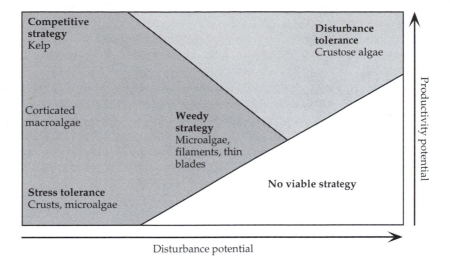

7 A generalized model of the occurrence of algal functional groups across disturbance and productivity gradients. (After Grime 1977 and Steneck and Dethier 1994.)

Algal crusts are especially well-defended. Fleshy crusts like *Ralfsia* are hard, and snail grazers suffer conspicuous tooth damage when forced to eat them. Calcified crusts, like *Lithothamnium*, are even harder, and are immune to most herbivores except limpets, chitons, and urchins. Bob Steneck (1983) has shown that the evolution of these coralline crusts parallels the evolution of scraping herbivores in the fossil record.

As described in Chapter 2, chemical defenses commonly complement structural defenses in seaweeds and other sessile organisms (for review see Hay and Fenical 1988; Hay 1996). Some seaweeds, such as *Desmarestia*, produce sulfuric acid that dissolves the calcium carbonate teeth of potential consumers.[8] Other seaweeds utilize secondary metabolites (chemicals derived from metabolic by-products), such as phenolic and halogenated compounds, for defensive purposes. Like many terrestrial plants, seaweeds often respond to grazing by increasing the production of defensive compounds. The common rockweed (*Fucus*) increases phenolic concentrations when grazed, limiting further grazing damage (Van Alsytne 1988). It also produces more phenolic compounds when nitrogen is in limited supply (Yates and Peckol 1993). This occurs because the availability of carbon increases. Phenolics are carbon-based compounds, and when nitrogen limits growth, algae have an excess of carbon, leading to phenolic production. When nitrogen is not limiting, most carbon is tied up in growth, leaving less for phenolic production.

Mark Hay and his colleagues (Hay et al. 1987; Duffy and Hay 1990; Duffy 1990) have suggested that the chemical defenses of most seaweeds

Desmarestia   Normal urchin tooth   Eroded urchin tooth

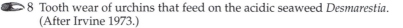 8 Tooth wear of urchins that feed on the acidic seaweed *Desmarestia*.
(After Irvine 1973.)

evolved primarily to limit the effectiveness of large herbivorous fishes and
urchins rather than smaller grazers such as amphipods and snails. They
have shown that small grazers often prefer to eat, and live in association
with, seaweeds that are heavily defended. Through this preference, the grazers are protected from their own consumers, either by sequestering ingested
defensive compounds or by living in habitats avoided by generalist consumers. Thus, the small crustaceans that live on and consume chemically
defended algae have lifestyles similar to many specialist insect herbivores
that live in close association with chemically defended host plants while
doing them minimal damage and gaining protection from their own consumers.[⬤ 9]

ROCKY INTERTIDAL HERBIVORES

The two sources of primary production that characterize rocky intertidal
food webs are harvested by different organisms using different means.
Benthic diatoms and seaweeds are eaten primarily by slow-moving resident
grazers that have mouthparts modified for scraping hard surfaces or excavating tissue from macroalgae. In New England and Canada, these grazers
include periwinkles, amphipod and isopod crustaceans, sea urchins, and
limpets. Unlike those in shallow-water tropical habitats (Randell 1967),
fishes in most cold temperate habitats are not typically important herbivores
(Vermeij 1978). A second trophic pathway links water column dynamics to
benthic shoreline populations. Pelagic microorganisms, including algae
(phytoplankton), cyanobacteria, and animals (zooplankton), are particularly
abundant in nutrient-rich coastal waters, and enter benthic food chains

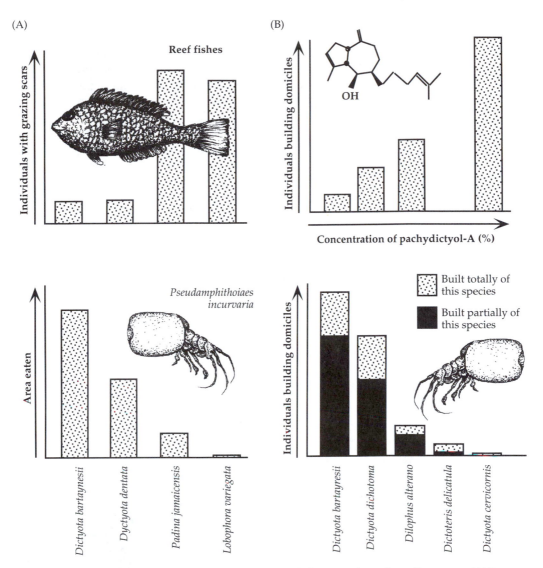

9 Relationships between chemically defended seaweeds and small grazers. (A) In an experimental test of feeding preferences of herbivorous reef fishes and amphipods, the fishes avoided the chemically defended algae that the amphipods preferred. (After Hay and Duffy 1990.) (B) The amphipod *Pseudamphithoiaes* preferentially builds its domiciles from seaweeds that either experimentally (top) or naturally (bottom) contain high levels of defensive chemicals. (After Hay 1996.)

through sessile filter feeders. On rocky intertidal shores of the Atlantic coast of North America, mussels and barnacles are the most conspicuous and abundant sessile filter feeders.

Grazers

The green sea urchin *Strongylocentrotus droebachiensis* is the most voracious scraping herbivore on New England shores, but its extreme vulnerability to physical stresses and bird predation restrict it for the most part to subtidal and deep tide pool habitats.[🐚 10] Sea urchins are also uncommon in bays and estuaries, where salinity may vary widely. Like other echinoderms, they have no physiological mechanism to deal with varying osmotic conditions.

Sea urchins feed with a movable jaw of five continually growing calcium carbonate teeth, called an Aristotle's lantern (see Chapter 2, 🐚 28). Urchins consume seaweeds by ingesting them whole, taking bites out of them, or scraping their surfaces. At high densities, they are capable of consuming all but heavily calcified coralline crusts (Sebens 1985). In subtidal habitats, groups of hundreds of thousands of urchins ("fronts") can move through an area, consuming all erect algae and scouring the substrate down to algal crusts and bare rock, creating "urchin barrens" (Witman 1985).

Both overharvesting and disease have taken a heavy toll on Atlantic coast sea urchin populations over the past two decades. Sea urchins are harvested for their eggs, which are an expensive delicacy in Japan. Overharvesting has become a serious problem in Canada and New England as well as

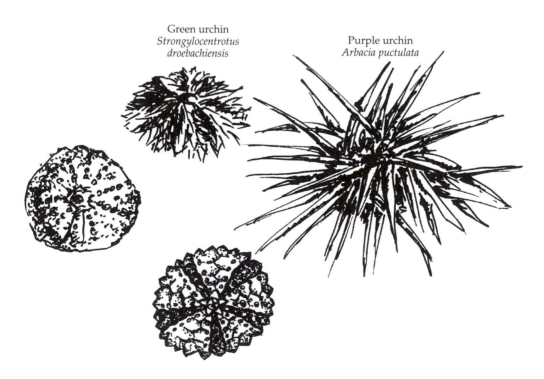

Green urchin
Strongylocentrotus droebachiensis

Purple urchin
Arbacia puctulata

🐚 10 Western Atlantic sea urchins of rocky shores.

on the west coast of North America. In the north Atlantic, diseases have also devastated urchin populations and led to catastrophic local extinctions. Similar diseases have been responsible for the large-scale die-offs of Caribbean sea urchins that have contributed to changing Caribbean reef habitats from coral-dominated to algae-dominated systems (Lessios et al. 1984). In the northern Atlantic, outbreaks of a water-borne amoeboid pathogen have been associated with unusually warm water temperatures, and have been responsible for shifting entire subtidal habitats from urchin barrens dominated by algal crusts to kelp beds (see Scheibling and Stephenson 1984; Scheibling 1986).

Higher on New England rocky shores, three species of herbivorous gastropods of the genus *Littorina* are the dominant grazers. Periwinkles feed with a tonguelike radula, but modified for use as a rake or file (Steneck and Watling 1982).[📀 11] They scrape diatoms, algal sporelings, and soft algal tissue into their mouths, where the food is entangled in mucus and pulled into the digestive system. Periwinkle radulae are incapable of digging deeply into algal tissue and tend to grind down if the animals eat structurally defended algae.

By far the most common periwinkle on rocky shores of New England is the common European periwinkle, *Littorina littorea*. On both open-coast and estuarine shorelines, this snail is often found at densities of 200–500 per square meter, and can play an exceedingly important role in the structure and organization of shoreline communities (Lubchenco 1978, 1983; Bertness 1984a).[📀 12] As its name implies, however, it is a relatively new introduction from Europe. *Littorina littorea* has planktonic development, and fossil evidence suggests that it may have established populations in northern

📀 11 The tonguelike radula of the periwinkle *Littorina* is used to rake the surfaces of diatoms and soft algal tissue.

Adult   Juvenile   Veliger
                                               larva

12 The common European periwinkle, *Littorina littorea*.

Canada prior to the eighteenth century (Vermeij 1982a). If so, its numbers remained small until the middle of the nineteenth century, when it began moving south, reaching Cape Cod by the turn of the century and Chesapeake Bay by the 1950s (Carlton 1982). It is more likely, however, that the common periwinkle was introduced intentionally or inadvertently by settlers in Nova Scotia in the middle of the nineteenth century and then quickly spread south.[13]

Littorina littorea is a habitat generalist found in all New England shoreline habitats that have at least some hard substrate. It commonly reaches 2–3 centimeters in length. Morphological variation in spire height and shape can be striking, and is largely due to variation in growth rates. Fast-growing snails develop more globose, higher-volume shells than slower-growing snails (Kemp and Bertness 1984).[14]

From salt marshes to wave-exposed rocky shores, the common periwinkle has become the most abundant intertidal herbivore throughout its biogeographic range, and has had a considerable effect on shoreline communities of the Atlantic coast of North America. In salt marsh habitats it has displaced the native mud snail *Ilyanassa* (Brenchley and Carlton 1983), and on open-coast rocky beaches it is responsible for limiting the distribution and success of palatable green algae and algal sporelings (Lubchenco 1978, 1983). It may have had its greatest effect, however, on the estuarine cobble beaches of Narragansett Bay and Long Island Sound, where its numbers can exceed 1000 per square meter (see the discussion of zonation on cobble beaches below).

Both the rough periwinkle, *Littorina saxatilis*, and the smooth periwinkle, *Littorina obtusata*, are smaller native species that, in contrast to the common periwinkle, have direct development and are never found at extremely high densities. *Littorina saxatilis* is a small snail rarely larger than 8 millimeters in length, found at high intertidal heights above where European periwinkles occur (Gosner 1978). It has received little experimental attention in North

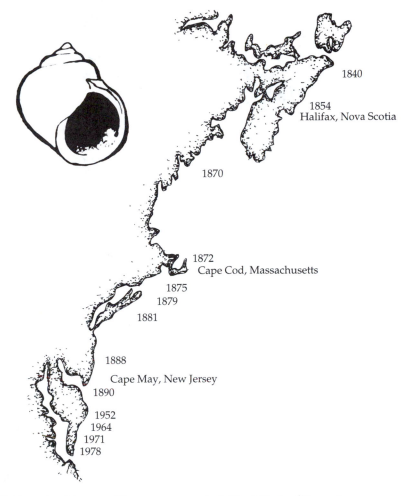

13 Invasion history of the common periwinkle, *Littorina littorea*, in North America. (After Carlton 1982.)

America, and can be difficult to distinguish from juvenile European periwinkles. In Europe, *Littorina saxatilis* grows faster and reaches larger sizes on protected shores than on wave-exposed shores; these differences are controlled by both environmental and genetic factors (Hughes and Roberts 1981; Janson 1983).

In New England, *Littorina obtusata* is almost always found associated with algal canopies of the knotted wrack, *Ascophyllum nodosum*. Shell shape and thickness in *Littorina obtusata* are highly variable. In wave-exposed habitats it develops a larger foot for resistance to increased wave stresses (Trussell et al. 1993). Shell thickness in this snail is also a plastic response to the presence of its predators. *Littorina obtusata* from wave-protected habitats

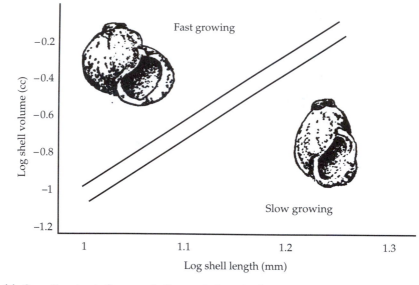

⬤▷ 14 Growth rates influence shell morphology in the common periwinkle, *Littorina littorea*. Fast-growing snails are more globose, or
round, than slow-growing snails. (After Kemp and Bertness 1984.)

have thicker shells than those from wave-exposed habitats because of phenotypic shell thickening in response to exposure to predaceous crabs
(Trussell 1996). Thin-shelled individuals transplanted to low wave energy
habitats with high predator densities develop thick shells, as do snails
reared with predaceous crabs in the laboratory. Average shell thickness in
New England *Littorina obtusata* has also increased over the last century
(Vermeij 1982b; Seeley 1986), apparently as a plastic response to the introduction of green crabs.

In contrast to periwinkles, whose radulae are designed for raking surfaces, limpets and chitons often have radulae with iron-capped teeth,
designed to excavate and shovel algal tissue into their mouths (Steneck and
Watling 1982). Neither limpets nor chitons, however, are very common or
important herbivores in the western Atlantic relative to other temperate
areas. Whereas there are ten common species of shallow-water limpets in
Puget Sound, and eight on European shorelines, there are only two species
on New England shorelines, and neither is particularly abundant. The
absence of a well-developed limpet and chiton fauna in the western Atlantic
is probably a consequence of Ice Age disturbances and more recent severe
winter ice scouring events that have driven most of these organisms to
extinction and prevented their reinvasion (Wethey 1985).

The most common limpet in New England, the tortoise shell limpet,
Acmaea testudinalis, is a feeding specialist that preferentially settles and feeds
on the coralline crust alga *Clathomorphum*.[⬤▷ 15] The relationship between

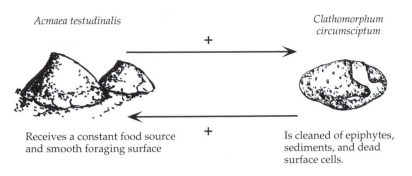

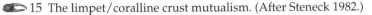

🐚 15   The limpet/coralline crust mutualism. (After Steneck 1982.)

Acmaea and *Clathomorphum* has been described as a coevolved mutualism (Steneck 1982). From this association, *Acmaea* gets a constant food supply and a smooth surface on which to forage. In return, *Clathomorphum* gets its surface cleaned of fouling epiphytes and sediments. Like most other crustose algae, *Clathomorphum* depends on grazers for its persistence. Without grazers, it is outcompeted for light by faster-growing algae or suffocates under accumulated sediment.

Chitons are also uncommon in the western Atlantic, particularly in comparison to the west coast of North America and the temperate shores of New Zealand and Australia. In the Puget Sound region of Washington State, five shallow-water chitons are common. They play an important ecological role in these communities, and reach sizes of up to 30 centimeters in length. In New England, only three small (generally less than 30 millimeters) chitons occur, none are common, and all are largely restricted to subtidal habitats.

Filter Feeders

The second major group of herbivores on New England rocky shores comprises the sessile filter feeders, of which acorn barnacles and blue mussels are the most conspicuous. Acorn barnacles dominate middle to high rocky intertidal habitats throughout New England, but become less dominant on Canadian shores as a result of increased ice damage and possibly the lower food supplies available in northern oligotrophic waters (Chapman, personal communication). Acorn barnacles were described by Louis Agassiz as shrimps glued to rocks by their heads, enclosed in a calcium carbonate house, kicking food into their mouths with their feet. Acorn barnacles actually do less kicking and more passive straining to catch their food than Agassiz suggested. Although they will beat their feet (cirri) in still water, actively straining particles from the water, in the field they are almost always passive feeders, relying on ambient currents to deliver food to their feeding appendages (Sanford et al. 1994; but see Trager et al. 1990).

Since barnacles are important fouling organisms on ships and pilings, they have received a great deal of attention, and as a result, we understand their biology and ecology probably better than that of any other shoreline organism. Both the American and British naval research offices have sponsored considerable basic research on the biology of barnacles.

Found throughout the northern Atlantic, the northern acorn barnacle, *Semibalanus balanoides* (also known as the northern rock barnacle), is the most common barnacle in New England.[ 16] Other common shoreline barnacles include *Chthamalus fragilis*, found at high intertidal heights on hard substrates south of Cape Cod, and *Balanus crenatus* and *Balanus eburneus*, common in subtidal and estuarine habitats, respectively.[17]

Acorn barnacles are simultaneous hermaphrodites (functioning at the same time as males and females). Internal fertilization is accomplished by a penis up to five times the animal's maximum body dimensions. This reproductive strategy makes it necessary for barnacles to live close together, and their larvae settle in response to specific chemical signals that indicate the presence of conspecifics (Crisp and Meadows 1962). In the northern acorn barnacle, fertilization occurs in the late summer, and the eggs are incubated in the mantle cavity until late winter, when they are released as nauplius larvae.[18] The larvae remain in the water column for 6–8 weeks, depending on temperature and food concentrations, where they move at the mercy

16 The ubiquitous northern acorn barnacle, *Semibalanus balanoides*.

Semibalanus balanoides
Northern acorn barnacle
North of Long Island Sound
Coastal

Chthamalus fragilis
Little gray barnacle
High intertidal
South of Cape Cod

Balanus crenatus
Crenate barnacle
Subtidal

Balanus improvisus
Bay barnacle
Estuarine

Balanus eburneus
Ivory barnacle
Estuarine

17 Common acorn barnacles of the Atlantic coast of North America.

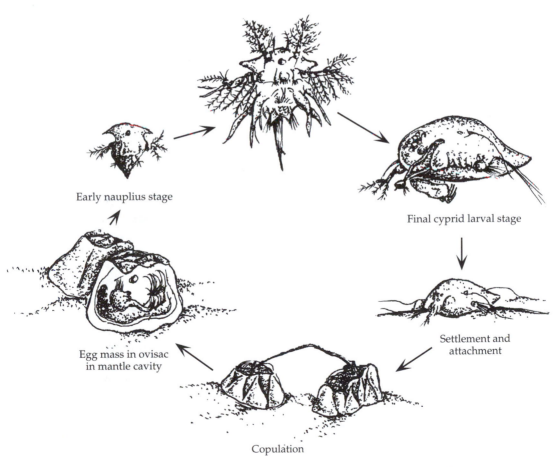

Early nauplius stage

Final cyprid larval stage

Settlement and attachment

Egg mass in ovisac in mantle cavity

Copulation

18 Life cycle of an acorn barnacle.

of local oceanographic conditions, largely as passive particles. They pass through four nauplius larval stages before metamorphosing into a terminal cyprid larval form that selects a habitat and settles, committing irreversibly to a permanent adult habitat. Settlement involves attachment to the substrate with adhesive antennal glands and the secretion of a calcareous external skeleton. Life span is highly variable; many recruits survive less than a day, while others can live up to 4–5 years.

Because acorn barnacles have a relatively long planktonic larval period and larvae that behave like passive particles, their populations have long been assumed to be genetically homogeneous. Recent evidence, however, has suggested that acorn barnacles show genetic variation at both large and small spatial scales. In New England, transplant studies have revealed that barnacles from warmer bays that retain larvae are more tolerant of thermal stress than barnacles from cooler open-coast habitats or bays that do not retain larvae (Bertness and Gaines 1992). At even smaller spatial scales, Paul Schmidt and David Rand have documented strong intertidal height zonation of genotypes that is the product of intense selection and mortality each settlement season (Schmidt and Rand 1999)

Blue mussels are the second most abundant and conspicuous rocky intertidal filter feeder in New England. Until recently, all blue mussels in the northern Atlantic were thought to belong to the same species. There are, however, two species of *Mytilus* that occur in the northwestern Atlantic: *Mytilus edulis*, which occurs from Cape Hatteras, North Carolina, to Newfoundland, Canada, and Iceland, and *Mytilus trossulus*, which occurs from Newfoundland to Nova Scotia (Koehn et al. 1984; Koehn 1991; McDonald et al. 1991). Hybrids between *M. trossulus* and *M. edulis* are found in Nova Scotia and Newfoundland. Neither the ecology of *M. trossulus* or that of the hybrid zone between *M. trossulus* and *M. edulis* has been well studied. Earlier reports of microgeographic variation among *M. edulis* populations in Canada (Gartner-Kepkay et al. 1983) are almost certainly due to the interdigitation of these two species and their hybrids (Hilbish 1996). The two species have different thermal limits, but are not visually distinguishable, and appear to play identical roles in the habitats they are found in. Like northern acorn barnacles, blue mussels become increasingly patchy and less abundant at higher latitudes as a result of ice disturbance (Stephenson and Stephenson 1971) and, probably, the inability of oligotrophic waters to support large populations of filter feeders. The southern limit of the distribution of *Mytilus edulis* at Cape Hatteras is known to be maintained by thermal stress. Blue mussel recruits are regularly transported south of Cape Hatteras, but persist only until summer temperatures reach about 27°C, when they die back (Gray 1960).

In contrast to barnacles, blue mussels are active filter feeders that pump a water current over their gills with cilia, and therefore their success is not as tightly linked to flow as that of barnacles (Wildish and Kristmanson 1979). Blue mussels also have external fertilization, with females releasing up to 25 million eggs per season. Fertilization is followed by a planktonic larval life of up to 3 months. In the plankton, mussel larvae are passively transported by ocean currents, leading to wide dispersal and gene flow (Scheltema 1971b, 1986).

Blue mussels initially settle as small (0.5–2.0 millimeters) plantigrade larvae in association with filamentous algae near adult mussel populations. After doubling in size, however, many juvenile mussels migrate into association with adults (Bayne 1964). The initial settlement of plantigrade larvae on algae probably minimizes competition with adult mussels while providing shelter from physical stresses and predators.[19]

Blue mussels attach to the substrate by means of proteinaceous byssal threads, produced by a byssal gland at the base of the foot and applied to the substrate by a groove in the foot.[20] Using their byssal threads like

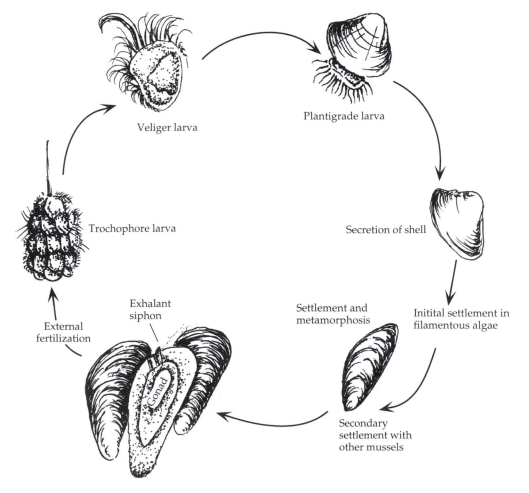

Veliger larva

Plantigrade larva

Trochophore larva

Secretion of shell

External fertilization

Exhalant siphon

Gonad

Settlement and metamorphosis

Initial settlement in filamentous algae

Secondary settlement with other mussels

19   Blue mussel life cycle.

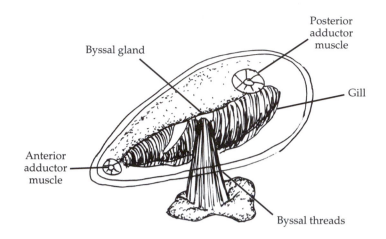

Byssal gland

Posterior
adductor
muscle

Gill

Anterior
adductor
muscle

Byssal threads

🐚 20 Byssal attachment in mussels.

climbing ropes, small mussels (less than 2 centimeters) are extremely mobile. For all practical purposes, mussels become sessile when they get too heavy for the strength of their byssal threads to move them, but they can reattach by byssal threads if dislodged from the substrate.

Blue mussels are typically found in dense beds in low intertidal and shallow subtidal habitats. As is the case for barnacles, group living in mussels has both costs and benefits. Dense mussel beds enhance external fertilization success (Denny et al. 1985), provide protection to small, vulnerable juveniles from crab and fish predators (Bertness and Grosholz 1985), protect intertidal individuals from heat and desiccation stresses (Bertness and Leonard 1997), and buffer individuals from wave stress (Denny et al. 1985). By providing a rough topography on the bottom, dense mussel beds may promote turbulent mixing of the water, which can limit the development of a food-depleted near-bed boundary layer (Fréchette et al. 1989). Conversely, competition among neighbors for food is intense in dense beds (Harger 1971; Bertness and Grosholz 1985; Okamura 1986).

Mussels are also important bioengineers that modify shoreline environments and provide habitat for other organisms (Suchanek 1986). By slowing water flow, trapping and binding sediments, and depositing particulate wastes, mussel beds accelerate sedimentation rates, prevent erosion, and limit the mobility of cobbles on rocky shores by binding them to the surface (Stephens and Bertness 1991). Moreover, by increasing the hard-substrate surface area, mussel beds serve as an important habitat for encrusting algae and animals, small mobile organisms that live within the mussel matrix, and soft-sediment tube-building organisms.

In spite of wide dispersal and high gene flow, adult mussels show strong genetic differentiation on relatively small spatial scales (Hilbish and Koehn 1985). In particular, mussels in bays and estuaries are better at coping with low salinities than mussels in nearby open-coast habitats, and this dif-

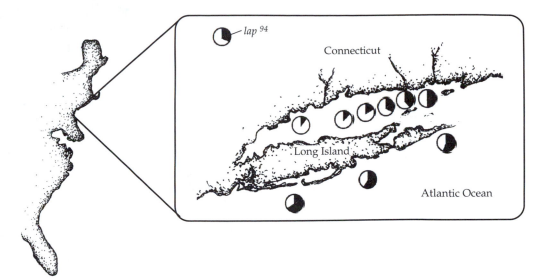

21 The frequency of the *lap*[94] allele (darkened area of pies) among blue mussels, *Mytilus*, along a salinity gradient in Long Island Sound. In spite of the wide dispersal of larvae, adult mussels in the sound show low *lap*[94] frequencies due to differential mortality of individuals with this allele. (After Koehn and Hilbish 1987.)

ference is due to genetic differentiation in their ability to regulate cell volume under changing osmotic conditions. Genetic differentiation in this trait on relatively small spatial scales is maintained by strong selection on mussel settlers every year. Wide dispersal annually leads to a random mixing of juvenile genotypes, but post-settlement differential mortality results in strong spatial genetic differentiation in adult mussels.[21]

In addition to the barnacles and mussels that dominate most New England rocky intertidal seascapes, low intertidal and shallow subtidal habitats also support a variety of filter-feeding clonal invertebrates. These include encrusting tunicates such as *Botryllus* and *Botrylloides*, solitary tunicates such as *Styela*, bryozoans such as *Bugula*, hydroids such as *Obelia* and *Hydractinia*, and shell-boring (*Cliona*) and encrusting (*Microciona*) sponges. Most of these organisms are active filter feeders that propagate asexually, which allows them to monopolize substrate rapidly and outcompete most solitary space holders. As a rule, these clonal subtidal dominants are precluded by heat and desiccation stresses from living in intertidal habitats (Jackson 1977).[22]

PREDATORS ON ROCKY SHORES

Predators on Atlantic rocky shorelines include shell-drilling snails, starfishes, crabs, fishes, and birds.[23]

Predaceous snails that use their tonguelike radulae to bore holes in mussel shells or the opercular plates of barnacles are common predators, partic-

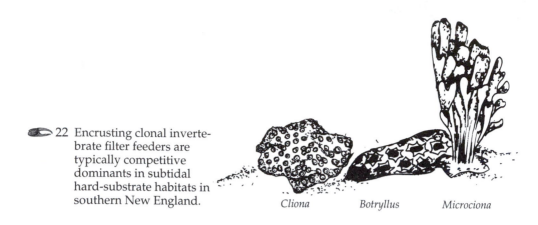

22 Encrusting clonal invertebrate filter feeders are
typically competitive
dominants in subtidal
hard-substrate habitats in
southern New England.

Cliona   *Botryllus*   *Microciona*

ularly in open-coast habitats. In New England and elsewhere in the northern
Atlantic, the dog whelk (*Nucella lapillus*) is the most abundant intertidal
predaceous snail in open-coast habitats. *Nucella* belongs to the thaid family
of snails, which are common predators in shallow, hard-substrate habitats
worldwide. In ancient times *Nucella* was a prized natural resource because it
was used by the Phoenicians to make royal purple dyes.

Dog whelks live at middle to low intertidal heights, feeding primarily
on mussels and barnacles. When common, they can sharply limit populations of their prey (Menge 1976). In New England, dog whelks are rare and
inefficient predators in highly wave-exposed habitats, since they are easily
dislodged by waves and must hide in cracks and crevices during extreme
weather (Menge 1978). They are abundant and important predators at open-
coast locations not exposed directly to waves, and are particularly effective
under algal canopies and on the undersides of boulders, which protect them
from waves and keep them cool and moist during low tides. Interestingly, in
Nova Scotia, where seaweed canopies dominate wave-exposed shores,
Nucella is abundant even in the most exposed habitats. Antony Chapman of
Dalhousie University has suggested to me that the ability of the snails to live
in these exposed habitats is the result of the presence of the seaweed canopy,
which is characteristic of the nutrient-poor waters of the northern Atlantic.
The canopy itself may dominate these shores because water column food
resources are not sufficient to support filter-feeder populations. In bays and
estuaries, low salinities, crab predators, and winter ice generally limit the
distribution of many drilling snails.

Nucella has direct development and crawl-away young. These traits,
combined with its limited mobility, lead to low gene flow, population isolation, and genetic adaptation to local conditions (Vermeij 1982a). *Nucella* is a
seasonal breeder and, like other thaid gastropods, aggregates in the late
summer or early fall in large breeding groups of tens to hundreds of individuals (Etter 1989). Copulation occurs in these breeding aggregations, and
females deposit eggs in semi-transparent proteinaceous egg capsules, which

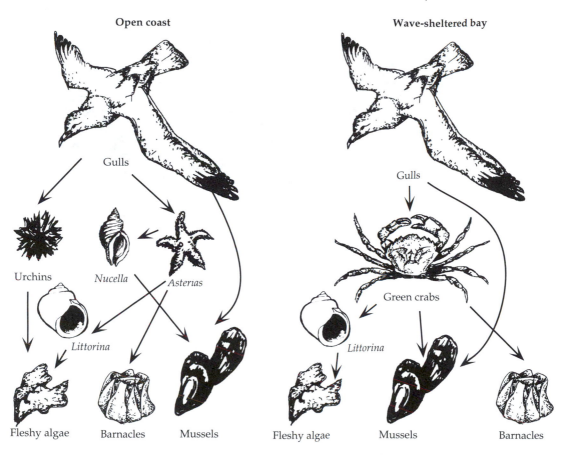

Open coast

Gulls

Urchins   *Nucella*   *Asterias*

Littorina

Fleshy algae   Barnacles   Mussels

Wave-sheltered bay

Gulls

Green crabs

Littorina

Fleshy algae   Mussels   Barnacles

23 Food webs for open-coast and protected bay sites in the Gulf of Maine.

they attach to the undersides of rocks. Crawl-away juveniles emerge from these capsules the following spring, and most return to breed at the same location throughout their 5- to 7-year life span.[24]

Like other thaid snails, *Nucella* are notoriously variable in morphology and color. Individuals living in wave-exposed locations have a larger foot and are more resistant to wave dislodgment than individuals from low wave stress areas. Foot area, however, is a plastic trait that is environmentally determined. Juvenile snails reared in the laboratory develop a small foot characteristic of low wave energy populations, whereas juveniles from low wave stress populations transplanted to high wave stress habitats develop a large foot characteristic of high wave stress populations (Etter 1988a).[25]

In contrast to the high plasticity of foot size, the color of thaids appears to be under strong genetic control. In New England, *Nucella* is found in a wide range of solid and striped color morphs.[26] White color morphs dominate protected habitats, while darker color morphs dominate wave-exposed

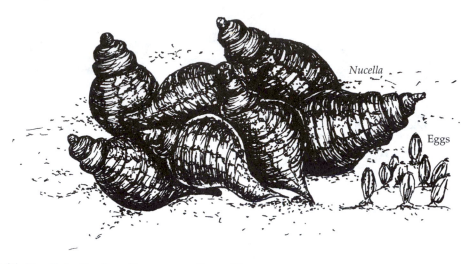

◔ 24 *Nucella lapillus* breeding aggregation with eggs.

habitats. Ron Etter (1988b) has shown that heat stress in protected habitats selects for reflective light-colored morphs (and against heat-absorbing darker morphs), and has suggested that cryptic coloration and protection from predation by birds may also play a role in maintaining these color morphs.

In bays and estuaries south of Cape Cod, the oyster drill, *Urosalpinx cinera*, replaces *Nucella* on hard substrates as a drilling predator. *Urosalpinx* is

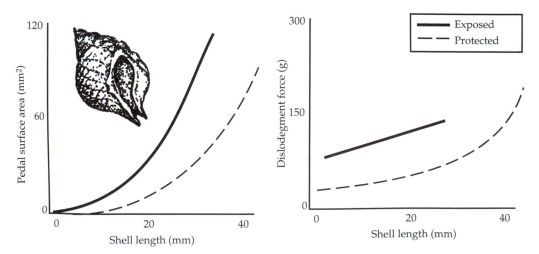

◔ 25 The dog whelk, *Nucella*, develops a larger foot and shell aperture at wave-exposed sites, resulting in the snails being able to withstand higher wave forces. (After Etter 1988b.)

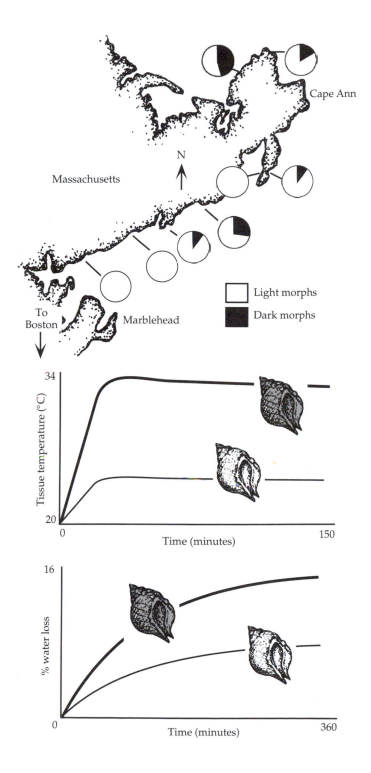

Cape Ann

Massachusetts

N
↑

□ Light morphs

■ Dark morphs

To
Boston

Marblehead

Tissue temperature (°C)

34

20

0   Time (minutes)   150

16

% water loss

0   Time (minutes)   360

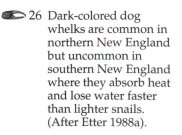

26 Dark-colored dog whelks are common in northern New England but uncommon in southern New England where they absorb heat and lose water faster than lighter snails. (After Etter 1988a).

best known as a pest of oyster beds, but it can also be an important predator of barnacles and mussels on rocky shores (Katz 1985). *Urosalpinx* is native to the Atlantic coast of North America, with a fossil record that goes back over 25 million years (Carriker 1955). Its original distribution appears to have been from Chesapeake Bay to Narragansett Bay, but during the nineteenth century it was spread as far north as Maine and as far south as Florida as a by-product of oyster farming. Oyster transplantation in the twentieth century has also introduced *Urosalpinx* to Europe and the west coast of North America.

Like *Nucella*, *Urosalpinx* is found only on hard substrates. It has a small foot, does not move well on soft substrates, and requires hard substrates on which to deposit eggs. Unlike *Nucella*, however, *Urosalpinx* is tolerant of low salinities (as low as 15 parts per thousand salt) and salinity fluctuations, allowing it to live in estuarine habitats inaccessible to *Nucella*.

Oyster drills have strongly seasonal activity patterns. Like dog whelks, they have direct development. They breed in the spring and summer and lay egg capsules, which they usually attach to the sides or bottoms of hard cobble or shell surfaces. The capsules hatch into crawl-away young in 30–45 days. Individuals can live for more than a dozen years. In the winter, *Urosalpinx* hibernates by moving into deep water and burying itself in the sediment.

Starfishes are also important predators on New England rocky shores, and often play a role in limiting the distributions of subtidal mussel beds. Since starfishes are extremely vulnerable to desiccation and have no mechanism to deal with varying salinities, they are usually found in low intertidal or subtidal habitats, especially on coastal shores with full-strength seawater. After open-coast storms, starfishes are commonly found thrown up onto the beach, which suggests that storm disturbances can play an important role in regulating their subtidal populations (Witman 1987). Starfishes move using a hydraulic vascular system, which is a radial network of water-filled tubules moved by both water pressure and muscles.

In low intertidal and shallow subtidal rocky habitats, starfishes are generalist predators of most sessile animals, but are particularly important mussel and barnacle predators (Lubchenco and Menge 1978). *Asterias forbesii* and *Asterias vulgaris* are the two most common shallow-water starfishes in the western Atlantic. They feed by everting the stomach, excreting enzymes to begin digesting their prey, and then sucking the resulting slurry into the digestive system. When small, *Asterias vulgaris* feeds primarily on bryozoans, hydroids, and small gastropods, but switches when larger to a diet of clams and mussels (Hulbert 1979). When eating a mussel, a starfish inserts its stomach between its slightly gaping (less than 1 millimeter) valves, and secretes digestive enzymes until the bivalve relaxes, opening the valves. When eating barnacles, it simply smothers them with digestive enzymes before slurping them into the stomach, where digestion is completed. Jon Witman and his colleagues recently found that starfish densities increase dramatically in response to massive recruitment of blue mussels, one of their favorite foods. They found that an unusually high mussel set in the Gulf of Maine in 1995 led to elevated starfish densities, and that when the mussel food supply was depleted, starfishes became less common and resorted to cannibalism in the absence of alternative food sources (J. Witman, personal communication).

Both *Asterias forbesii* and *Asterias vulgaris* have external fertilization, releasing eggs and sperm into the water column. Both of these starfishes are also good at regenerating lost body parts, and can regenerate an entire individual from little more than a couple of arms and a third of the central disk. [🐚27]

Crabs are another common and important predators on New England rocky shores, particularly in protected habitats. The two most common shallow-water predaceous crabs in the northern Atlantic are the introduced European green crab, *Carcinus maenas*, and the northern rock crab, *Cancer*

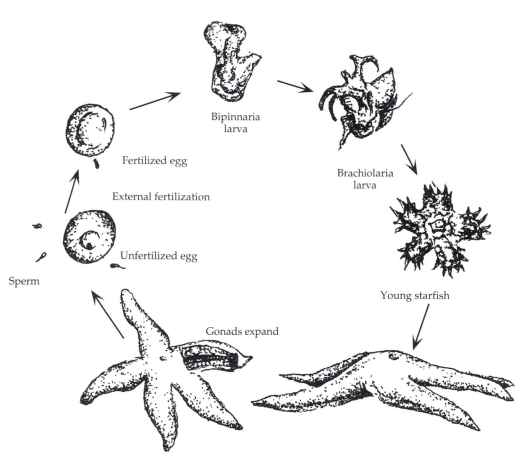

Bipinnaria
larva

Fertilized egg

External fertilization

Brachiolaria
larva

Unfertilized egg

Sperm

Young starfish

Gonads expand

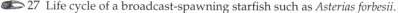

🐚 27 Life cycle of a broadcast-spawning starfish such as *Asterias forbesii*.

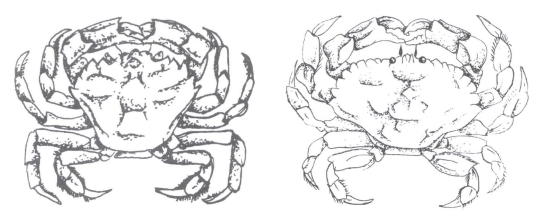

28 The green crab, *Carcinus maenus* (left), and the rock crab, *Cancer irroratus* (right), are common scavengers and predators of hard-bodied prey on rocky shores.

borealis.[28] Both are predators on mussels, barnacles, snails, and other hard-bodied prey. In contrast to starfishes, crabs are most abundant and have their greatest effects in protected bays and estuaries rather than in open-coast habitats. Crabs can generally cope with salinity fluctuations, but as relatively large mobile organisms, they have difficulty foraging in habitats with strong water movement. Rock crabs are found in low intertidal habitats in Maine, but are more common subtidally (Gosner 1978). Green crabs are the most abundant large intertidal crabs on rocky shores in New England, but were not known north of Cape Cod before 1900, and were probably introduced into Long Island Sound in colonial times (Glude 1955).

The reproductive behavior and growth of crabs are strongly affected by their hard external skeletons, which must be shed or molted for growth to occur. Both green crabs and rock crabs exhibit precopulatory courting behavior, in which males find premolt females and mount them in a copulatory embrace. Sometimes males carry around females for a number of days before molting occurs. When the female molts, her genital pore is exposed, and the crabs copulate. After copulation and fertilization, the female carries the fertilized eggs under her abdomen, aerating them by gently flexing her abdomen up and down under the carapace. The young are released as feeding larvae. Larval release by shoreline crabs often occurs during spring tides, which transport the larvae rapidly offshore, where they are exposed to fewer consumers (Morgan and Christy 1995). After developing in the water column for a number of weeks, crab larvae move back to shore by regulating their vertical position in the water column, moving to deeper water or into surface slicks that move them shoreward (see Morgan 1995).

Juvenile rock and green crabs are translucent and tend to be cryptic to their predators. Alvaro Palma, of the University of Maine, has shown that juvenile rock crabs preferentially settle on coarse sand substrates, where they are difficult for their predators to detect (A. Palma, personal communication). As the crabs grow larger, fewer predators are capable of consuming them, freeing them to live in other habitats. Molting always makes crabs vulnerable to predators and physical stresses, since immediately after shedding the old skeleton, they have soft shells and little protection. Molting green and rock crabs move into shallow water under rocks or seaweed canopies to minimize these risks.

Birds are more important predators on rocky shores than is often recognized. Herring gulls (*Larus argentatus*), black-backed gulls (*Larus marinus*), and purple sandpipers (*Calidris maritima*) commonly forage in rocky intertidal habitats during low tides. Gulls drop hard-bodied prey items from the air onto rocks to break them open (see Chapter 3). Their populations have increased over the last century due to human population growth and the proliferation of garbage dumps, which has probably led to their playing a larger role in coastal communities (Kadlec and Drury 1968). Herring gulls can be important predators of blue mussels, urchins, starfishes, and rock and green crabs (Dumas and Witman 1993; Marsh 1986), and may have important direct effects on rocky intertidal communities. For example, while gull predation has been shown to limit rock crabs to subtidal habitats, green crabs are less often eaten because they are cryptic and find refuge in algal canopies (Dumas and Witman 1993). Gull predation can also have important indirect effects on rocky shore communities by regulating population densities of crabs and urchins, which otherwise regulate prey abundance patterns. Gull predation on urchins in tide pools, for example, may enhance the persistence of palatable algae by removing these herbivores (Lubchenco 1978).

STRESSES IN ROCKY INTERTIDAL HABITATS

Most of the intertidal zone of rocky shores is subjected each day to aerial exposure, presenting problems of desiccation and extreme temperatures for the organisms that live there. Most organisms from other habitats have a relatively narrow window of about 10°C within which their biochemical systems will function. Any temperatures exceeding this range will result in death. For organisms living in rocky intertidal habitats, 10–20°C daily swings in temperature are routine. Rocky intertidal habitats on the east coast of North America are stressful even in comparison to other temperate zone rocky intertidal habitats, since they experience both relatively large tides and strong seasonality. In southern New England, intertidal organisms can be exposed to winter temperatures as low as –20°C (Kanwisher 1960a) and summer temperatures as high as 40°C (Bertness 1989). Daily variations in temperature are also impressive, with daily fluctuations of 20–30°C common in both the summer and winter. In the early spring in New England, for example, water temperatures may be 2–3°C, but intertidal rocks can heat up to 30–40°C within an hour during low tide exposures.

Desiccation

Desiccation, or water loss, during low tides is one of the most critical problems facing intertidal organisms. Mobile organisms that live in high intertidal habitats, such as crabs, can move into crevices, under seaweeds, and into tide pools at low tide to minimize dehydration problems (Kensler 1967).[🐚 29] Moving to safer habitats during low tide is not an option for sessile or slow-moving organisms, and the species found at high intertidal heights, such as barnacles and periwinkles, must be especially resistant to desiccation. The high intertidal barnacle *Chthamalus*, for example, can survive 28 days out of water, and the ribbed periwinkle *Littorina saxatilis*, which lives at extreme high intertidal heights, can survive over 42 days out of water (see Newell 1979 for other examples). Organisms characteristic of lower elevations on the shore are generally not capable of dealing with prolonged desiccation (Foster 1969).

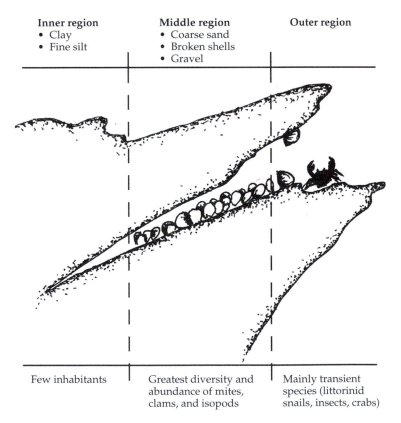

Inner region	Middle region	Outer region
• Clay • Fine silt	• Coarse sand • Broken shells • Gravel	

Few inhabitants	Greatest diversity and abundance of mites, clams, and isopods	Mainly transient species (littorinid snails, insects, crabs)

🐚 29 Crevice zonation. At high intertidal heights, cracks and crevices in rocks offer thermal and desiccation refuges for transient and specialized species. (After Carefoot 1977 and Kensler 1967.)

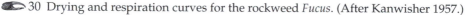

The problem of dessication can be reduced by increasing body water volume, limiting water loss, or simply tolerating large water losses. High intertidal snails (Vermeij 1971) and barnacles (Barnes and Barnes 1957) carry relatively large amounts of water in their shells. The globose shape of many high intertidal snails maximizes the water-holding capacity of their shells (Vermeij 1971). Water loss can also be limited by sealing off the outside environment. Snails, barnacles, and mussels all close their shells tightly when exposed to aerial conditions. Snails exposed to desiccating conditions can also attach to the substrate with a mucus thread and close the operculum to limit thermal contact with the rock surface and minimize water loss (Vermeij 1971).

Oxygen depletion can be a significant obstacle to limiting water loss. Barnacles address this problem by having a small opening between their valves (the pneumostome) that permits air movement while limiting water loss (Barnes and Barnes 1957). Mussels periodically open their valves and "airgape" during low tides to replenish their oxygen supply (Lent 1969). Some high intertidal organisms rely heavily on anaerobic metabolism, while many others are simply good at tolerating water loss. High intertidal seaweeds are particularly impressive in their ability to tolerate water loss. *Fucus, Ulva,* and *Enteromorpha* can all lose as much as 75–90 percent of their water during low tide, but rehydrate rapidly and resume active photosynthesis within an hour of being resubmerged (Kanwisher 1957; see Dring and Brown 1982 and Chapman 1986 for more recent work).[30]

High Temperatures

High summer temperatures are just as serious a problem as dessication for many intertidal organisms. During southern New England summers, water temperatures are only about 20°C, but both air and rock temperatures can reach 40°C during daytime low tides. As a result, intertidal residents can be

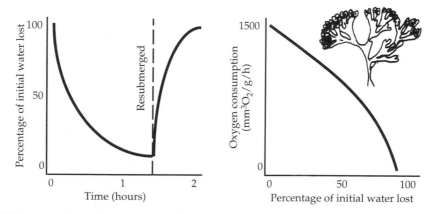

30 Drying and respiration curves for the rockweed *Fucus.* (After Kanwisher 1957.)

exposed to extreme high temperatures with large fluctuations. The metabolic rate of most cold-blooded organisms, which is normally temperature dependent, roughly doubles for every 10°C rise in temperature. While most intertidal invertebrates do not maintain a constant body temperature, as warm-blooded organisms do, many have relatively temperature-independent metabolic rates over the range of temperatures at which they usually live.[🐚 31] The mechanisms that permit this relatively temperature-independent metabolism are largely unstudied, although they are known to occur in cellular as well as whole-organism metabolism (Newell 1979).

Intertidal organisms avoid high summer temperatures and fluctuations with some of the same mechanisms they use to minimize water loss. Many live in thermal refuges such as crevices, tide pools, or algal canopies, thereby minimizing exposure to high temperatures. On protected shores in the Gulf of Maine, the algal canopy of *Ascophyllum nodosum* plays a major role in providing refuge for intertidal organisms from heat and desiccation (Menge 1978). Other sessile organisms, including many barnacles, mussels, and seaweeds, live in dense groups that buffer group members from water loss and high temperatures (Bertness 1989; Bertness and Leonard 1997). Nonetheless, the thermal limits of intertidal organisms are reflected in where they occur on the shore, with the most and least thermally tolerant species found at the highest and lowest levels of the shore, respectively.[🐚 32, 33, 34]

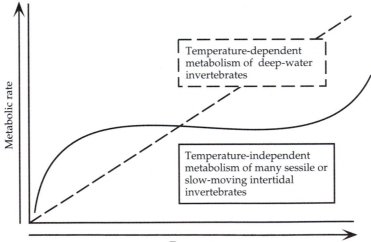

🐚 31 In contrast to most cold-blooded organisms, many intertidal invertebrates have metabolisms that are independent of temperature over the ranges they normally encounter. This allows them to remain active during normal intertidal temperature fluctuations. (After Newell 1969; Hochachka and Somero 1973.)

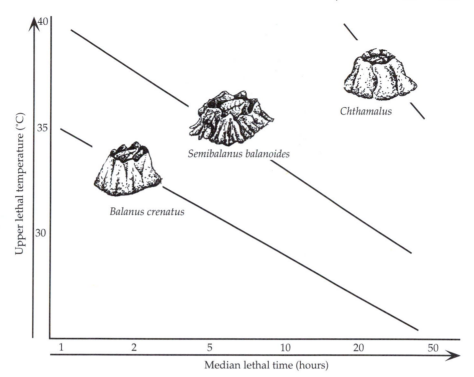

🦐 32 The thermal limits of Atlantic coast barnacles in Europe reflect their typical habitats on the shoreline. *Chthamalus* is the most tolerant of high temperatures and lives at the highest elevations. *Balanus crenatus* is the least tolerant and is found in subtidal habitats. (After Foster 1969.)

The effects of summer heating on rocky shore organisms are conspicuously influenced by rock color, composition, and size. Dark-colored rocks absorb rather than reflect heat, and thus get hotter than lighter-colored rocks. Sessile organisms such as barnacles may be unable to deal with the temperatures reached by dark-colored substrates (Raimondi 1988a). Rock size effects on heating can be particularly strong (Bertness 1989). Large boulders or rock benches have a large mass that buffers them from heating, while small, loose cobbles not embedded in sediment are not thermally buffered and heat up rapidly when exposed during summer daytime low tides. In Narragansett Bay, the differential heating of rocks of different sizes influences barnacle populations. Few barnacles survive to reproduce on small cobbles, while larger, thermally buffered boulders support large, age-structured barnacle populations (Bertness 1989). The rock benches common in open-coast habitats are probably important thermal refuges for many sessile invertebrates. Seaweed canopy cover can also buffer temperature fluctua-

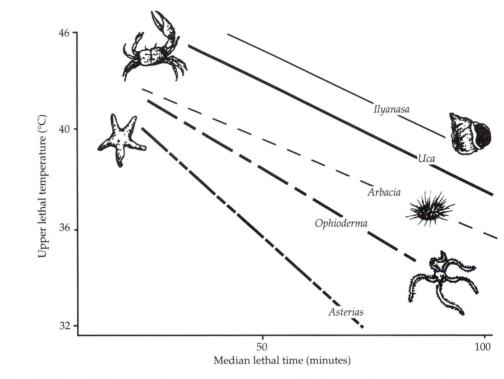

🦀 33 Thermal tolerances of some common western Atlantic shoreline invertebrates. (After Orr 1955.)

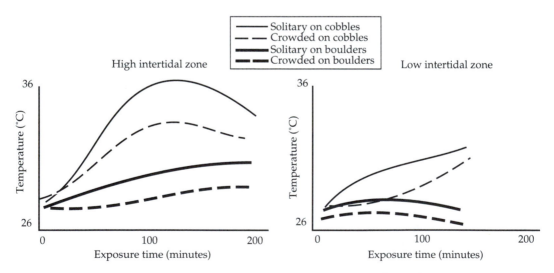

🦀 34 Acorn barnacle tissue temperatures during a summer low tide in Narragansett Bay as a function of rock size and crowding. (After Bertness 1989.)

tions on rock surfaces.[⬤ 35] Rock heating effects on the distributions of
rocky intertidal organisms, however, are probably not as important on
northern New England or Canadian shores, where summer temperatures
are not as high.

The importance of heat stress in determining both the local and biogeo-
graphic distributions of intertidal organisms was eloquently illustrated by
David Wethey's (1983a, 1984b) work with New England barnacles. In Long
Island Sound, two acorn barnacles are common on rocky shorelines.
Chthamalus fragilis occupies the high intertidal zone, while *Semibalanus bal-*

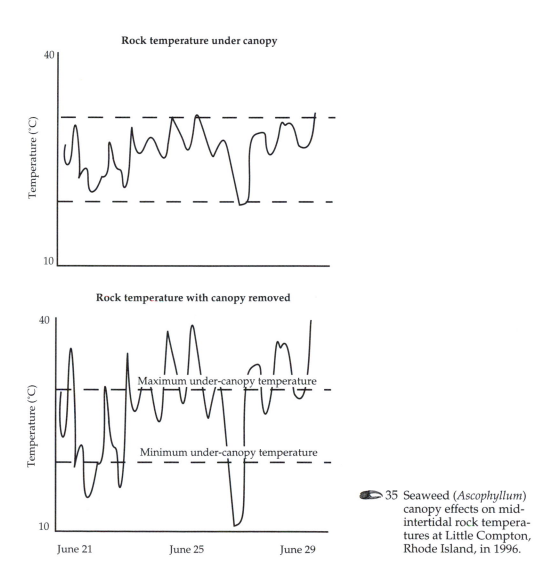

35 Seaweed (*Ascophyllum*)
canopy effects on mid-
intertidal rock tempera-
tures at Little Compton,
Rhode Island, in 1996.

anoides dominates the middle intertidal zone. As described in the last chapter, both of these barnacles are commonly found at high densities and settle widely in the intertidal zone, so that their distributions are often set by post-recruitment rather than recruitment processes. *Chthamalus* can tolerate higher temperatures, but *Semibalanus* is competitively dominant to *Chthamalus*, since it has shell walls filled with air passageways that allow it to grow faster and larger (Stanley and Newman 1980). Wethey (1984b) demonstrated that experimental shading reduced thermal stress and allowed *Semibalanus* to displace *Chthamalus* from high intertidal habitats. He also found that *Chthamalus* was restricted to high intertidal habitats by the competitive dominance of *Semibalanus*. He further suggested that the effect of thermal stress on the relationship between *Chthamalus* and *Semibalanus* has dramatic biogeographic consequences. North of Cape Cod, summer temperatures are not hot enough to exclude *Semibalanus* from high intertidal elevations. As a result, *Semibalanus* may displace *Chthamalus* from high intertidal habitats, setting the northern biogeographic limit of *Chthamalus*.[🐚 36]

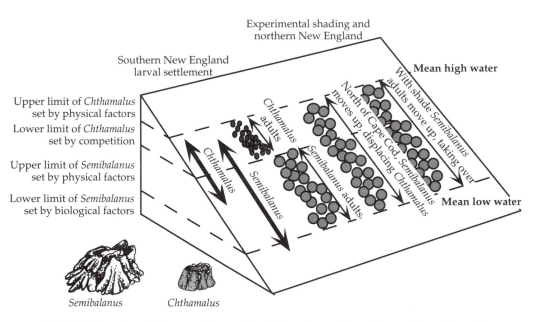

🐚 36 Zonation of the acorn barnacles *Chthamalus* and *Semibalanus*. In southern New England, *Chthamalus* is competitively displaced to high intertidal heights by *Semibalanus*. With experimental shading, *Semibalanus* moves to the high intertidal, displacing *Chthamalus*. This process occurs naturally at cooler northern latitudes, where *Chthamalus* is driven to regional extinction. (After Connell 1961; Wethey 1983a,b.)

Low Temperatures

Extreme low winter temperatures are also a problem for organisms on New England and particularly Canadian rocky shorelines. On wind- and wave-swept shorelines in January and February, ice sheets can cover high inter-tidal surfaces, leaving barnacles, mussels, and algae encased in ice for weeks at a time. Many mobile rocky shore organisms, such as snails (*Littorina, Nucella*) and crabs (*Carcinus* and *Pagurus*), migrate into subtidal habitats to avoid winter stresses. For most sessile organisms, however, extreme winter conditions are one of the costs of living in intertidal habitats in the northern Atlantic. Many barnacles and algae can tolerate extreme low temperatures remarkably well, although the mechanisms responsible are not always clear. Under typical New England winter conditions, as much as 50–70 percent of the water content of invertebrates and algae freezes (Kanwisher 1960a,b). Organisms survive freezing by limiting ice formation to regions between rather than within their cells (so that cells do not rupture) and by tolerating the intercellular hypersaline conditions that develop once ice forms (Kanwisher 1957). Additionally, some intertidal invertebrates probably uti-lize organic "antifreezes" analogous to those known to occur in arctic fishes (Devries 1971). Acorn barnacles, for example, have increased concentrations of glycerol (a possible organic antifreeze) in the winter, corresponding to increased resistance to cold (Newell 1979).

Differential freezing tolerances can play a leading role in the zonation of northern New England seaweeds. The rockweed *Fucus* is often encased in ice during the winter in high intertidal habitats in New England, but suffers little mortality (Kanwisher 1957). Similarly, *Mastocarpus stellatus*, a mid-intertidal red seaweed, shows no adverse effects when frozen daily for over a month. In contrast, Irish moss (*Chondrus crispus*), which dominates many subtidal New England habitats, dies when exposed to freezing conditions (Dudgeon et al. 1989). Cold winter exposure commonly kills intertidal *Chondrus*, and in north-ern New England, freezing appears to limit *Chondrus* to subtidal habitats (Dudgeon et al. 1989; Mathieson and Burns 1975).[🐚37]

Cold winter conditions are also important in determining the biogeo-graphic distributions of species on New England rocky shores. Cape Cod is a major biogeographic boundary for many marine taxa, forming the south-ern range limit of many boreal species and the northern limit of many south-ern species (Vermeij 1978). The importance of cold winters in maintaining the northern limit of southern species was demonstrated by W. C. Allee (1923). Allee documented that severe winters north of Cape Cod kill few boreal species, but southern species that successfully recruit north of Cape Cod suffer heavy mortality in severe winters.

As described in the first chapter, ice damage is an extremely important consequence of the cold winter conditions on Canadian rocky shores, partic-ularly north of Nova Scotia (Stephenson and Stephenson 1971). Ice can encase intertidal organisms and rip them from the shoreline (Mathieson et

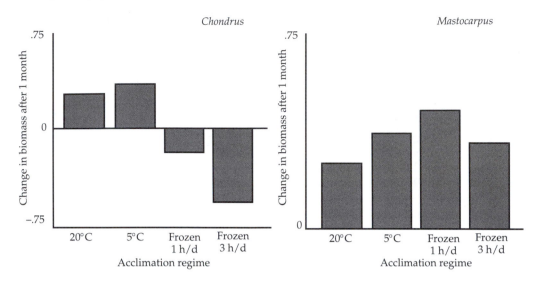

🦪 37 Freezing effects on the intertidal red seaweeds *Chondrus* and *Mastocarpus* result in *Mastocarpus* dominating the higher intertidal zone in colder habitats. When exposed to simulated low tide freezing, *Mastocarpus* grew, while *Chondrus* died. (After Dudgeon et al. 1989.)

al. 1982), leaving many northern rocky shores in a constant state of recovery from chronic disturbance. Ice scour from floating ice is also a major problem for organisms on many Canadian rocky shores. In the Gulf of Saint Lawrence, for example, heavy annual ice scouring restricts northern acorn barnacles to living in cracks and crevices. Bourget and Holm (1994) have suggested that this chronic ice scouring has led to selection for increased settlement in cryptic crevice habitats in barnacles in the Gulf of Saint Lawrence.

Wave Stresses

Wave stress in exposed open-coast habitats can play a major role in the design, abundance, and distribution of rocky intertidal organisms and acts as an important modifier of shoreline zonation patterns. Wave velocities can be measured with a simple, inexpensive device called a wave force **dynometer** (Bell and Denny 1994).[🦪 38] Such measurements show that wave velocities on exposed shorelines can reach 50 meters per second (Denny et al. 1985; Denny 1987).

One of the most conspicuous effects of wave forces on intertidal organisms is the limitation of organism sizes (Denny et al. 1985). Small, squat organisms, such as barnacles and limpets, live with much of their bodies within surface boundary layer conditions, which minimizes their exposure to wave forces, and have streamlined shapes that further minimize lift and drag. As organisms increase in size, their vulnerability to wave forces increases, and unless they compensate with increased attachment and/or

⌐⊃38 A simple wave force dynometer consists of a spring mounted in a plastic tube, attached on a swivel mount to the substrate. The instrument measures the force exerted on a drogue attached to the spring.

structural strength, they become more vulnerable to being ripped off the substrate. Wave forces may thus often set the upper size limit and optimal size of intertidal organisms in wave-exposed habitats (Denny et al. 1985; Gaylord 1998).[⌐⊃39]

A number of mechanisms limit the effects of wave stress in coastal habitats. Dense groups of organisms buffer individuals from wave stresses. Mussels in mussel beds, for example, are less exposed to wave forces than solitary individuals, allowing individuals in beds to exceed the calculated maximum solitary size (Denny et al. 1985). Algal canopies also reduce wave forces and permit organisms to live in habitats that they could not otherwise occupy (Holbrook et al. 1991). In particular, many of the mobile organisms associated with mussel beds and algal canopies (including carnivorous and herbivorous snails) are clearly dependent on canopy wave buffering in wave-exposed habitats.

The macroalgae on wave-exposed shorelines often are much larger than the sessile invertebrates. In contrast to invertebrates, which rely on small size, firm attachment, and rigid bodies to cope with wave stress, seaweeds

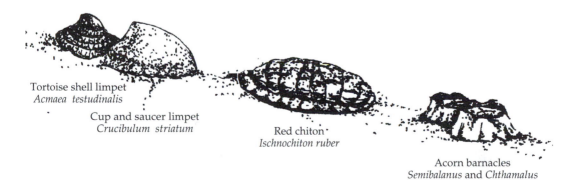

Tortoise shell limpet
Acmaea testudinalis

Cup and saucer limpet
Crucibulum striatum

Red chiton
Ischnochiton ruber

Acorn barnacles
Semibalanus and *Chthamalus*

⌐⊃39 Sessile and slow-moving invertebrates on wave-exposed rocky shorelines are small and have streamlined profiles to minimize the effects of wave stress.

often rely on flexibility. By bending, algae reduce their exposed surface area and move closer to the low-wave-energy boundary layer (Koehl and Wainwright 1977). Flexibility allows many seaweeds (such as the shallow subtidal kelp) to attain relatively large sizes while minimizing their exposure to waves.[40] Other soft-bodied organisms, such as anemones, sea squirts, and hydrozoans, also depend on flexibility to deal with wave forces (Koehl 1984a,b).

    Spatial variation in wave stresses can be important in generating distribution patterns both within and between seaweed species. A good example of wave stress effects on the morphology of an algal species can be found in *Chondrus crispus*. In protected bays and estuaries, *Chondrus* has a tall, bushy morphology, whereas on wave-exposed shorelines, it has a shorter, more heavily branched turf morphology. *Chondrus* from bays, with its high surface area, would likely be ripped off by waves in exposed habitats. Constant

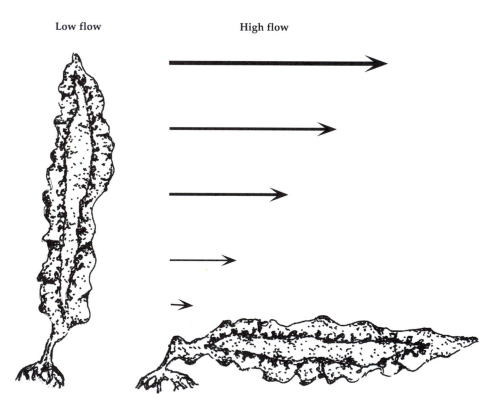

Low flow   **High flow**

40 Many seaweeds cope with high wave energy by bending into the flow, by which they reduce their profile and drag, and also move closer to the substrate, where they experience lower wave forces. This strategy allows them to achieve larger body sizes than if they had rigid bodies.

Protected   Exposed

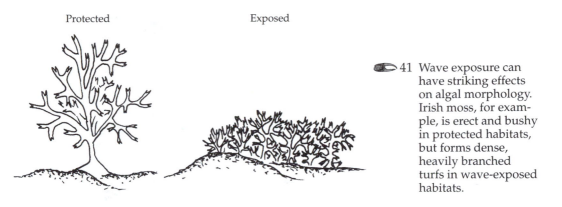

41 Wave exposure can have striking effects on algal morphology. Irish moss, for example, is erect and bushy in protected habitats, but forms dense, heavily branched turfs in wave-exposed habitats.

pruning and biomass loss due to waves generates the turf morphology found in exposed habitats by stimulating holdfast expansion and stipe regeneration and proliferation.[41] Like *Chondrus*, the knotted wrack *Ascophyllum* also decreases in height, but increases in stipe density, with increasing wave stress.

Morphological variation in seaweeds across wave stress gradients, however, must be interpreted with caution, since conspicuous morphological differences in flexible organisms do not necessarily influence their responses to hydrodynamic stresses (Carrington 1990). Morphological variation in soft-bodied sessile invertebrates in response to wave stress gradients is also probably common, but is not well studied. Steve Palumbi (1984, 1986) has shown that individual sponges can increase their investment in structural support and take on a more compact, rigid structure in response to wave stress exposure.

Species differences in morphology or resistance to wave damage commonly lead to species replacements across wave stress gradients. An example can be found in coastal habitats in the Gulf of Maine, where *Mastocarpus stellatus* replaces *Chondrus crispus* in wave-exposed habitats. Though these two seaweeds have similar breaking strengths, *Chondrus* fronds have a higher surface area, which increases drag and biomass loss (Dudgeon and Johnson 1992).

Vulnerability to wave dislodgement is often enhanced by epiphytic organisms that increase drag forces on their hosts. The horse mussel, *Modiolus modiolus*, for instance, commonly forms dense subtidal beds off rocky shores in the Gulf of Maine. It is much more vulnerable to wave dislodgement when kelps are attached to its shells (Witman and Suchanek 1984). Similarly, the introduced green alga *Codium fragilis* spp. *tomentosoides* has had a major effect on cobble shorelines because it increases drag on cobbles. This makes the cobbles more vulnerable to dislodgment, which kills associated organisms (Carlton and Scanlon 1985).[42]

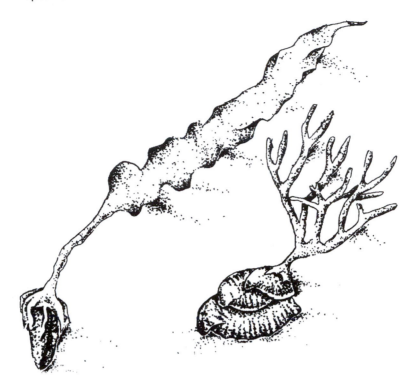

42 Examples of seaweeds growing epiphytically on other organisms, increasing the probability that those organisms will be dislodged by waves, include kelps on horse mussels (left) and *Codium* on slipper limpets (right).

Wave forces can also shift the distributions of organisms by reducing desiccation stress at high intertidal heights (Stephenson and Stephenson 1971). On exposed shorelines, waves crashing onto the shore splash higher than on more protected shores, permitting organisms whose upper intertidal limit is set by heat and desiccation to live at higher intertidal heights. Typically, this effect shifts the entire zonation sequence to higher intertidal heights.[43]

Wave forces also influence the distributions of intertidal organisms by abrading surfaces of settlers. Algal **whiplash**—the thrashing of seaweeds against the substrate, resulting in the death of recent settlers—can be an important source of mortality, particularly for sessile organisms associated with algal canopies in high wave energy environments (Menge 1976). On wave-exposed rocky shorelines in Maine, for example, surfaces adjacent to algal canopies often have a band of bare space, or whiplash halo. This band is caused by the physical removal of organisms by whiplash abrasion and the grazing of snails that are dependent on the algal canopy for protection from heat and wave stresses.[44]

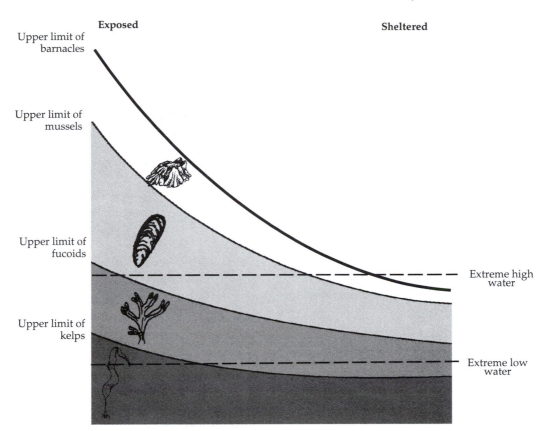

43 Relationship between the intertidal heights of distribution zones on rocky shores and wave exposure. (Modified for New England after Lewis 1964.)

44 Whiplash halos around rockweeds prevent most larval recruitment.

ZONATION

The zonation of rocky shores has long interested ecologists and has been well studied on the Atlantic coast of North America. Rocky intertidal shores have served as a model system for studying the zonation of shoreline organisms introduced in the last chapter. This pronounced vertical or elevational zonation is the product of strong gradients in physical and biological stresses. At high intertidal heights, only organisms capable of tolerating long periods out of water and extreme temperature fluctuations persist. At low intertidal heights, where habitats are not exposed to such intense disturbance, organisms face less stressful physical conditions, but more severe biological stresses. Consequently, only organisms capable of successfully dealing with consumers and competitors persist at low intertidal heights.

Wave-Exposed Shores

The vertical zonation of organisms on rocky shores on the east coast of North America has been studied most intensively on wave-exposed shorelines in the Gulf of Maine (Menge 1976; Grant 1977). Wave-exposed headlands are characterized by discrete zones of densely packed organisms and little (usually less than 10 percent) unoccupied (free) space.[🐚45] The acorn barnacle *Semibalanus balanoides* typically occupies a distinct high intertidal band, but is replaced at middle intertidal elevations by the blue mussel *Mytilis edulis*. The mussel zone usually extends into the shallow subtidal zone, where kelp beds occur.

The simplicity of these habitats is largely due to wave stress and shock. Few organisms can tolerate these stresses, and those that can (barnacles, mussels, and kelps) dominate these habitats because waves limit the abundance of their predators. Where algal canopies are absent, predatory crabs, fishes, and snails and herbivorous snails and urchins are rare and ineffective consumers (Menge 1978), leading to the dominance of only a few wave-tolerant species. As already mentioned, this scenario appears to break down in the oligotrophic waters of the Canadian Maritimes, where even extremely wave-exposed rocky shores support robust seaweed canopies that harbor dense snail populations.

In the absence of consumers, competition for space is usually an important determinant of the zonation of wave-exposed shores in New England. *Semibalanus* dominates the high intertidal zone because it is the only common primary space holder able to live at high intertidal heights. *Mytilus* typically replaces *Semibalanus* at intermediate elevations due to its competitive dominance. Recruitment densities, however, also influence zonation patterns on wave-exposed shores. Without sufficient recruitment to saturate the available substrate space, competition for space does not become a major force in determining organism distribution patterns.[🐚46]

Variation in recruitment conspicuously affects barnacle population dynamics and morphology. At low densities, acorn barnacles grow into short, squat individuals wider than they are tall. In contrast, at high recruit

Wave-exposed open coast zonation

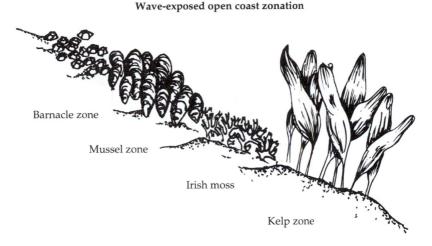

Barnacle zone

Mussel zone

Irish moss

Kelp zone

Protected coast zonation

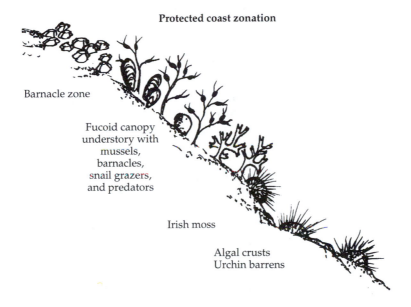

Barnacle zone

Fucoid canopy
understory with
mussels,
barnacles,
snail grazers,
and predators

Irish moss

Algal crusts
Urchin barrens

45 Some of the effects of wave exposure on rocky intertidal zonation are the result of low consumer densities. The relative lack of predatory snails, starfishes, and sea urchins on wave-exposed shores typically leads to conspicuous mussel and kelp beds.

densities, crowding causes barnacles to grow up rather than out, since they are constrained by neighbors. Crowding has interesting consequences for barnacle growth and reproduction, as well as population dynamics. Crowded barnacles invest less in structural support (shell material), since

High recruitment **Low recruitment**

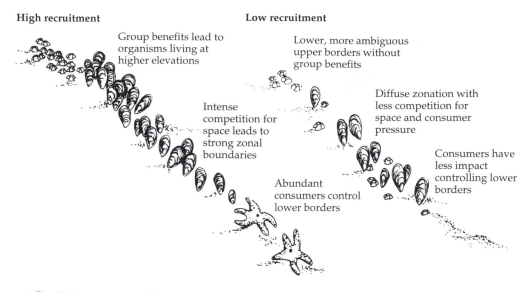

Group benefits lead to organisms living at higher elevations

Lower, more ambiguous upper borders without group benefits

Intense competition for space leads to strong zonal boundaries

Diffuse zonation with less competition for space and consumer pressure

Abundant consumers control lower borders

Consumers have less impact controlling lower borders

🐚 46 Recruitment effects on rocky intertidal shoreline zonation patterns.

they are supported by neighbors (Wu et al. 1977; Wu 1980). Moreover, since crowded individuals are elevated higher above the surface, they experience higher food fluxes and grow faster than solitary individuals (Pullen and LaBarbara 1991; Bertness et al. 1998). Crowded, columnar barnacles have larger mantle spaces, permitting them to produce larger clutches of eggs (Wethey 1984a). Crowded barnacles can also buffer one another from thermal stress and increase survivorship at high intertidal heights (Bertness 1989). Balanced against these advantages of crowding, however, are serious disadvantages. Overcrowding leads to intense competition for space and structurally weak individuals that are easily crushed by neighbors or external forces (Barnes and Powell 1950; Bertness 1989). The balance between these costs and benefits of crowding in barnacles is determined by environmental conditions. In habitats where barnacle growth rates are high (low intertidal and high-flow habitats), the disadvantages of crowding typically lead to massive mortality of crowded individuals. In habitats with low growth rates or harsh physical stresses, the advantages of crowding can outweigh the disadvantages (Bertness 1989; Sanford et al. 1994).[🐚47]

Mussel recruitment on wave-exposed shores is tightly linked to both the supply of recruits and the presence of barnacles. Barnacles enhance mussel recruitment, since mussels preferentially settle on and have higher survivorship on rough-textured surfaces (Menge 1976; Petraitis 1987). High rates of mussel settlement on barnacles and rapid mussel growth and mobility lead to mussels overgrowing and suffocating the barnacles. Thus,

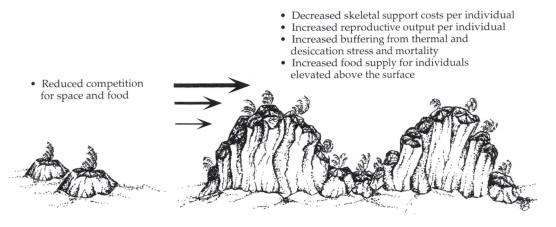

- Reduced competition for space and food

- Decreased skeletal support costs per individual
- Increased reproductive output per individual
- Increased buffering from thermal and desiccation stress and mortality
- Increased food supply for individuals elevated above the surface

47　Costs and benefits of crowding in acorn barnacles.

barnacles facilitate mussel settlement, which paves the way for their ultimate demise.[　48]

Physical disturbances, common in all rocky intertidal systems, play a particularly large role in creating spatial patterns on rocky shores exposed to heavy waves (Dayton 1971; Osman 1977; Sousa 1979a, 1985). Wave shock can rip organisms from the substrate (Menge 1976; Witman 1987), and waterborne objects such as driftwood and stones (Dayton 1971; Shanks and Wright 1986) can crash onto shorelines and destroy organisms. These physical disturbances kill sessile space holders, reduce competition for space, and stimulate the invasion of competitively subordinate rapid colonizers. On wave-exposed headlands in Maine, for example, winter wave exposure is often responsible for scouring rocky shores of mussel recruits in all but refuge habitats.

Protected Shores

Zonation on protected shores in New England is strikingly different from that on wave-exposed shores. The major similarity is the barnacle zone demarcating the upper intertidal border in both habitats. On protected shores, there is often considerable free space (often more than 50 percent) at lower elevations, few mussels, a profusion of seaweeds, and abundant mobile consumers. Farther north, in the Canadian Maritimes, protected habitats are strongly affected by ice, which is more common and heavier in bays and estuaries than in coastal habitats. Bays and estuaries experience more ice damage, since they are not as thermally buffered by the open ocean as coastal habitats are, and because their lower-salinity waters freeze more readily.

The intertidal community on protected rocky shores in New England differs markedly from that on shores exposed to heavy waves, since con-

Barnacle recruits increase surface
roughness

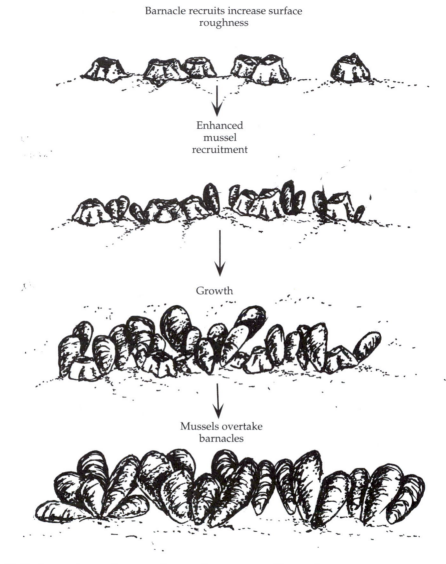

Enhanced
mussel
recruitment

Growth

Mussels overtake
barnacles

48 By increasing surface roughness, the presence of barnacles can
facilitate the recruitment and competitive dominance of mussels.
(After Menge 1976.)

sumers become more important and a seaweed canopy of relatively unpalatable larger seaweeds, like *Ascophyllum* and *Fucus*, buffers understory organisms from physical stresses. The algal canopy at middle elevations generally
supports a rich assemblage of organisms, many of which could not persist at
these intertidal heights without the cool, shaded, moist habitat provided by
the canopy. The herbivorous snails *Littorina littorea* and *Littorina obtusata*

limit the abundance of ephemeral algae and even the success beneath the canopy of the algal sporelings of canopy-forming seaweeds. The carnivorous snail *Nucella*, the starfish *Asterias*, and crabs limit the abundance of both mussels and barnacles (Ebling et al. 1964; Menge 1976).[🐚49]

Northern acorn barnacles, while sparse, are often the most common sessile understory animals. They live at much lower elevations on wave-protected shores than on exposed shorelines, since mussels (their major spatial competitors) are kept in check by predators under the canopy.[🐚50] Epiphytic algae (algae that grow on other algae), including ephemeral, opportunistic species like the sea lettuce *Ulva* and specialists like *Polysiphonia*, grow attached to canopy seaweeds. They escape their herbivores by living on unpalatable algae that are not heavily grazed (Harlin 1975, 1980). On extremely sheltered shores, sediment accumulation under the canopy precludes the successful recruitment and survival of most sessile organisms characteristic of hard substrates, including barnacles and other seaweeds.

In New England, the smooth periwinkle *Littorina obtusata* is almost always found associated with algal canopies of the knotted wrack *Ascophyllum nodosum*. This snail resembles *Ascophyllum* air bladders, and may be camouflaged from predators in association with *Ascophyllum*. In Europe, green crabs are important predators of this snail, and restrict smaller snails to high intertidal heights and algal cover (Williams 1992).[🐚51] The relationship between *Littorina obtusata* and *Ascophyllum*, however, has not been carefully examined in North America. Periwinkles appear to rely on *Ascophyllum* for cover and camouflage, but how much adult *Ascophyllum* they consume is unclear. Their grazing on diatoms and epiphytes that grow on the canopy could benefit *Ascophyllum* by limiting algal fouling, but this hypothesis has not been critically examined.

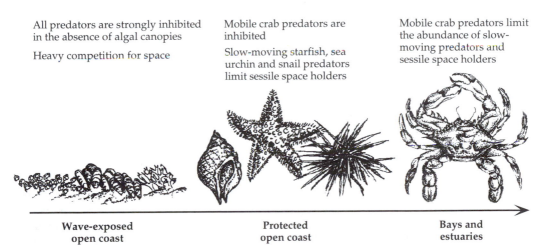

All predators are strongly inhibited in the absence of algal canopies

Heavy competition for space

Mobile crab predators are inhibited

Slow-moving starfish, sea urchin and snail predators limit sessile space holders

Mobile crab predators limit the abundance of slow-moving predators and sessile space holders

Wave-exposed open coast   **Protected open coast**   **Bays and estuaries**

🐚 49 The general influence of wave exposure on the predators and sessile space holders of rocky shores.

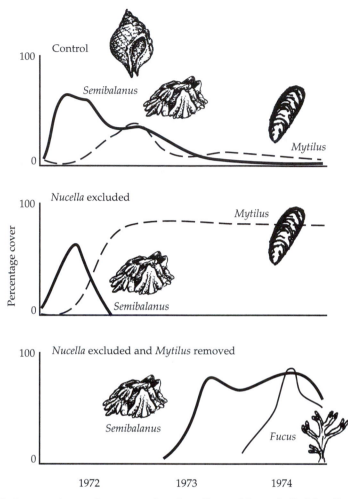

50 An experiment demonstrating the effects of dog whelk (*Nucella lapillus*) predation on protected New England rocky shores. (After Menge 1976.)

The presence of an *Ascophyllum* canopy on New England rocky shores is strongly correlated with wave stress. *Ascophyllum* canopies are absent from heavily wave-exposed shores. They are present in protected open-coast sites, but reach their most extensive development on the shores of protected embayments (Leonard et al. 1998a). *Ascophyllum* recruits attach very weakly to the substrate, precluding recruitment in wave-swept habitats (Vadas et al. 1990). Even if recruits were able to settle in wave-exposed habitats, adult *Ascophyllum* would probably be ripped from the substrate because of their large size. *Ascophyllum* in open-coast areas is relatively short in comparison to individuals in protected bays, further suggesting that wave stresses can

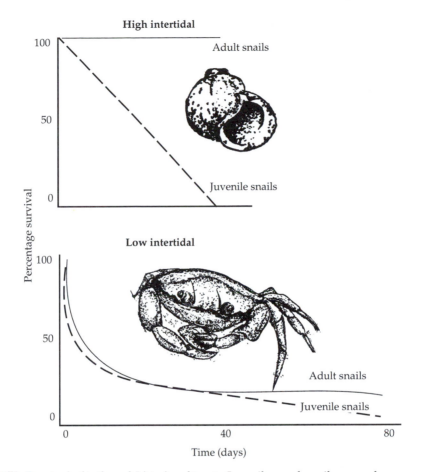

High intertidal

Adult snails

Juvenile snails

Low intertidal

Adult snails

Juvenile snails

Time (days)

51 Survival of tethered *Littorina obtusata*. Juveniles are heavily preyed upon by green crabs at both high and low intertidal heights. Adult snails are preyed upon only at low tidal heights, where larger crabs occur. (After Williams 1992.)

limit *Ascophyllum* success. Once a canopy is established, grazers limit the recruitment of most algal sporelings, including those of species that are immune to herbivores as adults (Lubchenco 1983; Bertness and Leonard, unpublished data). Adult *Ascophyllum* is eaten infrequently due to a tough outer cuticle and chemical defenses.

Direct positive associations among sessile organisms that result from neighbor amelioration of heat and desiccation stresses are important in protected rocky intertidal communities (see Chapter 4 for a general discussion of direct positive interactions). Dense groups of barnacles, mussels, and algae can all buffer group members from heat and desiccation stresses at high intertidal levels and permit them to live at higher intertidal heights than would be possible without neighbors (Hay 1981; Lively and Raimondi 1987; Bertness 1989; Bertness and Leonard 1997). The presence of dense aggregations of seaweeds, barnacles, and mussels also ameliorates stressful physical conditions for many associated organisms, such as snail and crustacean consumers, allowing them to live at higher intertidal heights than they could without these neighbors. Thus, positive associations among dense aggregations of mussels, barnacles, and seaweeds, in general, can set the upper intertidal limits of intertidal organisms.[🐚 52]

At the upper border of the *Ascophyllum* canopy, a strip of bare space may separate the canopy from the barnacle zone. This bare zone is the result of the whiplashing of algal fronds and the foraging of snail consumers that live

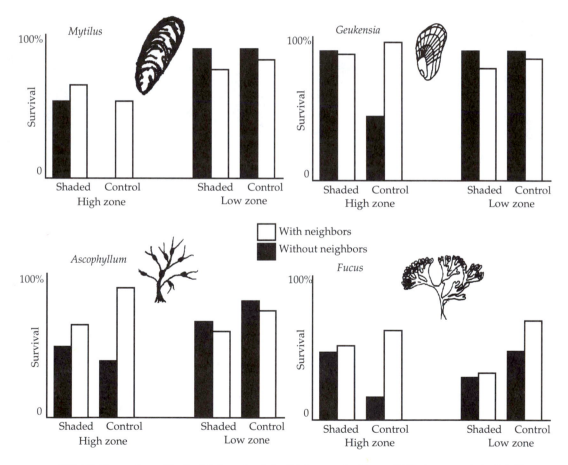

🐚 52 Group benefits for high rocky intertidal space holders. At their upper intertidal borders, most common sessile organisms are dependent on their neighbors for habitat amelioration and survival. Unless shaded, blue mussels, ribbed mussels, and common rockweeds and barnacles suffer heavy mortality without neighbors. (After Bertness and Leonard 1997.)

under the canopy (Menge 1978) (see Figure 44). At its lower border, the *Ascophyllum* canopy is usually replaced by a zone of free space, rockweeds (*Fucus*), and Irish moss (*Chondrus*). Occasional foraging by starfishes and urchins in this zone leaves bare space, which is colonized by these algae (Lubchenco 1983). Predation by crabs and snails prevents mussels from displacing *Chondrus*, so that grazer-resistant adult *Chondrus* is the functional dominant in this zone. Once established, *Chondrus* benefits from snail grazing, which cleans it of epiphytes, and even bounces back from moderate urchin grazing because its holdfast regrows new fronds when clipped back.

Tide Pools

The tide pools characteristic of New England shorelines have their own zonation. Pools at high intertidal heights heat up rapidly and become hypersaline in the summer, and may become hypotonic due to rains and freeze in the winter. These physical stresses limit the types of organisms that can live in these pools. Pools at lower elevations are frequently flooded by tides, which decreases these physical fluctuations. Large pools at lower elevations may even be thermally and osmotically stable enough to support sea urchins, which can limit fleshy algae, leading to a community of crustose algae and animals (limpets, urchins, and chitons) similar to that found in subtidal habitats.[53]

Algal abundance in tide pools is strongly influenced by herbivores. At intermediate intertidal heights, tide pools containing few periwinkle grazers (*Littorina*) are dominated by rapidly colonizing ephemeral green algae, such as *Ulva* and *Enteromorpha*, while pools with high periwinkle densities are dominated by grazer-resistant algae such as *Chondrus* and *Ralfsia*. Jane Lubchenco (1978) showed that these relationships are causal with manipulative field experiments at Nahant, Massachusetts. Removing periwinkles led to the invasion of *Ulva*, which ultimately dominated grazer-free tide pools, competitively excluding other algae. Conversely, adding adult periwinkles to pools dominated by ephemeral green algae led to only the most grazer-resistant algal forms persisting. Pools that initially lacked snail grazers were found to resist snail recruitment, since the algal canopy harbored crabs that ate all snail recruits (Lubchenco 1978).[54] Since periwinkles are relatively mobile, these differences in tide pool algal composition driven by snail densities probably occur at larger, between-habitat spatial scales, rather than within sites (Ron Etter, personal communication). At wave-exposed sites, periwinkles are often rare, leading to tide pools with algal assemblages not controlled by grazing. In contrast, at protected sites, periwinkles are often common, and their grazing appears to control algal abundance and species distribution patterns.

Maximum algal diversity in New England tide pools occurs at intermediate herbivore densities. This relationship between herbivore density and algal diversity is a consequence of herbivore feeding preferences. Periwinkles prefer ephemeral green algae (the competitively dominant algae in tide pools), and intermediate snail densities increase diversity by preventing

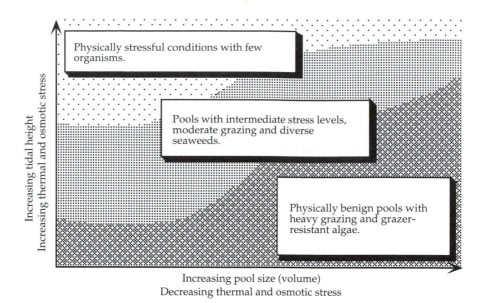

Increasing pool size (volume)
Decreasing thermal and osmotic stress

53 Effects of the size and intertidal height of tide pools on their occupants.

them from competitively excluding grazer-resistant algae. In contrast, on nearby emergent rock surfaces that dry during low tide exposure, the grazer-resistant rockweed (*Fucus*) is competitively dominant, since it is resistant to both grazing and desiccation, whereas *Ulva* and *Enteromorpha* are vulnerable to desiccation. As a result, high snail densities decrease algal diversity in these habitats, because increased herbivory hastens the dominance of *Fucus*. [55]

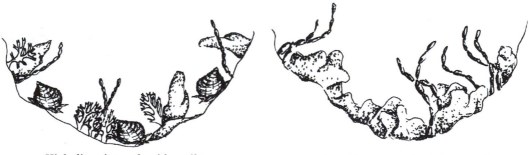

High-diversity pools with snails

Dominated by a mixture of palatable green seaweeds (*Ulva, Enteromorpha*) and slower-growing and well-defended species such as algal crusts (*Lithothamnium*) and Irish moss (*Chondrus*).

Low-diversity pools without snails

Dominated by fast-growing, competitively dominant green seaweeds such as *Ulva* and *Enteromorpha*. Green crabs colonize these pools and can limit the recruitment of herbivorous snails.

54 Jane Lubchenco's (1978) demonstration of the relationship between grazers and seaweed diversity in New England tide pools.

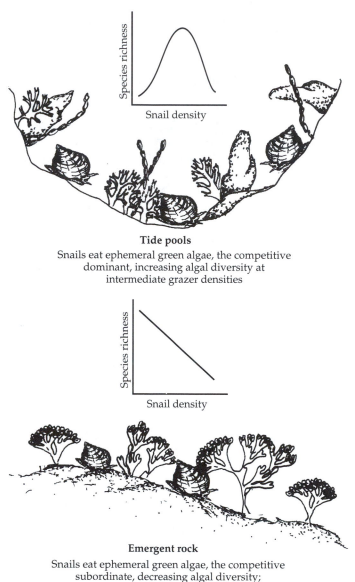

Tide pools
Snails eat ephemeral green algae, the competitive
dominant, increasing algal diversity at
intermediate grazer densities

Emergent rock
Snails eat ephemeral green algae, the competitive
subordinate, decreasing algal diversity;
Fucus dominates

🐚 55 The differential effect of herbivore grazing pressure on algal diversity in and out of tide pools. (After Lubchenco 1978.)

Subtidal Zonation

The vertical zonation of New England rocky shorelines extends well into subtidal habitats. Shallow subtidal habitats off sheltered rocky shores are usually characterized by heavy urchin and starfish grazing that eliminates most erect fleshy seaweeds. As a result, this zone is typically covered by

pink and red coralline crusts (*Lithothamnium* and *Phymatolithon*), the only algae capable of withstanding heavy urchin grazing. Dominance by coralline algae is clearly grazer-dependent, since without constant grazing, algal crusts are covered with sediment or overgrown and replaced by more palatable but competitively superior erect algae (Dethier 1981; Slocum 1980). Removal of urchins from shallow subtidal habitats leads to the establishment of a diverse algal assemblage dominated by kelps, and the successful recruitment sessile invertebrates (Himmelman 1983; Witman 1985).

On wave-exposed shores in New England, kelps dominate shallow subtidal hard substrates, but are usually replaced by horse mussel (*Modiolus modulus*) beds at depths greater than 10 meters.[🐚 56] The lower limit of kelps is set by urchin grazing (Witman 1985, 1987). In the shallow subtidal kelp zone, strong currents and storm surges limit the abundance of urchins. Below the kelp zone, reduced wave disturbance allows the persistence of urchins, which leads to the elimination of kelps, enabling horse mussel beds to persist without kelp-induced dislodgment.[🐚 57] Removing urchins from the horse mussel zone leads to kelps growing on horse mussel shells and the subsequent dislodgment of the mussels during storms due to the

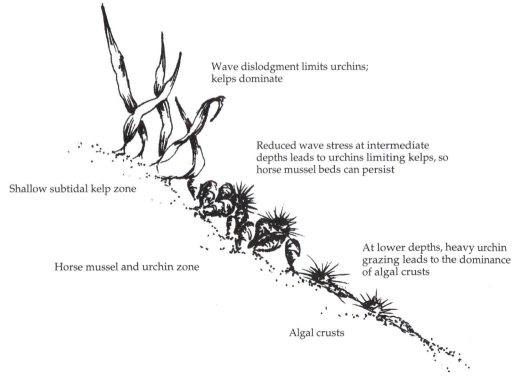

Wave dislodgment limits urchins; kelps dominate

Reduced wave stress at intermediate depths leads to urchins limiting kelps, so horse mussel beds can persist

Shallow subtidal kelp zone

At lower depths, heavy urchin grazing leads to the dominance of algal crusts

Horse mussel and urchin zone

Algal crusts

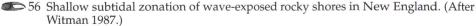

🐚 56 Shallow subtidal zonation of wave-exposed rocky shores in New England. (After Witman 1987.)

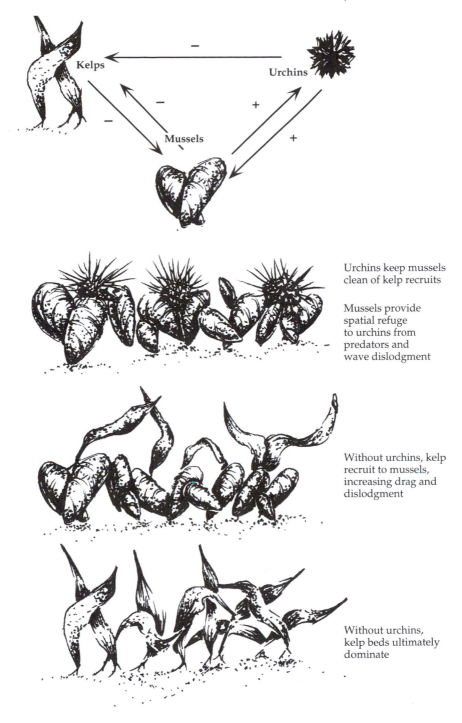

Urchins keep mussels
clean of kelp recruits

Mussels provide
spatial refuge
to urchins from
predators and
wave dislodgment

Without urchins, kelp
recruit to mussels,
increasing drag and
dislodgment

Without urchins,
kelp beds ultimately
dominate

57 Jon Witman's (1987) demonstration of the positive relationship
between horse mussel beds and sea urchins in shallow subtidal
rocky habitats.

increased drag. Consequently, while mussels benefit from urchins grazing on epiphytes, the urchins are hidden from their predators and protected from wave dislodgment by the horse mussel beds. This association between horse mussels and urchins is a good example of a facultative mutualism, in which both participants benefit from the association, but neither is totally dependent on it. Since horse mussel beds are pockets of high species diversity (Witman 1985), this mutualism indirectly maintains diversity.

One of the most marked features of subtidal habitats in general is the dominance of clonal organisms such as bryozoans, tunicates, and sponges. Jeremy Jackson (1977) has argued that organisms with solitary body plans, such as barnacles and mussels, dominate intertidal habitats due to their ability to colonize highly physically stressful and disturbed habitats. In contrast, clonal organisms dominate subtidal habitats due to their competitive dominance over solitary organisms and their ability to rapidly cover primary space with vegetative growth.

Protected Bays

In the Gulf of Maine, rocky shorelines in protected bays and estuaries differ markedly from coastal habitats. These shorelines are characterized by less wave energy, higher thermal stress, and increased crab predation relative to coastal habitats. Tidal currents influence communities in these habitats much as wave shock influences communities in coastal habitats. At the mouths of bays or constrictions in estuaries, tidal currents are strong, and intertidal communities resemble wave-exposed coastal habitats.[⬤ 58] A high intertidal barnacle zone gives way at intermediate intertidal heights to dense mussel beds dominated by small individuals, with shallow subtidal habitats often containing kelp beds. At nearby low-flow sites, a high intertidal barnacle zone gives way in the middle intertidal to an *Ascophyllum* canopy, which is replaced at low intertidal heights by a zone with considerable free space and patchy mussel beds made up mostly of large individuals.

One of the most striking effects of increased flow on the structure of estuarine communities is that the zonation of sessile invertebrates and seaweeds can be shifted higher by as much as a meter. High-flow sites are exposed to increased fluxes of planktonic larval recruits and the particulate food of filter feeders, leading to high barnacle, mussel, and herbivorous snail densities. These high densities buffer sessile organisms from high intertidal physical stresses, allowing these organisms to live at higher elevations than are possible at the low densities found in nearby low-flow habitats. At low-flow sites, recruit and particulate food delivery is low, and the effect of predators is much stronger than at high-flow sites. Intense snail grazing in low-flow habitats can lead to a lack of seaweeds other than grazer-resistant canopy species, and intense crab predation can restrict mussels to refuge habitats and large size classes that are immune to predators (Leonard et al. 1998a).

South of Cape Cod, the open coast is generally dominated by soft-substrate habitats, untouched by recent Ice Age scouring events. However, the shorelines of Narragansett Bay and Long Island Sound have characteristic

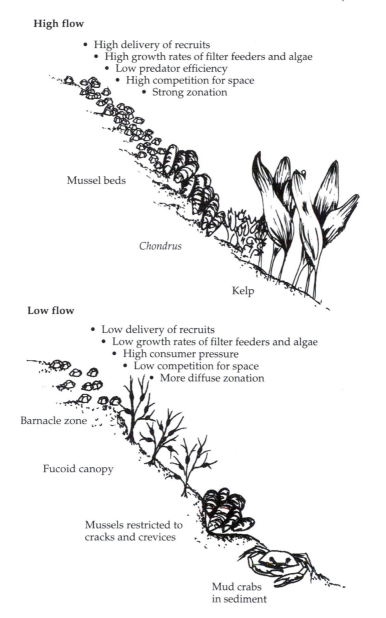

High flow
- High delivery of recruits
  - High growth rates of filter feeders and algae
    - Low predator efficiency
      - High competition for space
        - Strong zonation

Mussel beds

Chondrus

Kelp

Low flow
- Low delivery of recruits
  - Low growth rates of filter feeders and algae
    - High consumer pressure
      - Low competition for space
        - More diffuse zonation

Barnacle zone

Fucoid canopy

Mussels restricted to cracks and crevices

Mud crabs in sediment

58 High- and low-flow rocky intertidal habitats in tidal Maine estuaries. (After Leonard et al. 1998a.)

cobble shores left by receding glaciers. Intertidal communities on cobbles are simple for a number of reasons. First, they exist south of Cape Cod, a major biogeographic boundary excluding many species of the rich boreal Atlantic fauna. Second, these protected cobble beach habitats are thermally stressful. They are isolated from the thermal buffering of coastal habitats, and the smaller cobble substrate is thermally labile in comparison to large rocks. Finally, these inland estuarine habitats have been exposed to large-scale environmental insults for at least three centuries, and the effects of eutrophication, increased sedimentation, overfishing, and pollution on these communities are difficult to estimate with certainty.

Cobble beaches in Narragansett Bay and Long Island Sound have a high intertidal zone dominated by the cordgrass *Spartina alterniflora* or by bare cobbles. At high intertidal heights, summer thermal effects are strongest, and cobbles can be dislodged and abraded in winter, leaving them devoid of cover. Cobbles that are embedded in the root mat of cordgrass are shaded from heating, stabilized from rolling, and often encrusted with barnacles and mussels. Cobble beach cordgrass stands also buffer wave shock and facilitate the recruitment of other halophytic ("salt loving") vascular plants. *Atriplex*, *Suaeda*, *Salicornia*, and *Limonium*, all plants found in salt marshes, are common on estuarine cobble beaches, but are restricted to the zone immediately behind cordgrass stands.[⬤ 59] This pattern is the result of the cordgrass trapping the waterborne seeds of these plants as well as limiting the direct effects of waves on the plants during storms (Bruno and Rand, unpublished data). Over longer periods of time, probably measured in decades, the establishment of cordgrass on cobble beaches also appears to facilitate the successful establishment of other high marsh species (particu-

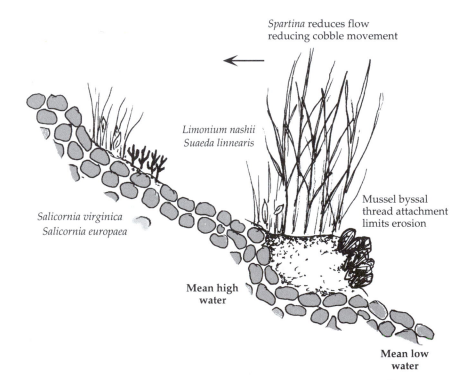

⬤ 59 Zonation of wave-protected cobble beaches in southern New England.

larly the marsh hay, *Spartina patens*). This enhances sedimentation and ultimately leads to the development of typical high marsh plant communities.
John Bruno has argued that cordgrass should be considered a keystone facil-
itator or bioengineer, since it fundamentally changes the cobble beach habitats that it colonizes through habitat modification.

The middle intertidal zone of cobble beaches in Narragansett Bay and
Long Island Sound typically consists of cobbles covered with barnacles,
fleshy algal crusts, and extremely high (600 to 1000 individuals per square
meter) densities of the herbivorous snail *Littorina littorea* (Bertness et al.
1983; Bertness 1984a; Petraitis 1987).[🐚 60] These extremely high herbivore densities, which are probably due to the lack of predators and competitors in these habitats, exclude all algae other than herbivore-resistant
leathery algal crusts (*Ralfsia* and *Hildenbrandia*). Experimental removal of
the European periwinkle from these cobble beaches has dramatic consequences, suggesting that its invasion has played a major role in changing

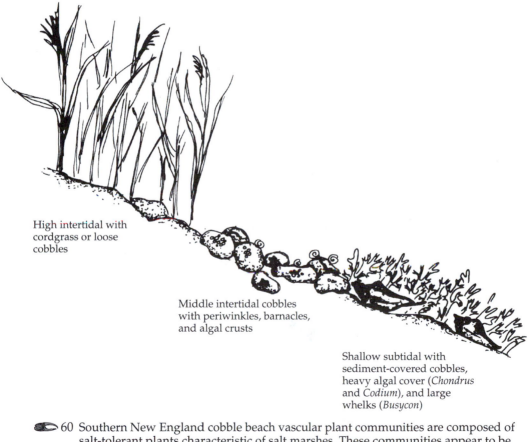

High intertidal with
cordgrass or loose
cobbles

Middle intertidal cobbles
with periwinkles, barnacles,
and algal crusts

Shallow subtidal with
sediment-covered cobbles,
heavy algal cover (*Chondrus*
and *Codium*), and large
whelks (*Busycon*)

🐚 60 Southern New England cobble beach vascular plant communities are composed of
salt-tolerant plants characteristic of salt marshes. These communities appear to be
strongly influenced by both seed supply and physical disturbance.

these shoreline landscapes. Snail removal leads rapidly to the development of a dense algal canopy, the accumulation of sediment, and the establishment of organisms typical of soft-sediment habitats (Bertness 1984a). Without the snails, the barnacles and algal crusts typically found in these habitats are covered with algae and suffocate under sediment. Thus, the presence of the snail may have been responsible for shifting communities in wave-sheltered habitats from soft-sediment assemblages dominated by tubiculous worms, crustaceans, mud crabs, and marsh grasses to assemblages more typical of hard-substrate habitats. The mechanism of these effects is simple. High periwinkle densities continually clear hard surfaces of diatoms, inadvertently preventing sediment accumulation and the development of an algal canopy. In the absence of snails, algal canopies become established, enhancing sedimentation and stabilizing sediments, and leading to the recruitment of tube-building organisms, which further stabilize the sediment.[61]

Why *Littorina* is restricted to the intertidal zone of cobble beaches is unclear. The predatory whelk, *Busycon*, or predatory crabs may play a role in limiting periwinkles to intertidal habitats, but this has not been experimentally tested. As a consequence of *Littorina* staying in the intertidal zone year-round, however, the shallow subtidal habitat has an algal canopy of Irish moss, dead man's fingers (*Codium*), and sea lettuce (*Ulva*). Without constant grazing, the subtidal habitat is covered with sediment and sediment-dwelling organisms (Bertness 1984a).

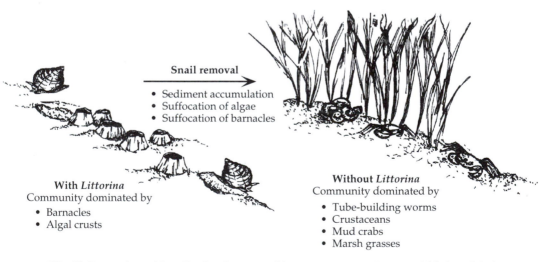

Snail removal

- Sediment accumulation
- Suffocation of algae
- Suffocation of barnacles

With *Littorina*
Community dominated by
- Barnacles
- Algal crusts

Without *Littorina*
Community dominated by
- Tube-building worms
- Crustaceans
- Mud crabs
- Marsh grasses

61 In southern New England, periwinkle grazing can maintain cobble beach habitats by preventing sediment accumulation and the establishment of a soft-sediment community. (After Bertness 1984a.)

CONSERVATION ISSUES

Rocky shores were thought historically to be relatively immune to severe environmental problems. Shorelines have been used for centuries for the disposal of sewage and industrial wastes under the assumption that the dilution of wastes in the vast ocean waters would render them harmless. The open populations of most shoreline organisms have also long been assumed to buffer them from overharvesting and local environmental contamination. Most shallow-water marine organisms have pelagic dispersal, which decouples adult benthic populations and opens them to heavy movement of recruits between populations. This metapopulation structure has been suggested to give local populations considerable resilience, or the ability to bounce back from perturbations.

In spite of this rationale and the historical lack of concern, shorelines are one of the most threatened environments on the planet. When Europeans began colonizing the east coast of North America 300 years ago, western Atlantic waters were teeming with cod, and lobsters were so common that they were dug into the ground as fertilizer. The shores of Narragansett Bay, Long Island Sound, and Chesapeake Bay were paved with dense oyster beds. Now all of these major shoreline features are gone, and their loss has probably changed the dynamics and structure of shoreline communities. The loss of cod and other nearshore fish predators to overharvesting has probably had major effects on benthic prey populations (Witman and Sebens 1992). Similarly, the loss of oyster populations and their ability to increase water clarity now prevents seagrass and seaweed beds from thriving in Narragansett Bay, Long Island Sound, and Chesapeake Bay (Ulanowicz and Tuttle 1992).

Jim Carlton (1993) has argued that changes in our coastal waters over the past few centuries have led to numerous local and global extinctions. Few of these have been documented, since the extinction of many rare organisms probably occurred before biologists had the opportunity to describe them. The extinctions we are most aware of have occurred when the habitats that specialist organisms depended on were destroyed. For example, the eelgrass limpet, *Lottia alreus*, went extinct on the Atlantic coast of North America with the demise of shallow-water eelgrass beds in the 1930s (Carlton et al. 1991; see Chapter 6). Organisms dependent on bay and estuarine habitats (such as oysters and scallops) are more at risk than those found in coastal habitats because they are more heavily harvested and affected by human activities. However, the vulnerability of even open ocean populations to human activities has recently been demonstrated with the unprecedented collapse of the Atlantic cod and flounder fisheries of Newfoundland and George's Bank and the Pacific coast salmon fisheries.

Overharvesting, pollution, and species introductions are the most pressing conservation issues confronting contemporary nearshore communities. These are all old problems that have affected shorelines for centuries, but have become particularly important over the last few decades. Overexploitation of marine resources can have substantial direct and indirect effects on marine communities. On rocky shores, harvesting of organisms by humans can deplete populations and change the size distributions of those organisms, with the potential for cascading effects on shoreline communities (Castilla and Duran 1989; Kingsford et al. 1991). Probably the best-known example of human overexploitation triggering shoreline community changes was the hunting of sea otters for fur. The decimation of Pacific Coast sea otter populations early in the twentieth century led to a sea urchin population explosion and the destruction of kelp beds from California to Alaska (Estes and Palmisano 1974; Estes et al. 1978). Since kelp beds are important nursery grounds for commercially important fish populations, there were serious economic repercussions. East coast examples of community shifts due to overharvesting include exploitation of sea urchins and oysters, as well as commercial fishing.

One problem with the depletion of marine organisms with external fertilization and pelagic larval dispersal is that populations can be reduced to levels so low that fertilization and successful recruitment may be limited. The potential importance of these **Allee effects** (see Chapter 3) is just beginning to be realized. For sea urchins, seaweeds, clams, and other organisms that release their gametes into the water column, fertilization may not occur at low densities due to gamete dilution. In addition, for sessile and slow-moving organisms that need to settle with members of their own species, densities below some threshold level may simply be too low for successful recruitment to occur (Quinn et al. 1993). Understanding the importance of these population success thresholds is vital in determining harvesting practices and designing refuge habitats. Reed (1990) has shown that with kelp, high densities of sexually reproducing spores (one spore per square millimeter or more) are necessary to trigger pheromone-induced sperm release. This entirely precludes kelp reproduction at low densities. Jim Quinn and his colleagues (1993) have shown that refuges from harvesting where Allee effects do not limit populations are needed for the persistence of heavily harvested shoreline organisms such as sea urchins. They also suggest that similar considerations are needed in designing marine preserves.[🐾 62]

Pollution is an equally important threat to shoreline communities, especially in shallow coastal bays and estuaries. Sewage runoff, which is often dumped directly into nearshore habitats, can influence these environments by causing eutrophication and blooms of nuisance seaweeds, which displace native plants and animals (Littler and Murray 1975; Valiela et al. 1992). Toxic paints applied to boat bottoms to prevent fouling by marine organisms are another pervasive problem in nearshore waters. They can leach into the water, killing many organisms. In Europe, fouling paint use has been linked to the decline of predatory snail populations in port cities (Gibbs et al. 1988).

Oil pollution, however, may be the most widespread threat to our shorelines (Suchanek 1993). It has been estimated that at least 4 billion liters of oil enter the oceans each year as a result of human activities. Surprisingly, only 12 percent of this occurs due to tanker spills, and only 8 percent

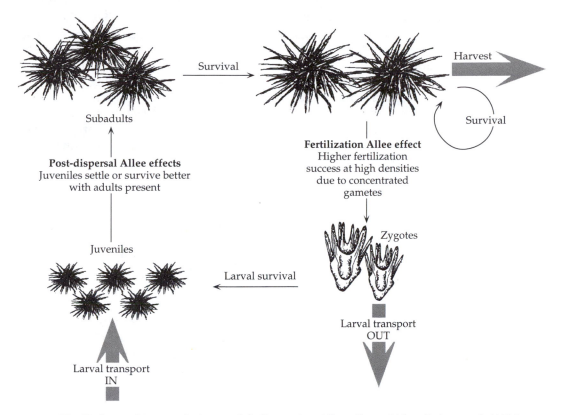

Subadults

Survival

Harvest

Survival

Post-dispersal Allee effects
Juveniles settle or survive better
with adults present

Fertilization Allee effect
Higher fertilization
success at high densities
due to concentrated
gametes

Zygotes

Juveniles

Larval survival

Larval transport
OUT

Larval transport
IN

62 Sea urchin population model, illustrating Allee effects. (After Quinn et al. 1993.)

of this oil is released in normal tanker operations. The majority of oil pollution (over 36 percent) comes from the runoff of municipal and industrial
wastes. Oil can clearly cause both chronic problems associated with its con-
sistent presence in marine systems and acute problems associated with
spills and natural disasters.

On rocky shores, not all organisms respond in the same way to oil.
Mobile organisms such as crabs, echinoderms, gastropods, amphipods, and
some long-lived seaweeds are extremely vulnerable to oil, and their loss can
lead to dramatic changes in shoreline populations, which may take many
years or decades to return to their original state (Suchanek 1993). To date,
however, the response of shoreline systems to chronic and/or acute oil pollution is not well understood.

Introduced species are another major concern on western Atlantic rocky
shores. Introduced species have become so common on temperate zone
shorelines that it is difficult to reconstruct what native communities were
really like before the invasions occurred. In the western Atlantic, the major
shell-crushing predator, *Carcinus maenus*, the herbivore *Littorina littorea*, and

some of the most dominant subtidal spatial competitors (*Codium fragilis* spp. *tomentosoides*, the lacy bryozoan *Membranipora membranacea*, and the encrusting sea squirt *Botryloides schleseri*) are all recent rocky shore introductions. Since introductions are continuing at a rapid rate, temperate zones around the globe are becoming a homogeneous mix of world-traveling hitchhikers that are displacing many indigenous organisms and entire native communities. Nearly half a century ago, Charles Elton called this homogenization of the world's biota the most dramatic biological change that the earth has ever experienced. Nowhere is this happening at a more rapid rate, or with greater consequence, than in nearshore marine environments.

SUMMARY

Rocky intertidal communities are one of the best-studied natural assemblages because of their conspicuous spatial patterns, aesthetic appeal, and relative ease of experimental examination. On the Atlantic coast of North America, rocky shores are restricted to New England and Canada, where recent Ice Age glaciers have scoured the shores of the sediments that cover the southern coastline.

The food chains of rocky intertidal habitats are dependent on two food sources. First, filter-feeding barnacles and mussels tap the plankton food supply in the water column, linking shoreline communities with water column processes. Second, herbivorous snails and urchins consume benthic algae ranging from diatomaceous films on hard surfaces to large seaweeds.

Rocky shorelines are characterized by striking elevational zonation, particularly where recruitment does not limit population densities and space for sessile organisms is limited. New England rocky shores are remarkably harsh physical habitats in terms of temperature fluctuations and desiccation, even in comparison with other temperate rocky shores around the world. As a general rule of thumb, the upper intertidal limits of rocky shore organisms are set by physical stresses and the mortality of recruits at higher elevations, whereas the lower intertidal limits are set by biological factors, such as consumer pressure and competition. In New England, drilling snails and starfishes are dominant predators in open-coast habitats. In bays and estuaries, rock and green crabs are dominant predators. Northern acorn barnacles, blue mussels, and seaweed canopies are dominant spatial competitors at middle to high intertidal heights on many New England shorelines, but are displaced by clonal invertebrates at lower elevations.

Most research on process and pattern on New England rocky shores has focused on the role of consumers and competition in generating distribution patterns. Positive associations among organisms and habitat-modifying bioengineers, however, are also clearly important organizing forces in these habitats. The upper intertidal limits of barnacles, mussels, and algal canopies are all strongly affected by habitat-ameliorating group benefits. Without neighbors, solitary individuals of all these organisms have lower survivorship due to heat and desiccation stresses. The algal canopies and mussel beds common on New England rocky shores are also important autogenic bioengineers that influence their habitats and the organisms that live there. Algal canopies and mussel beds both typically support a rich assemblage of organisms that are largely dependent on them for their persistence.

Whereas rocky shores have long been thought to be relatively immune to environmental problems due to the open populations of many shoreline organisms and the vastness of the sea, they are currently being seriously threatened. Pollution, overexploitation of commercially harvested species, and species introductions are dramatically changing our shorelines. It is a formidable challenge for us to understand them before they are irreversibly changed.

SUGGESTIONS FOR FURTHER READING

Chapman, A. R. O. 1986. Population and community ecology of seaweeds. *Advances in Marine Biology* 23: 1–161.

Connell, J. H. 1972. Community interactions on marine rocky intertidal shores. *Annual Review of Ecology and Systematics* 3: 169–192.

Little, C. and J. A. Kitching. 1996. *The Biology of Rocky Shores*. Oxford University Press, New York.

Lobban, C. S. and P. J. Harrison. 1994. *Seaweed Ecology and Physiology*. Cambridge University Press, Cambridge.

Moore, P. G. and R. Seed (eds). 1995. *The Ecology of Rocky Coasts*. Hodder and Stoughton, London.

Newell, R. C. 1979. *The Biology of Intertidal Animals*. Marine Ecological Surveys, Faversham, England.

Paine, R. T. 1994. *Marine Rocky Shore and Community Ecology: An Experimentalist's Perspective*. Ecology Institute, Oldendorf/Lohe, Germany.

Raffaelli, D. and S. Hawkins. 1996. *Intertidal Ecology*. Chapman and Hall, New York.

Stephenson, T. A. and A. Stephenson. 1971. *Life Between Tidemarks on Rocky Shores*. W. H. Freeman and Co., San Francisco.

Key References for Atlantic Coast Rocky Intertidal Communities

Bertness, M. D. and G. H. Leonard. 1997. The role of positive interactions in communities: Lessons from intertidal habitats. *Ecology* 78: 1976–1989.

Etter, R. J. 1988. Physiological stress and color polymorphisms in the intertidal snail *Nucella lapillus*. *Evolution* 42: 660–680.

Leonard, G., J. M. Levine, P. Schmidt and M. D. Bertness. 1998. Flow-generated bottom-up forcing of intertidal community structure in a Maine estuary. *Ecology* 79: 1395–1411.

Lubchenco, J. 1978. Plant species diversity in a marine intertidal community: Importance of herbivore food preference and algal competitive abilities. *American Naturalist* 112(983): 23–39.

Lubchenco, J. 1980. Algal zonation in the New England rocky intertidal community: An experimental analysis. *Ecology* 61(2): 333–344.

Mathieson, A. C., C. A. Penniman, P. K. Busse and E. Tveter-Gallagher. 1982. Effects of ice on *Ascophyllum nodosum* within the Great Bay Estuary System of New Hampshire-Maine. *Journal of Phycology* 18: 331–393.

Menge, B. A. 1976. Organization of the New England rocky intertidal community: Role of predation, competition, and environmental heterogeneity. *Ecological Monographs* 46: 355–393.

Steneck, R. S. and M. N. Dethier. 1994. A functional group approach to the structure of algal-dominated communities. *Oikos* 69: 476–498.

Steneck, R. S. and L. Watling. 1982. Feeding capabilities and limitation of herbivorous molluscs: A functional group approach. *Marine Biology* 68: 299–319.

Wethey, D. S. 1983. Geographical limits and local zonation: The barnacles *Semibalanus* and *Chthamalus* in New England. *Biological Bulletin* 165: 330–341.

Wethey, D. S. 1985. Catastrophe, extinction and species diversity: A rocky intertidal example. *Ecology* 66: 445–456.

Chapter *6*

Soft-Sediment Habitats

*I*ntertidal soft-sediment habitats, or **tidal flats**, are a common shoreline feature of the Atlantic coast of North America. The east coast of the continent has been geologically stable for the past 150 million years, and over this time, erosional sediments have accumulated on the continental margin, leaving sandy or muddy shores. Except for Canada and New England, where glaciers have scoured shorelines of sediments, leaving behind exposed rocky beaches and glacial debris, Atlantic shorelines of North America are dominated by dynamic soft-substrate habitats. Soft sediments exposed to heavy wave energy are highly mobile, which can prevent many organisms from living in them. On wave-exposed sandy beaches, waves continually scour the surface, removing settling larvae and preventing the establishment of a stable benthic assemblage. On these highly disturbed shorelines, the only organisms that flourish are those that can deal with these disturbances, such as surf clams and mole crabs.[ 1]

In soft-sediment habitats that are sheltered from waves, the muddy shores also appear at first sight to be depauperate, but they are actually biologically very rich. These habitats are more stable and can host a diverse assemblage of organisms, including microscopic organisms that live

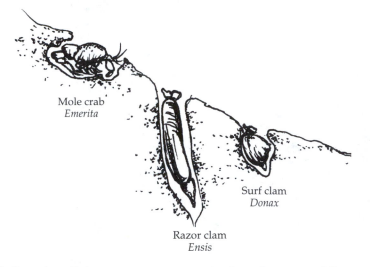

Mole crab
Emerita

Surf clam
Donax

Razor clam
Ensis

1 Organisms living on wave-swept sandy shores burrow rapidly to escape the surf, like razor clams, or use the surf to move and feed, like mole crabs and surf clams.

between sand grains, clams and worms that continually turn over the sediment with their feeding and burrowing, and predators ranging from snails to birds.

Soft-sediment habitats often support large predator populations. At low tide, when mud flats are exposed, shorebirds with beaks designed for probing sediments forage on worms, clams, and small crustaceans. At high tide, fishes and crabs forage in the same habitat, often taking the same prey as the shorebirds. In spite of the high biomass in soft-sediment communities, they are considered to be energetically subsidized assemblages because they are dependent on adjacent marsh and seagrass habitats for their primary production. Thus, soft-sediment communities are tightly linked to water column processes, because wave energy determines food and nutrient levels for the filter-feeding and deposit-feeding secondary consumers that dominate these habitats, as well as sediment supply.

The critical feature of tidal flats that differentiates them from rocky shores is that many of the conspicuous organisms living in soft substrates are much larger than the particles of the sediments they live in. These sediments are potentially considerably more unstable than the substrate on rocky shores. Furthermore, the organisms that live in soft sediments play a major role in the stability of their habitat. Burrowing and feeding on tidal flats can reduce sediment stability and thus the survivorship of larvae. In contrast, biogenic structures in soft sediments, such as worm tubes, grass stems, and mussel beds, can increase substrate stability and enhance the deposition of sediments and larvae. This dynamic interaction between soft-sediment organisms and the sediments in which they live makes community processes on tidal flats very different from those on rocky shores.

This chapter examines soft-sediment communities. I start by describing the physical habitat and the organisms of soft substrates. I then discuss the biological and physical processes that are important in shaping these common habitats. I end with a discussion of seagrass and sand dune plant communities and the role that vascular plants play in stabilizing and modifying soft-sediment environments.

THE PHYSICAL HABITAT

The sediments on tidal flats are the combined product of sediment supply and transport processes. Tidal flat sediments range from fine muds, composed of particles less than 0.05 millimeters in diameter, to coarse sands, composed of particles ranging from 0.5 to 1.0 millimeters in diameter. The distribution of mud and sand sediments on shorelines is largely a product of water movement. Rapidly moving water can carry larger, heavier particles than slowly moving water, so rapidly moving currents and waves erode the smaller particles from the bottom. In slower-moving water, sediment particles are deposited. Thus, the distribution of sediment types within and among habitats is a dynamic product of near-bed flow patterns. Tidal flats with little water movement are typically dominated by fine, muddy sediments, whereas tidal flats exposed to strong tidal currents or waves are dominated by coarser sands.[2] This relationship between water movement and sedimentation explains many large-scale patterns in the distribution of sediment types. On barrier islands, for example, wave-exposed shores are typically sandy, while substrates on their inland shores are variable. Estuaries can dump large amounts of sediments into the shallow water between the mainland and barrier islands, leaving the leeward coasts of the islands muddy and covered with salt marshes. When barrier islands are well offshore and experience strong wind-driven waves, the shoreward coasts are sandy.

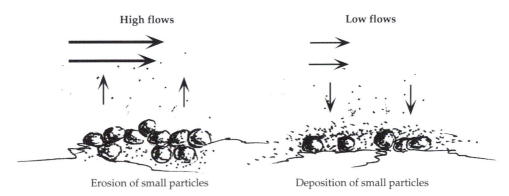

High flows   Low flows

Erosion of small particles   Deposition of small particles

 2 Flow effects on sediment size in soft-substrate habitats.

Water movement effects are also conspicuous at smaller spatial scales. Sediment sizes on tidal flats generally increase with increasing intertidal height, reflecting a parallel gradient in wave forces with intertidal height. Sediment size is also strongly affected by local topographic features that influence flows. Emergent vegetation, such as seagrasses or marsh grasses, slows water velocities and enhances sediment deposition (Nowell and Jumars 1984). Flows also leave characteristic small-scale topographic features on soft sediments. In low flow areas, sediment surfaces are usually smooth, but on tidal flats with higher water movement, near-bed flows carve regularly spaced grooves, sand waves, and ripples of size-sorted sediments in the substrate.

The slopes and shapes of soft-sediment shores are a complex product of the size of available sediments and local wave energy. As waves break on beaches, sediment is moved shoreward and deposited. The backwash of breaking waves, however, has the opposite effect, moving sediment back toward the ocean. How this shoreward and seaward movement of sediment balances out to give a shoreline its shape is largely a product of sediment size and wave energy. If a beach is made up of coarse material, such as pebbles, water moving up the beach drains into the beach face rather than moving back offshore as backwash. In this situation, sand or pebbles are carried up the beach, but not back again, leading to a steep slope. When sediment size is smaller, the beach face remains waterlogged, limiting subsurface percolation of water and causing a strong backwash. As a general rule, increasing wave energy and decreasing sediment size lead to beaches with flatter slopes. [🐚3]

Beaches with steep slopes and low wave energy are referred to as **reflective beaches**. On these beaches, sediment is deposited high on the beach, and the minimal wave energy present is reflected off the steep beach face. As

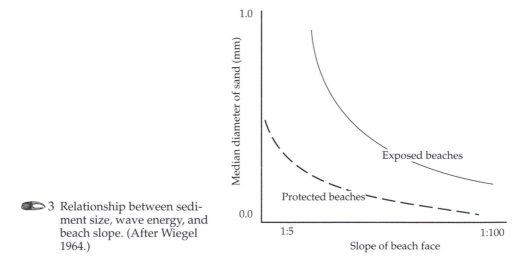

🐚3 Relationship between sediment size, wave energy, and beach slope. (After Wiegel 1964.)

wave energy increases, larger waves erode beach sediments, creating a **surf zone** where wave energy is dissipated. Under extreme high wave energy conditions, beaches can be maximally eroded, with all their sediment stored in a broad surf zone, rather than on the landward border of the beach. In these **dissipative beaches**, wave energy is dissipated in the surf zone before reaching the beach. Intermediate sediment size and wave energy conditions yield a variety of beach shapes between these reflective and dissipative extremes (see Brown and MacLachlan 1990).[🐚4]

Sediments on tidal flats can show marked vertical chemical gradients with substrate depth. The surface of muddy sediments in contact with the air during low tides is generally well oxidized. Below the surface, however, oxygen is depleted by the organisms in the sediment, and poor exchange with the surface leads to **anoxic** (without oxygen) sediments. Usually the boundary between oxygenated surface sediments and anoxic subsurface

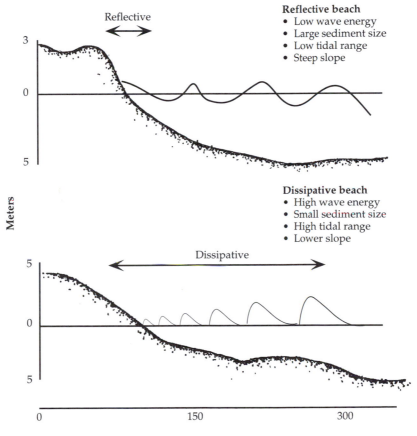

🐚 4 Reflective and dissipative beach characteristics.

sediments is sharp, and is demarcated by a black stain from the deposition of ferrous sulfide and the smell of rotten eggs from the production of hydrogen sulfide by anaerobic organisms. The depth of this boundary, called the **redox potential discontinuity**, varies with sediment grain size, organism oxygen input, and the feeding and burrowing activities of organisms living in the substrate (redox potential is a measure of the reduction/oxidation state or oxygen content of sediments). On muddy sediments where water and air do not readily move through the substrate, anoxic sediments can start only 1–2 millimeters from the surface. [🐚5] This is particularly the case in sediments with a high organic content, which increases microbial respiration and oxygen depletion. In coarser-grained, sandy sediments, water easily percolates through, and sediments drain during low tide exposure, allowing oxygen to diffuse 10–20 centimeters below the surface.

Oxygen availability in tidal flat sediments dictates sediment metabolic processes. In oxygenated surface sediments, aerobic decomposition pro-

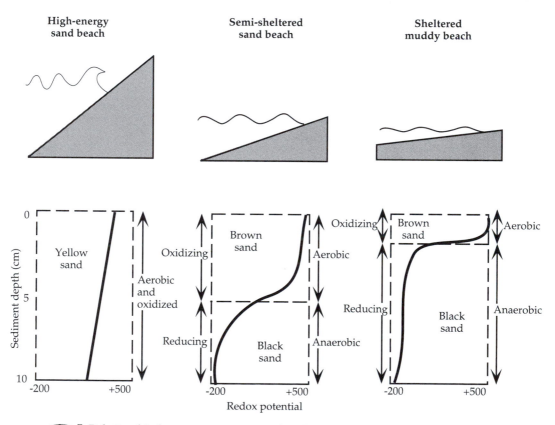

🐚5 Relationship between wave energy, beach sediments, and sediment oxygenation, or redox potential (Eh).

cesses prevail, but give way to **anaerobic** (without oxygen) processes below the redox potential discontinuity. Chemotrophic bacteria that utilize chemical energy for their food source play a major role in metabolism in these anoxic sediments (Howarth and Teal 1980).

The organisms that live in soft sediments often influence sediment chemistry. Burrowing crabs and worms oxidize sediments and deepen the oxygenated zone by reworking sediments and building and maintaining burrows that enhance drainage during low tides (Gray 1974). Burrows are often surrounded by conspicuous halos of oxygenated sediments and the microorganisms that are dependent on them. By processing sediments for food, deposit-feeding organisms remove organic material, which also increases sediment oxidation by decreasing microbial respiration. In addition, particle sorting by deposit feeders influences particle size distribution and can have strong indirect effects on sediment chemistry. In contrast to these primarily positive effects of macroorganisms on sediment oxidation, their waste products enhance sediment nitrogen levels, increasing metabolism and the depletion of oxygen by nitrogen-limited bacteria.

TIDAL FLAT ORGANISMS

Tidal flats host a rich assemblage of organisms, ranging from bacteria and protists to microscopic organisms that live between sediment particles and feed on bacteria and protozoans, to burrowing worms and bivalves, to swimming crabs and wading birds that forage on clams and worms. [⬤▷6] Tidal flat food chains are largely dependent on outside sources of fixed carbon and are thus tightly linked to water column processes. Whereas some deposit-feeding snails (such as the mud snail *Ilyanassa*) consume benthic diatoms, most tidal flat deposit feeders on fine sediments eat microorganisms associated with the decay of detritus from marsh plants and seagrasses. Moreover, the filter-feeding clams that dominate many tidal flats are dependent on plankton in the water column for their food. This coupling of water column and soft-sediment assemblages, however, is not a one-way street, but a dynamic interface. The water column supplies primary production in the form of plankton and detritus to soft-sediment assemblages, while the filtering and sediment processing of benthic consumers resuspends, recycles, and releases nutrients and dissolved organic material back to the water column. This strong positive feedback between benthic inhabitants of soft substrates and water column processes is an important consideration in understanding tidal flat community dynamics.

Meiofauna

One of the least conspicuous but most abundant groups of soft-sediment organisms are the microscopic (42–500 µm) invertebrates called **meiofauna** that live between sediment grains (**interstitially**) , feeding on bacteria and single-celled protozoan grazers.[⬤▷7] Typical meiofaunal communities consist of organisms that spend their entire lives as meiofauna as well as the

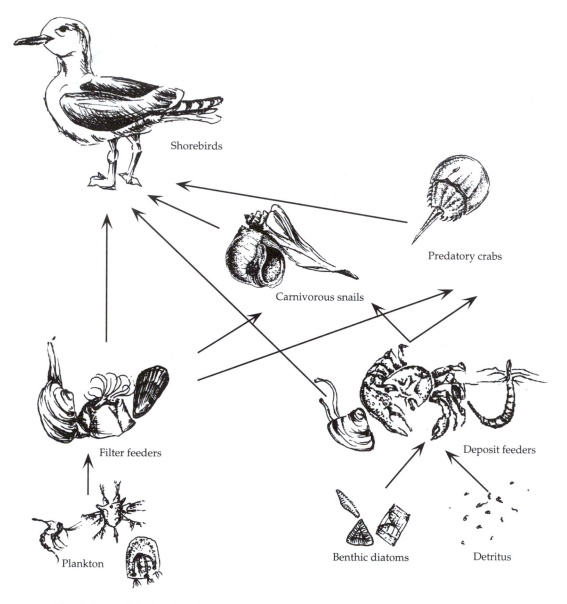

 6 A simplified soft-sediment food web.

juveniles of larger soft-sediment macroorganisms. Nematodes, benthic (harpacticoid) copepods, and ostracods are common permanent members of the meiofauna, whereas segmented worm and crustacean larvae are common transient members. Meiofauna are exceedingly abundant, particularly in well-oxygenated muds, where a single square meter of sediment can contain over 2 billion individuals. Sandy sediments, which have less surface area for bacterial growth, typically have lower densities of meiofauna.

Annelids

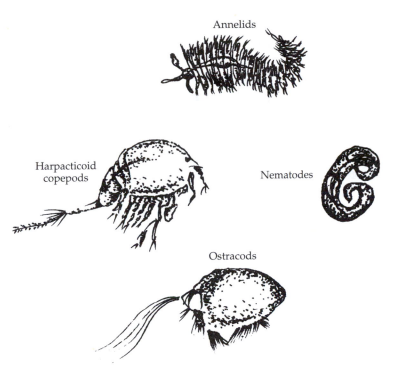

Harpacticoid copepods

Nematodes

Ostracods

7 Meiofauna that live between sand grains (interstitially) in soft sediments.

Meiofaunal densities are greatest in surface sediments, and usually decline precipitously with depth because of the decreased food availability and harsh anoxic conditions in deeper sediments. [8]

Meiofauna play an important role in soft-sediment food webs as a link between bacterial food sources and larger consumers. Meiofaunal population densities are controlled by food supplies, physical stresses, and meiofaunal predators (Coull 1985). Rich organic, oxygenated muds at low intertidal heights, not exposed to harsh physical conditions, usually harbor large meiofaunal populations. These populations, however, commonly are kept in check by predators and physical disturbance of the sediment (Bell and Coull 1978; Bell 1980; Hoffman et al. 1984; Ellis and Coull 1989).[9]

Filter Feeders

Filter-feeding bivalves are characteristic of soft-sediment shorelines. They serve as a critical link between water column primary production, benthic secondary production, and sediment metabolism. They are also, themselves, important prey of snails, crabs, fishes, and birds.[10]

Filter-feeding bivalves make their living by pumping water over their gills and selectively removing organic food material, especially pelagic

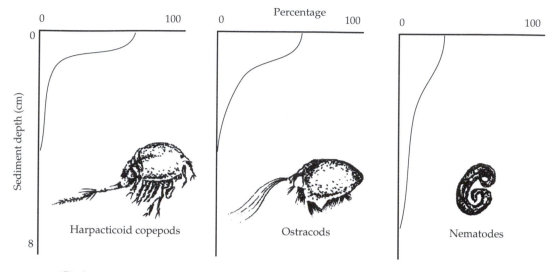

8 Depth distributions of common meiofauna, showing that most are found in well-oxygenated surface sediments. (After Whitlatch 1981.)

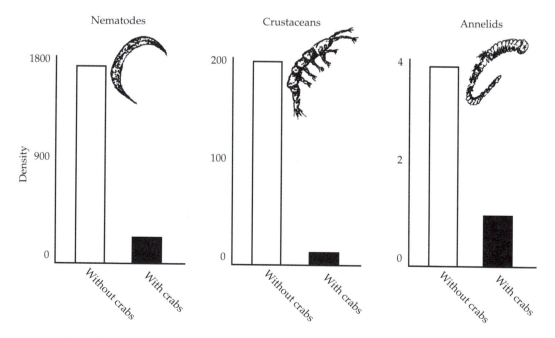

9 Fiddler crab foraging effects on meiofaunal populations in surface sediments (top 2 centimeters) of a New England salt marsh. (After Hoffman et al. 1984.)

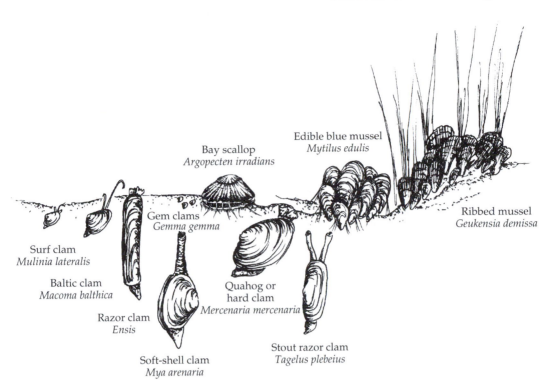

Edible blue mussel
Mytilus edulis

Bay scallop
Argopecten irradians

Ribbed mussel
Geukensia demissa

Gem clams
Gemma gemma

Surf clam
Mulinia lateralis

Baltic clam
Macoma balthica

Quahog or
hard clam
Mercenaria mercenaria

Razor clam
Ensis

Stout razor clam
Tagelus plebeius

Soft-shell clam
Mya arenaria

10 Common bivalves of Atlantic seashores.

diatoms. Primitive bivalves were deposit feeders that picked organic material from sediments with oral palps, while constantly cleaning their gills of fouling sediments. Modern bivalves have turned this gill sanitation problem into a way of obtaining food. The gills of most bivalves are enlarged to enhance their efficiency as pumps and particle filters.[11] Pumping rates are dependent on body size, temperature, and food concentrations, and are often impressive. An adult soft-shelled clam, *Mya arenaria*, can filter 4 liters of water per hour, and an adult American oyster, *Crassostrea virginica*, can pump as much as 14 liters per hour through its system. Particles are sorted by size on the enlarged gill surface. Larger particles, not usually good sources of nutrition, are rejected as **pseudofeces** without entering the digestive system. Smaller particles are conveyed by cilia on mucus strands through the digestive system, and nitrogen-rich waste products are released into exhalant currents (see Jorgenson 1966).

Where filter-feeding bivalves are abundant, they may play an important role in the flux of materials between the water column and the benthos (Officer et al. 1982; Dame, 1976, 1993). At natural densities in laboratory experiments, for example, the common quahog *Mercenaria* can increase system primary production, sedimentation, and benthic mineralization without

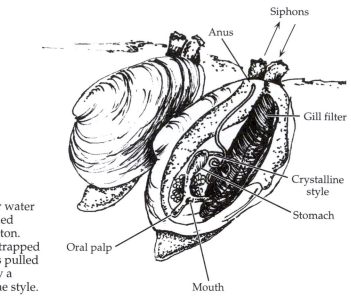

11 Most bivalves filter water through an expanded gill to collect plankton. Particulate food is trapped on mucus, which is pulled into the stomach by a revolving crystalline style.

depleting plankton through its effects on particle fluxes (Doering et al. 1986, 1987).[🐚12] Whether or not this occurs under field conditions, however, is unclear. Bivalves may also serve as important habitat bioengineers. In Chesapeake Bay, the loss of oyster beds and the water column filtering they provided may have contributed to an increase in suspended sediments, increased dominance of microbial and planktonic food webs, decreased water clarity, and decreased production of benthic algae and seagrasses (Newell 1988; Ulanowicz and Tuttle 1992). [🐚13]

Most soft-sediment bivalves are external fertilizers that seasonally release eggs and/or sperm into the water column, where fertilization occurs. [🐚14] The young typically develop in the water column for a number of weeks as

12 Possible bivalve water column production feedback. (After Doering et al. 1986, 1987.)

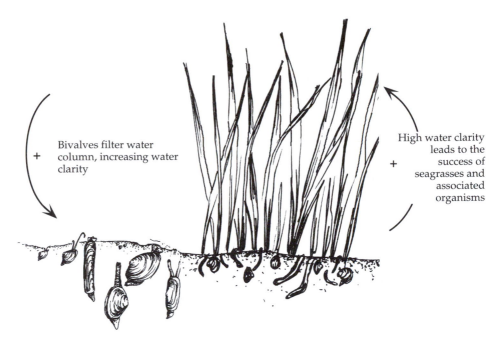

Bivalves filter water
column, increasing water
clarity

+

High water clarity
leads to the
success of
seagrasses and
associated
organisms

+

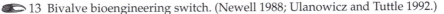

13 Bivalve bioengineering switch. (Newell 1988; Ulanowicz and Tuttle 1992.)

feeding larval stages before reaching the veliger stage, at which they settle. Settlement of bivalve larvae is influenced by hydrodynamic forces, which can dictate larval supply patterns and larval settlement preferences, which in turn determine the microhabitats where larvae choose to settle (see Butman 1987). Veligers initially attach to the substrate with proteinaceous byssal threads. Byssal thread attachment is a larval trait of many burrowing bivalves that is lost in adults. Byssal attachment is retained in adult mussels and juvenile scallops, allowing them to live **epibenthically** (on the surface of the sediment).

While some soft-sediment bivalves (scallops, oysters, and mussels) live on the surface, most are **infaunal** (live within the sediment). Burrowing in bivalves is accomplished with a muscular **foot**. Mechanistically, it is a very simple process. [15] When a clam relaxes the adductor muscles that serve to keep its shell closed, the valves open slightly (5–10 mm), since they are attached by a springlike proteinaceous hinge. With the valves slightly open, the clam is anchored in place, allowing the foot to be extended farther into the substrate. Once the foot is extended, the adductor muscles close the shell. This releases the shell as an anchor and forces fluid from the shell into the sediment, thereby loosening it. Closing the shell also increases vascular pressure in the foot, which swells to form an anchor. Foot retractor muscles then pull the shell down to the foot, and the process is repeated. The shell sculpture or ribbing that is common in many small bivalves is thought to increase friction and burrowing efficiency (Stanley 1970).

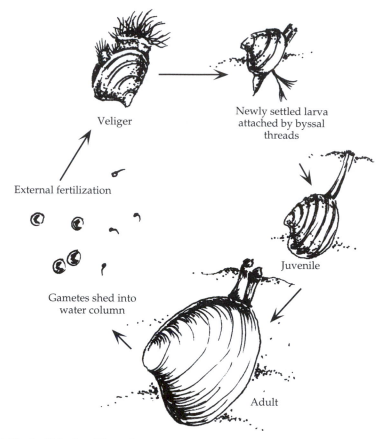

Veliger

Newly settled larva
attached by byssal
threads

External fertilization

Juvenile

Gametes shed into
water column

Adult

14 Typical bivalve life cycle.

Burrowing ability varies widely among bivalve species. Some bivalves, such as the quahog, *Mercenaria*, are sluggish and ineffective burrowers. *Mercenaria* completes fewer than 5 burrowing cycles a minute and moves very slowly. In contrast, the razor clam *Ensis* can complete up to 60 burrowing cycles a minute and moves relatively quickly through semi-permanent mucus-lined burrows. Rapidly burrowing clams have streamlined bodies, large muscular feet too large to retract into the shell, and soft parts tightly sealed into a closed mantle space, which increases their ability to generate pressure for foot extension.[⌐16] In contrast, slow burrowers, such as *Mercenaria* and *Mya*, have small, retractable feet, bulky shells, and unsealed mantle cavities (Stanley 1975).

Much of bivalve shell morphology is designed for protection against predators. The heavy shells of bivalves arose in the fossil record at the same time shell-crushing predators appeared, suggesting a major role for the predators in their evolution (Stanley 1970). Other shell features thought to

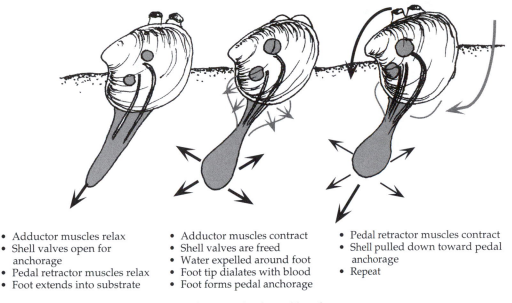

- Adductor muscles relax
- Shell valves open for anchorage
- Pedal retractor muscles relax
- Foot extends into substrate

- Adductor muscles contract
- Shell valves are freed
- Water expelled around foot
- Foot tip dialates with blood
- Foot forms pedal anchorage

- Pedal retractor muscles contract
- Shell pulled down toward pedal anchorage
- Repeat

🐚 15 Burrowing mechanisms of a typical infaunal bivalve.

reduce susceptibility to predators include fat, inflated shells, minimization of shell gapes, and reinforcing shell ribs and sculpture (Boulding 1984).[🐚17] Less obvious are bivalve adaptations for living deep in the substrate. Filter-feeding bivalves clearly need access to the sediment surface to feed, but they are easy prey for birds, crabs, and snails close to the surface (Blundon and Kennedy 1982). Largely in response to these predators, bivalves evolved siphons and rapid burrowing, which allow them to move deeper into the sediment as they grow.

Deposit Feeders

Members of many taxonomic groups, including segmented worms, amphipods and shrimps, sea cucumbers, and clams and snails, make their living extracting detritus or organic material from tidal flat sediments. Most of these species process large volumes of sediment and extract a variety of potential food sources, including bacteria, protozoans, diatoms, fungi, meiofauna, and nonliving particulate organic material. The role of deposit feeders in tidal flat communities is enormous. They serve as a critical link in marine food chains by converting detrital food sources into macroorganism biomass that is available to higher trophic levels. They are also important **bioturbators** that disturb the substrate when feeding, and their activities fundamentally influence the nature of tidal flat sediments.

Surface deposit feeders, such as the mud snails *Ilyanassa* and *Hydrobia*, fiddler crabs, pelagic fishes, and shrimps, feed on the sediment surface, which is rich in microalgae and diatoms. They are often capable of regulat-

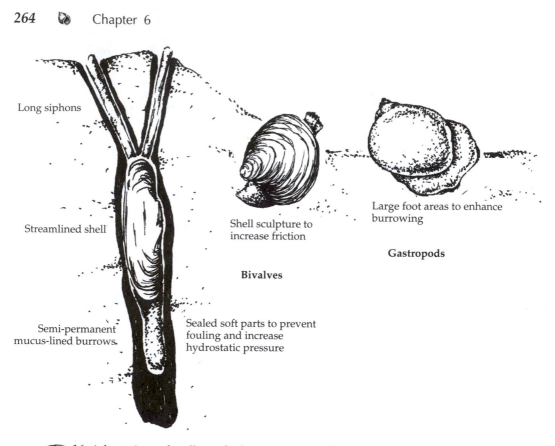

Long siphons

Streamlined shell

Semi-permanent mucus-lined burrows.

Shell sculpture to increase friction

Sealed soft parts to prevent fouling and increase hydrostatic pressure

Bivalves

Large foot areas to enhance burrowing

Gastropods

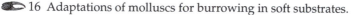 16 Adaptations of molluscs for burrowing in soft substrates.

ing primary production in surface sediments (Levinton and Lopez 1977; Hoffman et al. 1984), but their effects are usually limited to the top 2 or 3 centimeters. Burrowing deposit feeders, in contrast, both live in and feed on sediments, and can have important effects on sediments up to 30 centimeters deep.[18]

The actual food of deposit feeders has been the topic of considerable debate. Since most deposit feeders cannot break down the cellulose of plant material, it was long thought that bacteria and fungi capable of breaking down plant debris were the major food of most of these organisms. The "microbial stripping hypothesis" proposed that plant debris was made available to deposit feeders by bacterial and fungal decomposers (Odum and de la Cruz 1967).[19] More recently it has been recognized that microbe populations in most sediments are not large enough to support the observed growth of deposit feeder populations. Current models of deposit feeding characterize the microbial stripping hypothesis as overly simplistic and emphasize the utilization of a wide range of food sources by most deposit feeders (Lopez and Levinton 1987).

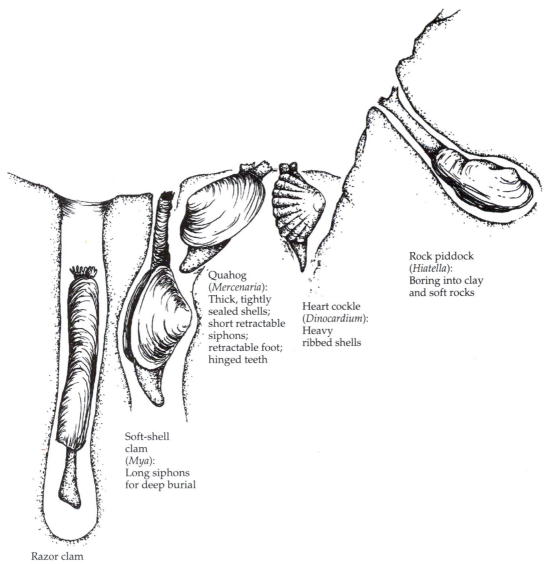

Rock piddock
(*Hiatella*):
Boring into clay
and soft rocks

Quahog
(*Mercenaria*):
Thick, tightly
sealed shells;
short retractable
siphons;
retractable foot;
hinged teeth

Heart cockle
(*Dinocardium*):
Heavy
ribbed shells

Soft-shell
clam
(*Mya*):
Long siphons
for deep burial

Razor clam
(*Ensis*):
Semi-permanent
deep burrows

🐚 17 Antipredator defenses in burrowing bivalves.

As a consequence of the poor nutritional quality of sediments, most deposit feeders process large volumes of sediment. Typically, deposit feeders process at least their own body weight in sediments daily, and some lugworms and protobranch bivalves can process over ten times their body weight daily (Lopez and Levinton 1987). Rhoads (1973) determined that the deposit feeder populations of Barnstable Harbor, Massachusetts, are capable

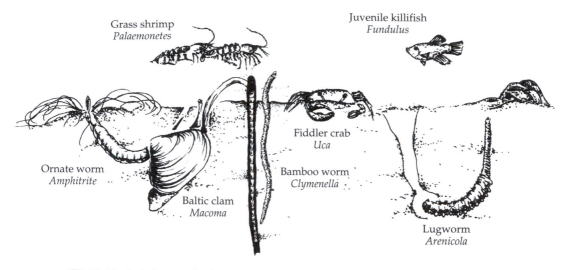

Grass shrimp
Palaemonetes

Juvenile killifish
Fundulus

Fiddler crab
Uca

Ornate worm
Amphitrite

Bamboo worm
Clymenella

Baltic clam
Macoma

Lugworm
Arenicola

🐚 18　Typical deposit feeders.

of entirely turning over all the sediments in Barnstable Harbor, Massachu-setts in the harbor every year and a half.

The effects deposit feeders have on tidal flat sediments are the result of bioturbation, selective removal and transport of particles of different sizes, and fecal deposition. The polychaete *Amphitrite*, for example, lives in semi-permanent J-shaped burrows and is a surface feeder.[🐚20] It feeds with modified head appendages, which are grooved and ciliated to convey food particles to the mouth. *Amphitrite* is a size-selective feeder, picking smaller particles from the substrate and rejecting larger ones without passing them through its digestive system. As a result of its feeding activity, mounds of size-sorted sediment particles accumulate at both ends of its burrow. The lugworm, *Arenicola*, lives in a U-shaped burrow and, like an earthworm, indiscriminately ingests large volumes of sediment, leaving large mounds of processed substrate at the back end of its burrow.[🐚21]

Many other polychaetes live head down in the substrate and transport deeper sediments to the surface. Both *Clymenella* and *Pectinaria* are common head-down polychaete deposit feeders that selectively remove and process small particles from below the surface, and then deposit processed particles and feces on the surface.[🐚22] This activity leads to size sorting of tidal flat sediment particles, leaving coarser sediments below the surface and fine sediments on the surface, where they are often resuspended.

Members of another interesting group of soft-sediment organisms, including some clams and polychaetes, can change their mode of feeding depending on environmental conditions. The primitive protobranch bivalves *Macoma* and *Yoldia* are facultative (optional) deposit feeders that utilize oral palps to remove sediment from the surface in low flows, but

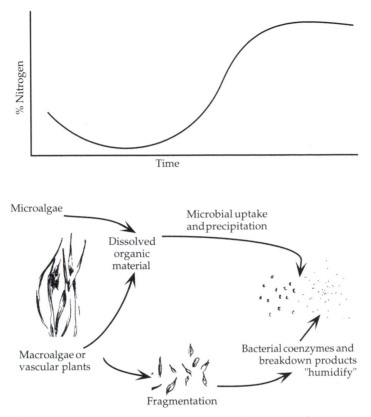

19 The microbial stripping hypothesis. Dead plant material is attacked by bacteria and fungi. This increases its food value to detritivores, which feed on these microbes and their by-products. (After Odum and de la Cruz 1967; Levinton and Lopez 1977.)

switch to filter feeding in high-flow habitats with high particle fluxes (Olafsson et al. 1994).[23] Some tentacle-feeding polychaetes also switch from filter feeding to deposit feeding depending on particle fluxes (Taghon et al. 1980). This plasticity in feeding method allows these organisms to adjust to local flow conditions, and they can have varying effects on habitats as a result.

Sediment reworking by deposit feeders can have enormous repercussions for virtually all aspects of sediment dynamics. Deposit feeders resuspend sediments, and so are an important link between benthic and sediment processes. Many deposit feeders move small, easily suspended particles to the surface, generate surface topography with their fecal mounds, and move gill-ventilating currents through their burrows. All of this activity enhances sediment resuspension and nutrient exchange with the water column. The constant reworking of sediments by deposit feeders may also play an impor-

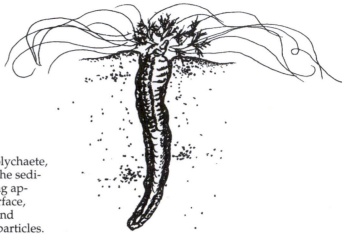

 20 The mop-topped polychaete, *Amphitrite*, lives in the sediment with its feeding appendages on the surface, where they collect and process small food particles.

tant role in increasing productivity by increasing oxygen and nutrient levels in the sediments; this idea has been called the "microbial gardening hypothesis" (Hylleberg 1975; Woodin 1977; Levinton and Lopez 1977; Connor et al. 1982). This hypothesis, however, has turned out to be difficult and complicated to evaluate critically.[24]

Predators

Predators in soft-substrate habitats include snails, crabs, worms, fishes, and birds. Tidal flat predators can be extremely numerous, and have a strong effect on the distribution and abundance of prey organisms. Trophic relationships in soft substrates, however, can be complicated by the fact that consumers often compete with, as well as eat, one another.

Snail predators are common in tidal flat habitats. Some, such as the oyster drill, *Urosalpinx cinera*, forage on the sediment surface, but are also common on rocky shores. Others, such as moon snails and whelks, are soft-sediment specialists that use a large foot like a plow to burrow through the

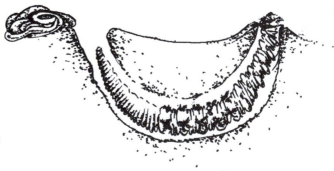

21 The lugworm, *Arenicola*, lives in a U-shaped burrow and, like an earthworm, ingests sediments. It leaves large castings of processed sediments at the posterior end of its burrow.

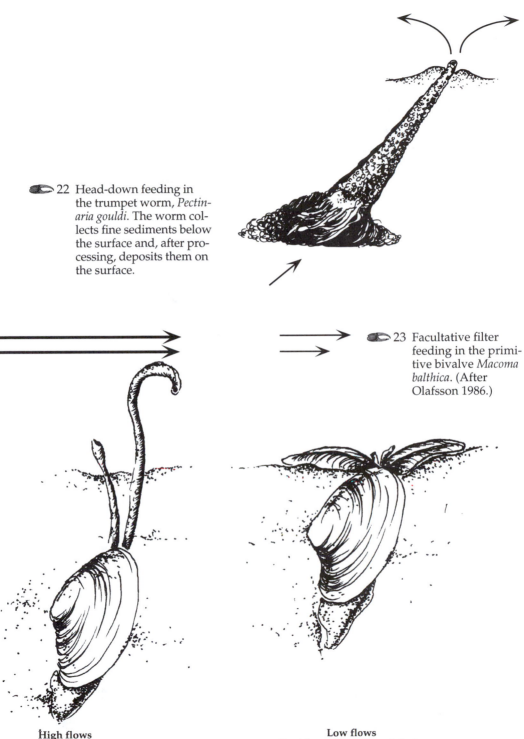

🐚 22 Head-down feeding in the trumpet worm, *Pectinaria gouldi*. The worm collects fine sediments below the surface and, after processing, deposits them on the surface.

🐚 23 Facultative filter feeding in the primitive bivalve *Macoma balthica*. (After Olafsson 1986.)

High flows
High particle fluxes allow filtering of food

Low flows
Particles settle on the substrate, allowing deposit feeding

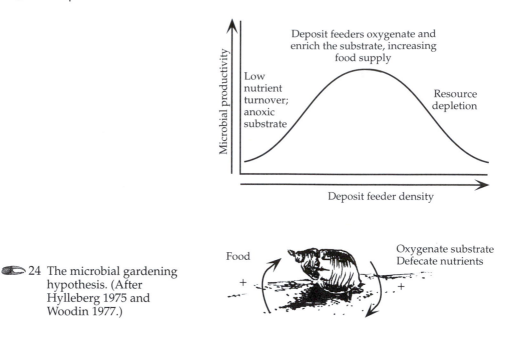

Microbial productivity

Low nutrient turnover; anoxic substrate

Deposit feeders oxygenate and enrich the substrate, increasing food supply

Resource depletion

Deposit feeder density

Food

Oxygenate substrate
Defecate nutrients

+　+

⬿ 24 The microbial gardening hypothesis. (After Hylleberg 1975 and Woodin 1977.)

substrate in search of bivalve prey. Moon snails (*Polinices*) and predatory whelks (*Busycon*) both rely on chemoreception (the ability to sense chemical cues) to locate prey, but have different means of killing and eating their victims. Like dog whelks on rocky shores, moon snails have a radula modified into a drill, which can rasp a small (2–3 millimeter) hole in clam shells. The hole is typically made by alternately applying an acid secretion to dissolve the calcium carbonate of the shell and rasping at the shell surface with the filelike radula. Drill holes are often made near the clam's adductor muscle, which holds the shell valves together; once the hole is complete, the moon snail secretes enzymes that relax the adductor muscle and spring the valves open. In contrast to moon snails, whelks pry their bivalve prey open. The whelk holds a bivalve with its foot and slowly wedges the edge of its shell between the two valves. Once the bivalve is opened, the whelk uses a rakelike radula to rasp away the prey's flesh.[⬿25]

The egg cases of moon snails and whelks are often found on tidal flat shorelines. Both of these predatory snails have separate sexes. After copulation, females lay eggs in protective casings, where they develop. Moon snails embed their egg capsules in a flexible sand collar cemented together with mucus. Whelks lay their eggs in tough, translucent egg capsules strung together in a long strand and attached to the substrate.

Crabs are major predators of both shelled and soft-bodied prey. The blue crab, *Callinectes sapidus*, is the most abundant and important crab predator in soft sediments ranging from Cape Cod to the Caribbean.[⬿26] Blue crabs are successful, voracious predators with broad diets, but their preferred prey

Moon snail
Polinices

Whelk
Busycon

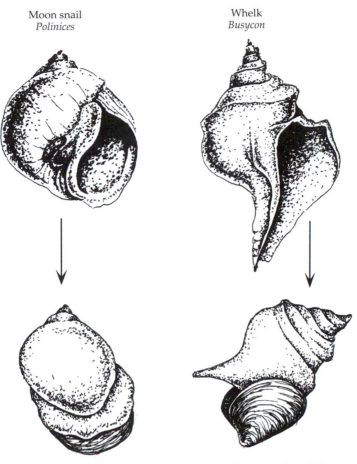

Drills hard-shelled prey

Pries shells open with shell lip or
smothers prey with large foot

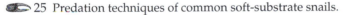 25   Predation techniques of common soft-substrate snails.

is molluscs. Blue crab populations are large enough throughout much of their range to support commercial harvesting, but their numbers have declined recently due to overexploitation. In the mid-Atlantic (from Delaware to the Carolinas), blue crabs are a good example of a predator whose foraging is a major structuring force in soft-sediment communities (Micheli 1997).

Blue crabs are highly mobile predators. Their posterior walking legs are modified into paddlelike propellers, which allow them to swim like other portunid crabs. *Callinectes* locate prey by chemoreception, digging through sediments with their walking legs and claws to extract prey, and using their strong claws to crack open shells or rip soft-bodied prey into pieces small enough to eat. When common, blue crabs can entirely deplete populations of clams in favored habitats, often restricting their prey to refuge habitats.

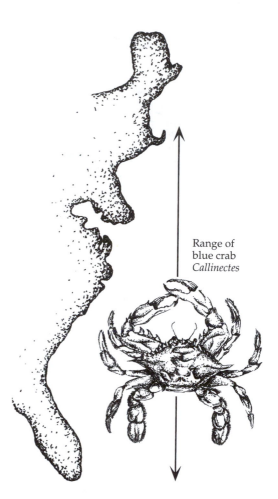

Range of
blue crab
Callinectes

26 Throughout much of its range, the blue crab *Callinectes sapidus* is a dominant predator in soft-sediment habitats.

Among the most important refuges from blue crab predation are habitats with high water movement. High flows inhibit crab foraging by making movement difficult and by interfering with chemoreceptive location of prey. Even when prey are abundant, turbulent flow breaks up their odor plumes, and the crabs are unable to detect them (Weissburg and Zimmer-Faust 1993). [27]

Blue crabs reproduce in the spring. After copulating, females migrate, up to 200 kilometers, to high-salinity waters near the mouths of bays to incubate their eggs. Larvae are released during extreme spring tides at high tide, and are flushed over the continental shelf, where they develop in the plankton for approximately 2 weeks. When ready to settle, they migrate to the surface, where they are passively transported back onshore by internal waves (Shanks 1983). Juvenile blue crabs utilize vegetated shallow-water habitats such as seagrass beds and marshes as nursery grounds, but as they

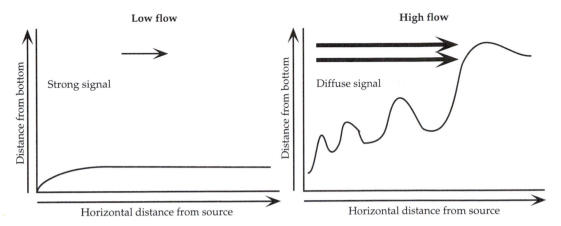

27 Flow speed effects on odor signals used by blue crabs to locate prey. Odor plumes remain intact and lead the crabs to prey in low flows, but turbulence in high flows breaks up odor signals, effectively hiding prey from predators. (After Weissburg and Zimmer-Faust 1993.)

grow large enough to escape their predators (fishes and other crabs) they move into deeper water. Shallow-water habitats, however, continue to be used by adults when molting and during reproduction, when they are more vulnerable to predators (Hines et al. 1987). As in other crabs, mating occurs only when females are molting. Molting exposes their genital openings, and males will often carry premolt females for a number of days in anticipation of a mating opportunity.

Although blue crabs are the dominant crab predator in Atlantic coast soft-sediment habitats, other crabs are also important predators. North of Cape Cod, the introduced European green crab (*Carcinus maenus*) is an abundant predator in soft sediments. Mud crabs (*Panopeus* spp.), while smaller in size, have strong claws for breaking hard-shelled prey and are also important predators in soft-sediment habitats. The horseshoe crab, *Limulus*, is also a common predator in soft sediments (see Botton 1984b; Woodin 1981).[28]

Particularly south of Cape Cod, rays and fishes are prominent predators in intertidal soft-sediment habitats. Rays forage over tidal flats at high tide, eating large worms, crustaceans, and molluscs. Their digging for prey can cause widespread mortality among nonprey organisms and trigger secondary succession in disturbed substrates (Van Blaricom 1982). The Atlantic stingray, *Dasyatis saina*, and the cow-nosed ray, *Rhinplera bonasus*, are the most common rays on southern Atlantic shorelines. The Atlantic stingray is the more abundant of these, but is not a pit digger, and so may not affect the physical structure of its habitat to the same extent as the larger cow-nosed ray. Cow-nosed rays can severely disturb substrates in the process of digging for large molluscan prey, and are even known to dig through seagrass beds while foraging (Orth 1975). In southern New England, the little skate,

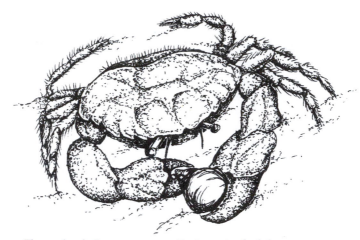

The mud crab, *Panopeus*, can use its strong, toothed claws to open hard-shelled prey.

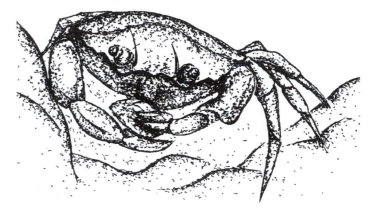

The introduced European green crab, *Carcinus,* has become an important predator in soft-substrate habitats throughout New England.

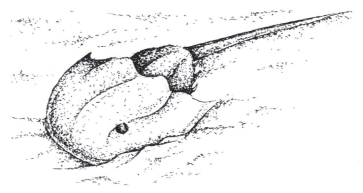

The horseshoe crab, *Limulus,* digs in shallow substrate for crustaceans, worms, and small clams.

28 Common predaceous crustaceans of western Atlantic soft sediments.

Raja erinacea, and the winter skate, *Raja ocellata*, are common predators on mud and sand flats.

A variety of predatory fishes forage on intertidal mud and sand flats. The cryptically colored flounders are "sit and wait" predators that feed on small shrimps and crabs when young and switch to a diet of fish as adults. They sit motionlessly, partially buried in the sediment, as chromatophores (pigment-bearing cells) in their skin allow them to change color to match their background. Hogchoker (*Trinectes maculatus*), spot (*Leistomus xanthurus*), and Atlantic croaker (*Micropogonias undulatus*) are all common predators of infaunal organisms on tidal flats south of Cape Cod.[◖➤29]

Soft-bodied segmented and ribbon worms are inconspicuous and often overlooked predators in soft-sediment habitats. They are particularly interesting since they are both prey of and competitors with crabs, birds, and

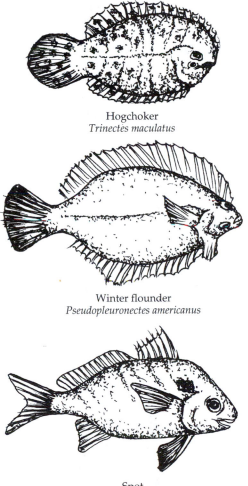

Hogchoker
Trinectes maculatus

Winter flounder
Pseudopleuronectes americanus

Spot
Leiostomus xanthurus

◖➤29 Common predatory fishes of soft-substrate habitats.

fishes. Experimental manipulation of predaceous worm densities has shown that they are often important predators in soft-sediment assemblages (Ambrose 1991, 1984a,b; Commito 1982; Commito and Shrader 1985).[🌢⊃30]

Shorebirds are also major predators in intertidal soft-substrate communities, often depleting populations of soft-bodied invertebrates near the sediment surface. In particular, seasonal or migrant shorebirds, which arrive in large numbers with depleted energy reserves, can strongly affect intertidal populations on mud flats (Schneider 1978; Schneider and Harrington 1981). Gulls and other year-round residents of tidal flats can also have significant effects on populations of soft-bodied prey (Ambrose 1986).

ZONATION IN INTERTIDAL SOFT-SUBSTRATE HABITATS

One of the most interesting features of intertidal mud and sand habitats is their lack of conspicuous vertical zonation. Whereas rocky shores and salt marshes are characterized by strikingly repeated bands of organisms paralleling the shoreline, vertical zonation in soft-substrate habitats, while present, is much more subtle. This contrast between intertidal soft-substrate habitats and shorelines inhabited by vascular plants or sessile invertebrates points to fundamental differences in the relative importance of biotic and physical forces in these communities (Peterson 1992). Zonation on rocky shores and in salt marshes is the product of strong gradients in physical

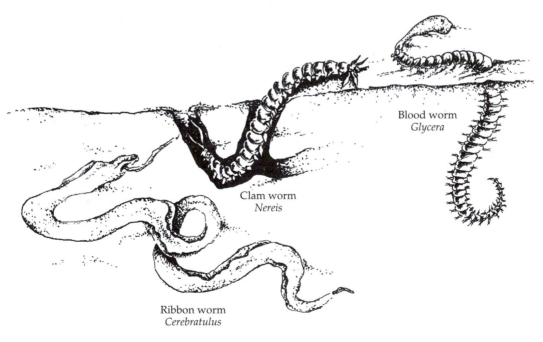

Blood worm
Glycera

Clam worm
Nereis

Ribbon worm
Cerebratulus

🌢⊃ 30 Common Atlantic coast soft-bodied predators.

stresses imposed by tides. In these habitats, physical stresses limit the distributions of organisms at higher intertidal elevations, while biotic forces (competition for space and predation) typically limit the distributions of organisms at lower elevations. Zonation schemes based on predictable elevational variation in physical stresses have also been proposed for intertidal soft-sediment shorelines (see Dahl 1952; Brown and McLachlan 1990; Raffaelli et al. 1991). In soft sediments, however, the stress gradient and spatial patterns in competition and predation are much more diffuse, leading to less defined zonation than on rocky shores.

The sediment itself buffers most of the residents of intertidal soft-substrate habitats from physical stresses.[⬤▷31] Whereas organisms on rocky shores routinely experience large swings in temperature and intense desiccation, organisms buried in only a few centimeters of sediment are largely buffered from these stresses. Even mobile organisms such as fishes, snails, and crabs bury themselves in sediments during low tide exposures. As a result, the elevational gradient in these physical stresses is more diffuse in soft sediments than on rocky beaches.

Competition for resources is also different for soft-sediment organisms than for the biota of rocky shores. On rocky shores, competition for substrate space is an important organizing force, particularly at low intertidal heights where space is often limited (Connell 1961; Jackson 1977). Competition for space is not as intense in soft-substrate communities because these habitats are three-dimensional, and because of the complicated interactions between soft-sediment organisms and the sediments they live in. On two-dimensional hard substrates, both larval recruitment and asexual clonal growth often lead to complete occupancy of available space. The ability of organisms to live at different depths in soft substrates relaxes these spatial considerations. Moreover, the strong effects that the organisms themselves have on the substrate often kick in at relatively low densities, precluding intense competition for space.[⬤▷32]

Predation pressure on organisms in intertidal soft sediments also does not show as strong an elevational pattern as is found on rocky shores. On temperate rocky shores, the dominant consumers are usually slow-moving invertebrates (such as snails, starfishes, and sea urchins) whose foraging is constrained by tides, focusing predation pressure at lower intertidal elevations. In contrast, many of the common predators of soft-sediment organisms, such as blue crabs, are highly mobile and are not restricted to low intertidal heights. Thus, while predation gradients with elevation probably occur in both habitats, these gradients are much less dramatic in soft substrates. Predation by birds is lower with decreasing intertidal height as a result of decreased foraging time during periods of submersion, but this decrease may be offset by fish predation, which usually increases with decreasing intertidal height.

There are also a number of more subtle effects on the intertidal zonation of soft-sediment organisms that are not important on rocky shores. Substrate grain size can influence where organisms settle (Butman et al. 1988), as well

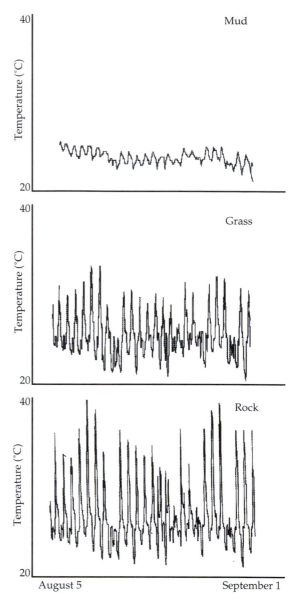

31 Intertidal temperature fluctuations on an open rock surface, within a marsh grass stand, and 10 centimeters under the surface of mud over a 30-day period in Narragansett Bay, Rhode Island.

as the success of organisms with different feeding methods (Rhoads 1974). Thus, tidal sorting of sediments to different elevations probably has important consequences on the vertical distribution of soft-sediment organisms. Reduced feeding times at high elevations and the depletion of suspended food particles by the large number of filter feeders in many tidal flat habitats may also be important considerations (Peterson and Black 1987). High elevations on many soft-sediment beaches are uninhabitable because of lack of

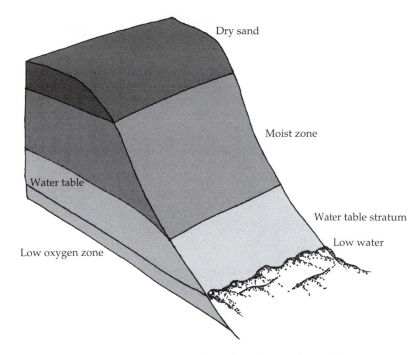

Dry sand

Moist zone

Water table

Water table stratum

Low water

Low oxygen zone

32 A simple vertical zonation scheme for soft-sediment shores. (See Brown and McLachlan 1990; and Raffaelli and Hawkins 1996 for discussion.)

food, rather than because of intolerable physical conditions, which are usually thought to set upper intertidal limits on rocky shores.

The Role of Competition

Competition within and among species plays a different role in soft-substrate than in hard-substrate habitats. In soft-substrate habitats, the three-dimensional utilization of the habitat, the dynamic relationship between the sediments and the organisms that live in them, and interactions among adults and juveniles all act to weaken the effects of direct spatial competition and change the way in which soft-substrate organisms compete with one another.

The best examples of competitive interactions among soft-sediment organisms that affect spatial distributions involve organisms that share access to the sediment surface, are relatively mobile, and are not strongly influenced by the three-dimensional nature of soft sediments. In Long Island Sound, Jeff Levinton and his colleagues (Levinton 1981; Levinton et al. 1985) have shown that the distributions of mud snails are affected by competition for space. In Europe, the mud snail *Hydrobia totteni* occurs at both high and low intertidal heights on mud flats, whereas on the Atlantic coast of North

America, *Hydrobia* is found exclusively at high intertidal heights. By experimentally removing a second mud snail, *Ilyanassa obsoleta*, found only in North America, Levinton demonstrated that *Hydrobia* is displaced to higher intertidal heights by the competitively superior *Ilyanassa*. *Hydrobia* avoids contact with *Ilyanassa*, leading to its displacement to high intertidal heights not inhabited by *Ilyanassa*. Similarly, the distribution of *Ilyanassa* is affected by the introduced European periwinkle, *Littorina littorea* (Brenchley and Carlton 1983). At lower tidal heights, *Ilyanassa* actively avoids contact with *Littorina*, because *Littorina* destroys *Ilyanassa* egg cases while foraging. This behavior leads to spatial segregation between *Ilyanassa* and *Littorina*. As a consequence of these competitive interactions, the distributions of *Littorina* and *Ilyanassa* are segregated by substrate type, while the distributions of *Ilyanassa* and *Hydrobia* are segregated by intertidal height.

The distribution of **tubiculous** (tube-building) amphipods, which live in small tubes made of sediment and mucus near the sediment surface on tidal flats, is also strongly influenced by competition for space. On sandy shores in Georgia, amphipod species and size classes partition the habitat by intertidal height and sediment depth (Croker 1967), and their distributions appear to be maintained by competition.[🐚33] Density-dependent migration of juvenile amphipods can lead to spatial segregation of large and small individuals (Wilson 1989). Experimental manipulations have also shown that the depth distribution of amphipods can be driven by competitive inter-

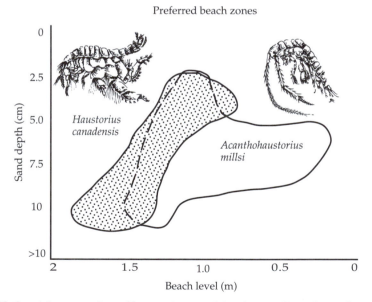

Preferred beach zones

🐚33 Spatial segregation of burrowing amphipods on a Georgia sand beach. (After Croker 1967.)

actions. In South Carolina, the amphipods *Acanthohaustorius* and *Pseudo-haustorius* are commonly found together on sandy shores, but at different depths in the substrate. *Acanthohaustorius* lives in oxidized surface sediments, whereas *Pseudohaustorius* lives in deeper anoxic sediments. By experimentally manipulating densities of these amphipods, Grant (1981a) showed that both prefer living in oxidized surface sediments, but that in the presence of *Acanthohaustorius*, *Pseudohaustorius* is displaced to deeper anoxic sediments.[🐚34]

The nonrandom distribution of polychaetes that share access to the sediment surface is also commonly the result of density-dependent migration (Levin 1981; Ambrose 1984a). The spionid *Polydora ligni*, for example, feeds with its tentacles on the surface. The regular spacing of *Polydora* individuals at high population densities is the result of individuals avoiding contact with one another, as well as the decreased feeding times and reduced growth of closely spaced individuals (Levin 1981).

In contrast to competition among mobile and sedentary soft-sediment organisms that share surface space, competition among larger, more sluggish infaunal organisms such as bivalves is often in the form of exploitative competition for food. Even at artificially high densities, competition among burrowing bivalves is reflected in reduced individual growth rates and fecundity, rather than the physical displacement or death of individuals.

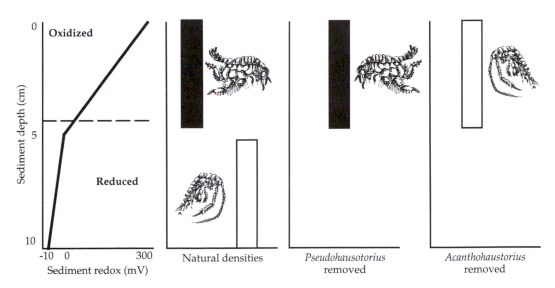

🐚 34 Competitive displacement of burrowing amphipods. Under field conditions, *Acanthohaustorius* (solid bars) occupies oxidized surface sediments and *Pseudo-haustorius* (open bars) occupies deeper, anoxic sediments. Experimental removals demonstrated that both prefer to live in oxidized surface sediments, but when the species occur together, *Acanthohaustorius*, the competitive dominant, displaces *Pseudohaustorius* to deeper levels. (After Grant 1981a.)

Pete Peterson at the University of North Carolina has documented this general pattern in bivalve assemblages in North Carolina, southern California, and Australia (Peterson 1977, 1979, 1982b, 1992; Peterson and Black 1987). [◖◗35] These results are strong evidence that competition for space is not a major driving force in the distributions of large infaunal organisms, and implies that competition for food among filter feeders is generally an important factor in soft-substrate habitats.

      The importance of food supply in influencing the outcome of competitive interactions among filter-feeding bivalves is elegantly illustrated by density-dependent growth in the Baltic clam, *Macoma balthica* (Olafsson 1986). In muddy substrates, the growth of *Macoma* is reduced at high densities, but in sandy substrates such density-dependent growth is not observed. This interesting result is due to the facultative feeding behavior of *Macoma*. In sand, *Macoma* is primarily a filter feeder, while in mud it relies more on deposit feeding. For *Macoma*, substrate food resources are apparently in limited supply at high clam densities, whereas water column resources are not. For filter feeders, resource renewal by currents is rapid and prevents local resource depletion, while for deposit feeders, local resource depletion is common.

Trophic Group Amensalism

A unique aspect of soft-substrate habitats is that many of the conspicuous organisms that live in soft sediments are much larger than the sediment grains they live in. Consequently, the movement of organisms through soft substrates strongly influences substrate stability, and animal/substrate interactions can play a major role in influencing organism distribution and abundance patterns.

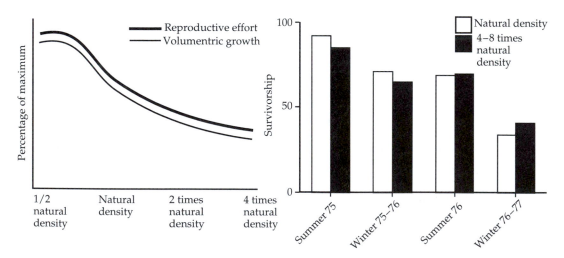

◖◗35 Peterson's finding that competition among filter-feeding clams leads to depressed growth and reproduction, rather than mortality. (After Peterson 1982b.)

Early studies in soft-substrate habitats suggested that filter-feeding organisms were more characteristic of sandy substrates than muddy substrates, while deposit feeders were more prevalent in muddy substrates (Sanders 1956). Rhoads and Young (1970, 1971) proposed that this trophic group habitat segregation was the product of the differential success of filter and deposit feeders in sand and mud habitats, respectively, and the adverse effects of deposit feeders on filter feeders (**trophic group amensalism**).[36] In their original articulation of the trophic group amensalism hypothesis, they proposed that filter feeders do better in sand than in mud, since fine sediments clog filtering devices and thus lead to low growth rates and high mortality. Conversely, they suggested that deposit feeders live in mud rather than sand substrates because muddy substrates often contain more organic material and have higher surface areas for bacterial growth, offering a better food source for deposit feeders. They tested the potential adverse effects of deposit feeders on filter feeders by examining the growth of the hard clam *Mercenaria* elevated over surfaces with and without deposit feeders, and found that the presence of deposit feeders lowered clam growth.

The initial trophic group amensalism hypothesis was elaborated on by Sally Woodin (1976), who emphasized the potential effects of adults on successful larval establishment. She recognized three functional groups—tube builders, bioturbators (primarily deposit feeders), and filter feeders—and noted that each of these groups is often found in dense assemblages that are dominated by single age classes. She suggested that each of these groups has negative effects on the others.[37] Tube builders were proposed to limit the success of both bioturbators and filter feeders by monopolizing space and by interfering with movement and larval recruitment. Bioturbators, by reworking sediments, were thought to adversely affect adult filter feeders

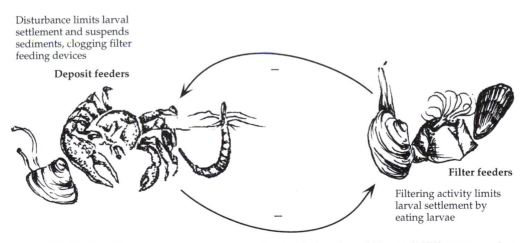

Disturbance limits larval settlement and suspends sediments, clogging filter feeding devices

Deposit feeders

Filter feeders

Filtering activity limits larval settlement by eating larvae

36 The trophic group amensalism hypothesis of Rhoads and Young (1970) suggests that deposit feeders and filter feeders adversely influence habitat conditions for one another.

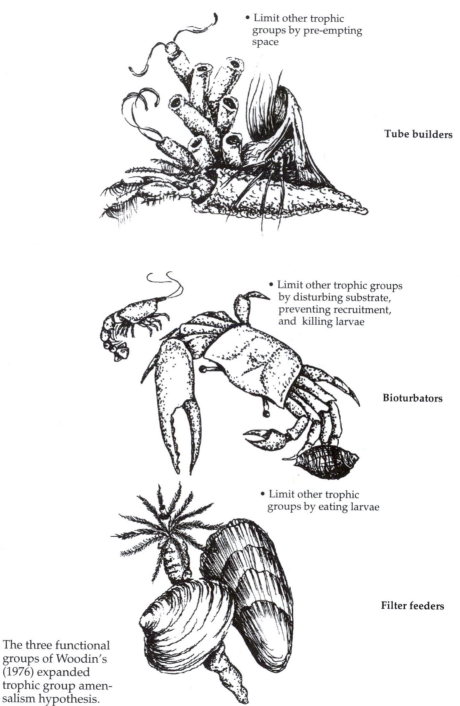

• Limit other trophic
groups by pre-empting
space

Tube builders

• Limit other trophic groups
by disturbing substrate,
preventing recruitment,
and killing larvae

Bioturbators

• Limit other trophic
groups by eating larvae

Filter feeders

🐚 37 The three functional
groups of Woodin's
(1976) expanded
trophic group amen-
salism hypothesis.

and to limit filter feeder recruitment by consuming settlers or burying them in sediment. Filter feeders were thought to adversely affect the recruitment of all groups by ingesting larvae (but see Commito and Boncavage 1989). Each of these groups was thought to strongly influence the age structure of assemblages, since recruits of one functional group were precluded by dense assemblages of any other functional group.

Woodin's expanded trophic amensalism hypothesis, particularly the role of adults in limiting larval recruitment, has received considerable attention since it was proposed nearly 20 years ago (for reviews see Wilson 1991; Olafsson et al. 1994). Many aspects of this hypothesis have received considerable support, but others have been found to be overly simplistic. Bioturbators have been found to limit the establishment of other organisms, but with exceptions. The deposit-feeding bivalve *Macoma balthica*, for example, limits the distribution of some worms while having no influence on others (Reise 1983). Strong deposit feeder effects on the recruitment and success of other trophic groups have also been documented, but only at very high deposit feeder densities (Ahn et al. 1993). Tube builder effects on the recruitment and success of other trophic groups have received the least support. Most studies have found that tube builder densities are not high enough to preclude movement or recruitment. Moreover, the potentially negative effects of tube builders on recruitment may commonly be offset by increases in larval supply through water flow modification by their tube structures (Eckman 1979, 1983; Jumars et al. 1981; Nowell and Jumars 1984). Thus, while the trophic group amensalism hypothesis provides a good general framework for thinking about the organization of soft-substrate communities, many exceptions have been found.

Physical Disturbance in Soft Sediments

Soft-sediment habitats are strongly influenced by physical disturbances, which disrupt sediments, kill infaunal organisms, and leave uninhabited sediments available for colonization. Pervasive physical disturbances in many soft-substrate habitats may keep population densities low enough to limit the role of competition. Both abiotic and biotic disturbances are important in soft-sediment habitats. Storms and severe weather can routinely move large volumes of sediment. In New England, shoreline scouring by ice can also transport large volumes of sediment and kill infaunal organisms. In many soft-substrate habitats the influence of these abiotic disturbances far outweighs the influence of biological disturbances (see Grant 1981b). For example, sediment movement on wave-exposed sandy beaches makes these habitats unsuitable for most organisms. Seasonal variation in wave energy can also dramatically change beach topography, killing some organisms by eroding sediments and burying others.

Biological factors can both increase and decrease the effects of abiotic disturbances. Deposit-feeding organisms rework sediments, increase sediment water content, often move small sediment particles to the surface, and resuspend sediments by defecating on the surface. These activities make sediments more vulnerable to abiotic disturbance and erosion (Rhoads and Young 1970). In contrast, seagrasses, marsh grasses, mussel beds, and tube-building invertebrates stabilize the substrate by binding sediments together and buffering the bottom from eroding currents.[🐚38]

A major consequence of disturbance in soft sediments is that most soft-sediment communities are best envisioned as a mosaic of patches in different stages of recovery from disturbance (McCall 1977). Disturbance creates patches with reduced densities of inhabitants, triggering immigration and secondary succession (Zajac and Whitlatch 1985). Recently disturbed habitats are typically characterized by "weedy" species—small worms and bivalves that are usually short-lived, fast-growing, rapidly reproducing, and easily dispersed. Less recently disturbed soft-substrate habitats are typically characterized by longer-lived climax species, which disperse less easily (see Oliver et al. 1985; Van Blaricom 1982). Large worms and clams with seasonal reproduction and relatively long life spans are typical of established soft-sediment habitats.

The mechanisms of secondary succession in soft-sediment communities are not clear. Early colonizers are usually rapid dispersers or organisms with nearby source populations. The effects of these early colonizers on one another and on later colonists, however, are more complicated (Gallagher et al. 1983; Zajac and Whitlatch 1982, 1985).[🐚39] Experiments have revealed that early colonizers can have positive, neutral, or negative effects on later colonizers (Gallagher et al. 1983). Early colonizers can improve conditions for

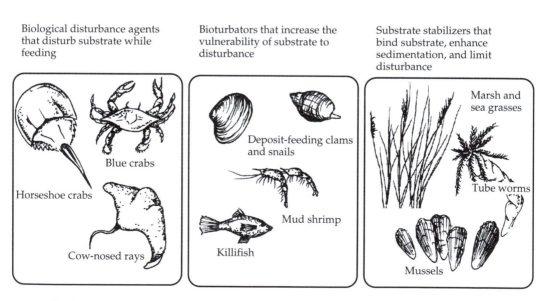

Biological disturbance agents that disturb substrate while feeding

Blue crabs

Horseshoe crabs

Cow-nosed rays

Bioturbators that increase the vulnerability of substrate to disturbance

Deposit-feeding clams and snails

Mud shrimp

Killifish

Substrate stabilizers that bind substrate, enhance sedimentation, and limit disturbance

Marsh and sea grasses

Tube worms

Mussels

🐚38 Organisms that influence the effects of physical disturbance on soft-substrate communities.

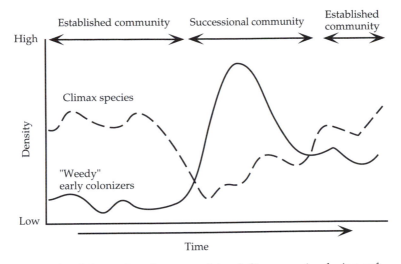

39 Generalized dynamics of opportunist and climax species during soft-sediment secondary succession. (After Zajac and Whitlatch 1985.)

later ones by modifying sediment stability, chemistry, and microflora, thus influencing larval settlement and survival (Meadows and Campbell 1971). Eckman (1979) found that tube worms that protrude as little as 5 millimeters from the sediment surface can enhance larval settlement by influencing local hydrodynamics and increasing the passive deposition of larvae.

Other studies have found strong negative effects of early colonizers on later ones and little support for facilitated secondary succession. Early colonizers can negatively affect later recruits by consuming larvae, depleting food resources, or harming recently settled larvae by bioturbation (McCall 1977; Watzin 1983, 1986; Zajac and Whitlatch 1982; Ambrose 1984a). The roles of facilitation and inhibition in soft-sediment secondary succession may vary over time and under different environmental conditions, as has been found in many other marine and terrestrial habitats (Whitlatch and Zajac 1985).[40] Facilitated succession may be more common early in colonization, before densities of consumers or bioturbators are high enough to inhibit settler success. Facilitated succession may also be more prevalent in higher flows, where larvae are vulnerable to erosional transport without the substrate stabilization of earlier recruits. Most studies of soft-sediment succession, however, have not systematically examined the process across environmental gradients or the effects of consumers on succession, so our understanding of these relationships is entirely speculative.

The Role of Predation

Predation by epibenthic predators, such as crabs, fishes, and birds, plays a major role in shaping the distributions and abundances of organisms in soft-

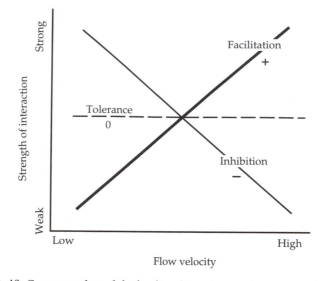

40 Conceptual model of soft-sediment secondary succession predicting the effects of early settlers on later ones as a function of flow
velocity. (After Zajac and Whitlatch 1985.)

sediment habitats. Experimental caging studies that excluded predators
from shallow soft-sediment habitats have been carried out over the last 20
years. These experiments have suggested that epibenthic predators commonly limit the distributions and population sizes of their prey in unvegetated sediments (Peterson 1979). While the importance of predators is
widely accepted, the difficulties of performing such experiments in soft-sediment habitats have raised questions about the generality of this finding (see
Hall et al. 1991; Olafsson et al. 1994). Soft substrates are dynamic, and cage
structures can increase sediment deposition and the passive deposition of
larvae and food by influencing local hydrodynamic conditions (Hulberg and
Oliver 1980). Ecologists have tried to solve these problems by controlling for
potential cage artifacts with cage controls (cages with missing sides and/or
tops to allow predators access) and by combining manipulative field experiments with sampling programs and laboratory work (for example, see
Lipcius and Hines 1986). On balance, these studies have supported the general finding that epibenthic predators play an important role in shallow, soft-
sediment habitats that are unvegetated.

The role of predators in soft-sediment habitats on the Atlantic coast of
North America has received considerable attention, but most studies have
focused on communities south of Cape Cod. The Cape is a major biogeographic boundary for many Atlantic species. As already mentioned, blue
crabs, which are found only south of Cape Cod, have been repeatedly shown
to have a strong effect on benthic communities (see Virnstein 1977; Lipcius

and Hines 1986). Other potentially important predators, including fishes (spot and hogchoker), are also found primarily south of Cape Cod. Woodin (1976) has argued that the dense infaunal assemblages found north of Cape Cod and the relative lack of these dense assemblages south of Cape Cod may be the result of the lack of dominant mobile soft-sediment predators, especially the blue crab, north of Cape Cod. Unfortunately, this important hypothesis has not been tested experimentally.

South of Cape Cod, the role of predators in limiting infaunal prey populations is well established. In the mid-1970s Bob Virnstein did a seminal series of experimental studies in which he manipulated blue crabs and predatory fishes on soft substrates in Chesapeake Bay. His results were dramatic. When access to the bottom by predatory crabs and fishes was limited, the density of shallow-burrowing bivalves and worms increased by as much as two orders of magnitude.[41] The only soft-sediment organisms not strongly affected by blue crab predation were those that burrow deeply (see also Blundon and Kennedy 1982).

More recently, Tuck Hines and his associates, in an elegant series of field and laboratory experiments at the Smithsonian Environmental Research Center, have refined our understanding of the role of blue crabs in dictating the abundances and distributions of bivalves. Both the baltic clam, *Macoma balthica*, and the soft-shelled clam, *Mya arenaria*, settle annually at high densities in mud and sand substrates in Chesapeake Bay. Few of these bivalves, however, survive their first year due to intense predation by *Callinectes*. Blue crabs dig in both mud and sand for their prey, but their foraging success is strongly influenced by prey density and substrate type. At high prey densities, blue crabs are extremely efficient foragers on clams, but their efficiency drops at lower clam densities.[42] This density-dependent foraging efficiency gives clams a refuge from blue crab predation, and clams therefore can coexist with blue crabs at low densities. *Mya* has a low density refuge from blue crabs in sand, but not mud, since it is more costly for blue crabs to forage in coarser, heavier substrates. In contrast, *Macoma* has a short siphon, is less detectable to crabs in both mud and sand, and thus persists at low densities in both substrates. Quahogs (*Mercenaria mercenaria*) have a similar refuge from crab predation in coarse substrates (Sponaugle and Lawton 1990).[43]

In addition to large epifaunal predators that live on the surface, predators that live within the substrate (infaunal predators), such as polychaete and ribbon worms, moon snails, and large whelks, can also play an important role in regulating infaunal populations (Peterson 1982a; Ambrose 1984a,b; Commito 1982). In the Gulf of Maine, common large infaunal polychaetes can limit the densities of smaller organisms such as burrowing amphipods (Ambrose 1984a,b; Commito 1982). The little work that has been done on infaunal predators suggests that they may be particularly important north of Cape Cod, where large mobile predators are less conspicuous, though this biogeographic aspect has yet to be examined. North of Cape Cod, large populations of these soft-bodied infaunal predators are common

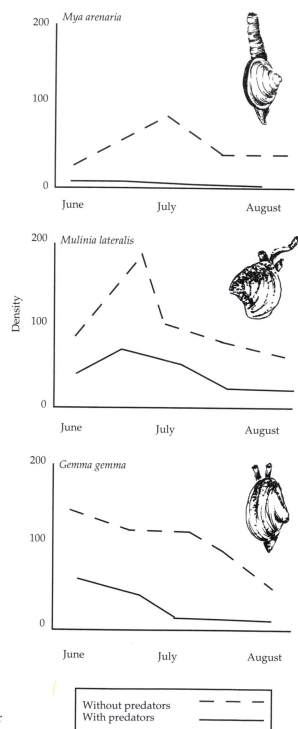

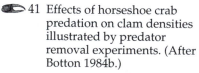 41 Effects of horseshoe crab
predation on clam densities
illustrated by predator
removal experiments. (After
Botton 1984b.)

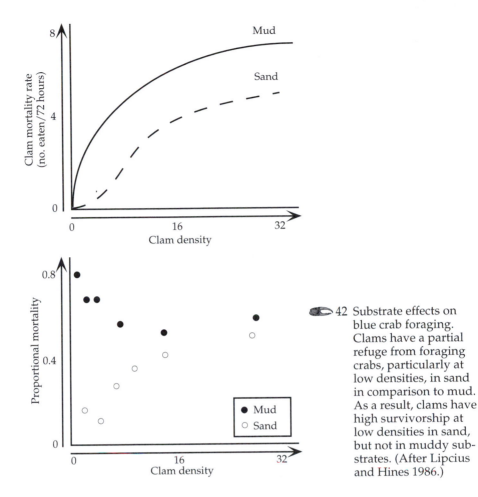

42 Substrate effects on blue crab foraging. Clams have a partial refuge from foraging crabs, particularly at low densities, in sand in comparison to mud. As a result, clams have high survivorship at low densities in sand, but not in muddy substrates. (After Lipcius and Hines 1986.)

enough that Maine supports a bait industry exporting large polychaetes to cities all along the east coast of North America. South of Cape Cod, dense populations of large burrowing polychaetes disappear, presumably as a result of increased predation by crabs and fishes.

Meiofaunal nematodes and benthic (harpacticoid) copepods, while less conspicuous and more difficult to study than larger infaunal predators, may be important recruitment filters. Predation by meiofauna (especially flatworms) on settling larvae may play an important role in soft-sediment communities by regulating the recruitment success of large infaunal organisms (Watzin 1983).

Complex interactions that result from organisms feeding at more than one trophic level are common in soft-sediment communities and often lead to predators having strong indirect effects on prey populations. Such indirect effects are particularly common in soft-sediment habitats because biotic

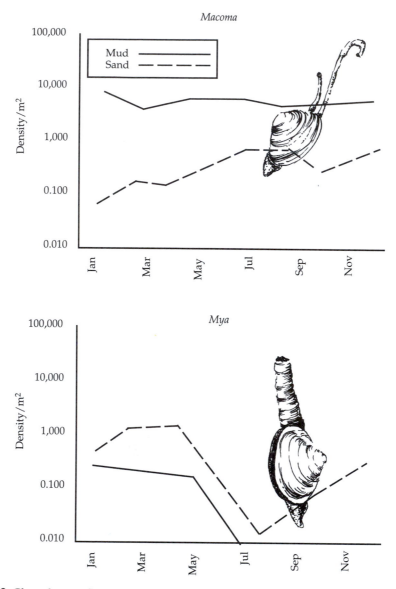

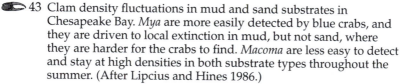

43 Clam density fluctuations in mud and sand substrates in Chesapeake Bay. *Mya* are more easily detected by blue crabs, and they are driven to local extinction in mud, but not sand, where they are harder for the crabs to find. *Macoma* are less easy to detect and stay at high densities in both substrate types throughout the summer. (After Lipcius and Hines 1986.)

structures often provide refuges for prey, and infaunal consumers commonly feed at more than one trophic level. Marsh grasses and seagrasses commonly provide refuges for prey (Vince et al. 1976; Peterson 1982a), leading to positive indirect effects of vegetation on prey populations.[44]

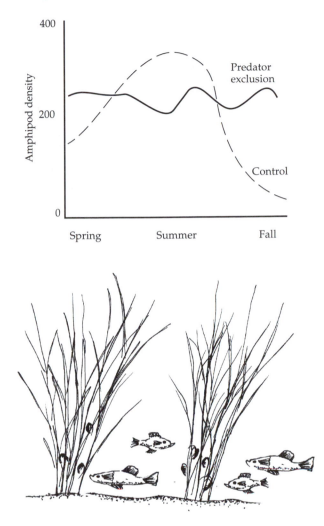

44 In Chesapeake Bay marshes, amphipod densities are kept low by killifish predation during most of the year, and killifish restrict amphipods to living in cordgrass culms for protection. (After Van Dolah 1978.)

More subtle indirect effects in soft-substrate communities are a consequence of trophic complexity. By preying on intermediate predators, large predators can often have positive effects on lower trophic levels. Ambrose (1984a) has shown that glycerid polychaetes can limit the population densities of nereid polychaetes, which regulate the densities of smaller infaunal invertebrates. Thus, glycerids can indirectly regulate populations at lower trophic levels. Predators of grass shrimp have also been shown to have strong positive effects on grass shrimp prey populations (Kneib 1988, 1991; Hines and Posey 1989).[45]

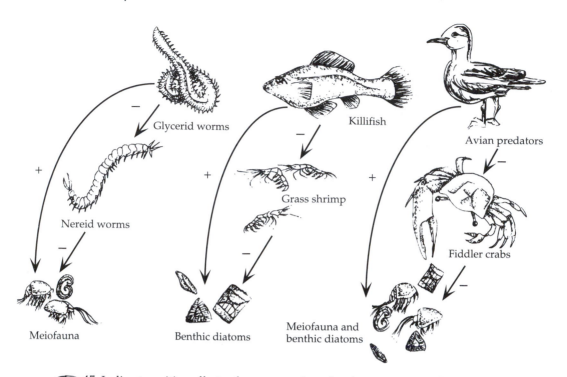

45 Indirect positive effects of consumers in soft-substrate systems. By controlling populations of their prey, predators stimulate lower trophic level densities. [After (from left to right) Ambrose 1986; Kneib 1981; and Hoffman et al. 1984.]

Predators in soft-sediment habitats can also influence the success of other predators (Hines and Posey 1989; Skilleter and Peterson 1994). Clam predators south of Cape Cod are a good example. The Baltic clam, *Macoma balthica*, is a common prey item of blue crabs from Chesapeake Bay to Florida. Blue crabs can strongly limit *Macoma* densities, particularly of smaller clams, which live close to the surface (Virnstein 1977; Lipcius and Hines 1986). Because of predation, only large *Macoma* can coexist with blue crabs. Large clams live deeper in the substrate, giving them a spatial refuge from blue crab predation. In addition to blue crabs, a number of fishes, particularly spot and hogchoker, also prey on *Macoma*. These fishes do not eat whole clams, but consume the siphons of feeding clams. Siphon nipping can be very common, and has two important indirect effects. First, since clams with nipped siphons are forced to live at shallower sediment depths, they are more vulnerable to predation by blue crabs than are clams with intact siphons. Second, siphon nipping affects the feeding strategy of *Macoma* and thus the effect of *Macoma* on the infaunal community. *Macoma* is a facultative deposit feeder (Olafsson 1986), which filter-feeds when the flux of suspended food is high, and deposit-feeds when little food is available in the water column. Since long siphons are required to deposit feed, siphon nipping restricts *Macoma* to filter feeding (Skilleter and Peterson 1994).[46]

BIOENGINEERING BY VASCULAR PLANTS IN SOFT-SEDIMENT HABITATS

Among the most important components of soft-substrate habitats are the vascular plant assemblages that commonly colonize and dominate sandy beach habitats in both the intertidal and shallow subtidal zones. Vascular plant colonization of soft sediments plays an important role in the ecology of soft-substrate communities by stabilizing substrates against erosion, stimulating further sedimentation, and providing prey refuges and food resources to the organisms that live within the plant canopies. In other words, vascular plants are important bioengineers on soft-sediment shores. Below, I introduce the vascular plants that modify seagrass meadows and sand dune plant communities. I devote the following chapter entirely to an examination of salt marsh plant communities, which are the most widespread and extensive vascular plant assemblages on Western Atlantic shorelines.

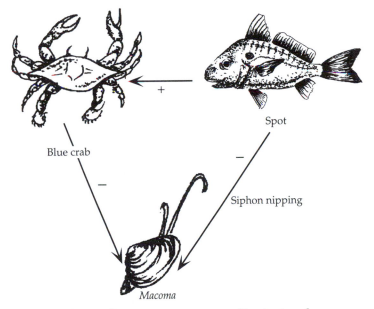

Blue crab

Spot

Siphon nipping

Macoma

46 Positive indirect effects among consumers. By nipping clam siphons, spot decrease the burrowing depths of clams and increase their vulnerability to blue crab predation. (After Lipcius and Hines, unpublished data.)

Seagrass Communities

Seagrass beds are a common feature of shallow subtidal and lower intertidal soft-substrate habitats on the east coast of North America. A good case can be made for considering seagrasses key bioengineers of these habitats, which they can potentially dominate. They play an exceptionally important role in both the geomorphology and ecology of shorelines. They influence the shape and stability of shorelines by enhancing sedimentation and binding sediments against erosion. Seagrass beds also dictate biological interactions in shallow-water habitats by serving as a source of fixed carbon for adjacent marine communities and food webs and as a refuge from predation for many organisms. Seagrass beds, however, are also sensitive to harsh environmental conditions and are particularly vulnerable to anthropogenic disturbances.[🐚47]

Seagrasses are rooted vascular plants of terrestrial origin that have successfully returned to the sea with a number of adaptations that allow them to live submerged. Since the diffusion of gases is slow in water, seagrasses have thin, permeable cell walls to enhance external gas exchange, and gas vacuoles and aerenchyma tissue to facilitate internal gas exchange (Penhale and Thayer 1980; Penhale and Wetzel 1983). Aerenchyma tissue, also common in the halophytic plants that live in salt marshes, forms gas conduits that run from the leaves to the roots. In waterlogged, anoxic sediments,

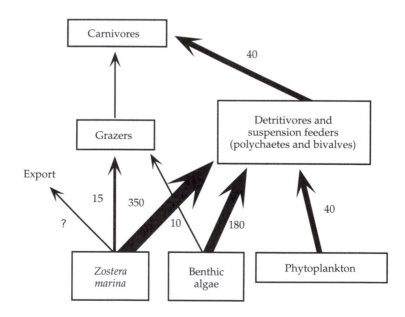

🐚 47 Energy flow in a seagrass bed (in grams of carbon per square meter per year. (After Thayer et al. 1975; Raffaelli and Hawkins 1996.)

aerenchyma allows the passage of photosynthetically produced oxygen to belowground roots and rhizomes, allowing aerobic respiration in an anaerobic environment. In some seagrasses, root and rhizome respiration is aerobic under daylight conditions but anaerobic under dark conditions, when photosynthetic oxygen cannot be provided by aerenchyma tissue (McRoy and McMillan 1977).

Living in seawater eliminates the need for large investments in structural support, and seagrasses have taken a flexible approach to their environment, much like large seaweeds. Gas vacuoles keep them erect and maximize light interception when they are submerged, and their flexible design allows them to bend rather than break when exposed to terrestrial conditions or water movement. By being flexible, seagrasses bend closer to the sediment surface when exposed to high current speeds or waves, minimizing their exposure to hydrodynamic forces (Koehl and Wainwright 1977).

A number of seagrass species are common on the Atlantic coast of North America. Eelgrass, *Zostera marina*, occurs exclusively at temperate latitudes. It is the dominant seagrass from the Maritime provinces of Canada to the Carolinas, and also dominates northern latitudes on the eastern Atlantic coast of Europe and in the Pacific Ocean (den Hartog 1970). South of the Carolinas and in the Caribbean, turtle grass, *Thalassia testitudinalis*, replaces eelgrass as the dominant seagrass. A third seagrass, *Halodule wrightii*, or shoalgrass, also occurs from the Carolinas to the Caribbean, usually living at shallower depths than *Thalassia* or *Zostera*.

Seagrasses are rhizomatous grasses that propagate both vegetatively and sexually. Vegetatively, seagrasses colonize by means of belowground rhizomes and nutrient-absorbing roots. Established seagrass beds rely primarily on vegetative propagation for their persistence (Tomlinson 1974). Sexual reproduction or fragmentation is necessary for colonizing new habitats (Orth et al. 1994) and for the local persistence of populations that experience heavy annual mortality (Keddy and Patriquin 1978; Gagnon et al. 1980). Seagrass pollen is carried in water, and is designed so as to maximize interception by receptive flowers (Ackerman 1997), but low seed set suggests that pollination efficiency is low. Seed dispersal, while poorly studied, appears to be passive, localized, and strongly affected by current patterns (Orth et al. 1994).[🐚 48]

Seagrass distributions are extremely sensitive to variation in light. Seagrasses cannot live at depths below where photosynthetic benefits outweigh respiratory costs. Since light transmission is a function of water column production and sediment load, which influence turbidity, seagrass distributions are strongly affected by eutrophication and other anthropogenic effects that influence water clarity. In Chesapeake Bay, seagrass beds historically occurred at depths of over 10 meters, but are restricted today to depths of less than 1 meter (Orth and Moore 1981). Runoff from organic fertilizers has increased plankton production in the water column and limited light transmission. Moreover, the loss of large oyster beds in Chesapeake Bay has probably resulted in the loss of natural filtration of the bay water

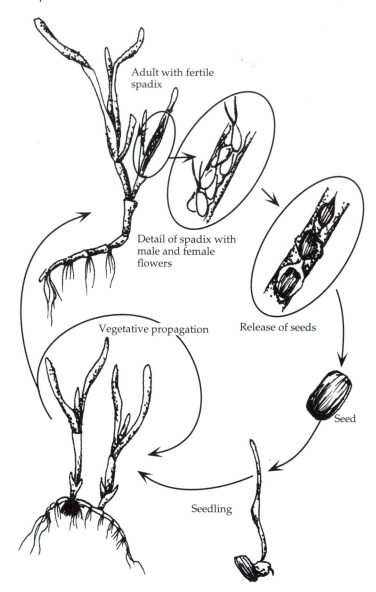

Adult with fertile
spadix

Detail of spadix with
male and female
flowers

Release of seeds

Vegetative propagation

Seed

Seedling

 48 Typical seagrass life cycle.

and an increase in the success of epiphytic algae and planktonic food webs. Contemporary seagrass beds in Chesapeake Bay cover less than 10 percent of the area they covered a century ago (Orth and Moore 1981, 1983).[49]

Eelgrass populations throughout the northern Atlantic Ocean in both Europe and North America experienced a catastrophic die-off in the early 1930s, which focused worldwide attention on the role of seagrass beds in marine food webs. The cause of eelgrass "wasting disease" is still uncertain,

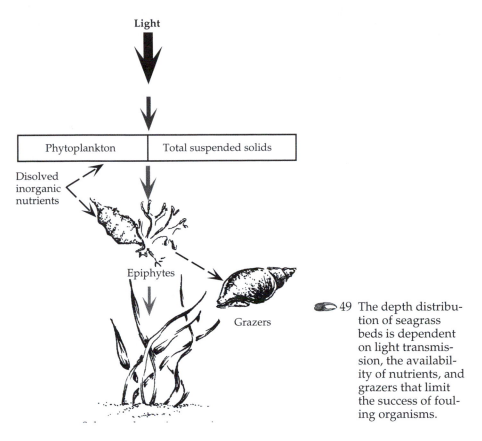

Light

Phytoplankton	Total suspended solids

Disolved inorganic nutrients

Epiphytes

Grazers

49 The depth distribution of seagrass beds is dependent on light transmission, the availability of nutrients, and grazers that limit the success of fouling organisms.

but its consequences were dramatic. Between 1930 and 1933, it destroyed 90 percent of the eelgrass in the northern Atlantic (Tutlin 1938). After much controversy, scientists attributed the disease to infection by a slime mold pathogen, *Labyrinthula macrocystis* (Rasmussen 1977). However, this pathogen was later found to be present in healthy stands of eelgrass, but restricted to the leaves of dead plants (Porter 1967; Phillips 1971). Rasmussen (1977) examined water temperature records in the northern Atlantic and found that eelgrass wasting disease occurred during years with abnormally high summer and winter water temperatures. High temperatures either directly or indirectly appear to have weakened eelgrass on a large scale, leading to infection by *Labyrinthula* and widespread death.

Eelgrass wasting disease had dramatic consequences on northern Atlantic shorelines. Sandy and muddy habitats that once supported extensive seagrass beds were eroded away, leaving rocky bottoms (Rasmussen 1977).[50] This shift in habitats from soft substrate to coarse cobble bottoms was accompanied by dramatic shifts in the benthic assemblage. Deposit feeders, common in the fine sediments of seagrass beds, were replaced by filter feeders. Fishes (Milne and Milne 1951) and birds (Cottam

Before wasting disease
Eelgrass-dominated shallow subtidal
zone with associated organisms

After wasting disease
Cobble-dominated shallow subtidal
zone with low species diversity

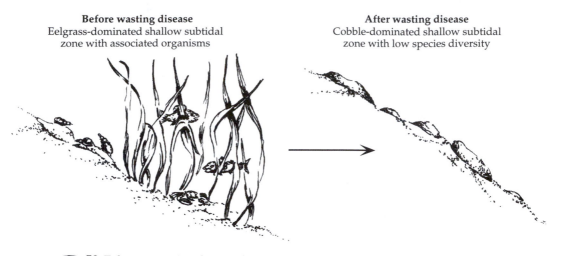

🐚 50 Eelgrass wasting disease dramatically reduced the dominance of eelgrass in the
northern Atlantic during the 1930s. The eelgrass die-off led to large shifts from
diverse seagrass communities to relatively simple cobble-bottom communities.

and Munro 1954) that depended on eelgrass for food or nursery grounds
suffered serious population crashes. The bay scallop, *Argopecten irradians*,
showed a particularly well documented population crash with the loss of
eelgrass beds (Thayer and Stuart 1974). Bay scallops depend on eelgrass not
only for their detritus-based diet, but in addition, scallop recruits attach to
eelgrass blades until they are large enough to assume a benthic existence.
After the eelgrass die-off, scallop harvests in North Carolina fell to less than
10 percent of their previous levels and showed no signs of recovery for
nearly 30 years (Thayer and Stuart 1974).

The consequences of eelgrass wasting disease serve as a striking example of the role rooted vascular plants play as bioengineers of soft-sediment
habitats. Like the marsh plants that often dominate high elevations on tidal
flats, seagrass beds at lower elevations clearly play an important role in
shoreline geomorphology, community development, and stability. Seagrass
beds increase the deposition of sediments and decrease the resuspension
and erosion of small sediment particles by slowing the flow of currents near
the bottom (Orth 1977; Gambi et al. 1990). Seagrass root mats also stabilize
sediments and limit disturbance by burrowing deposit feeders (Peterson
1982a; Ringold 1979). In Chesapeake Bay, the most conspicuous consequence
of the decline in eelgrass over the last few decades has been a marked
decrease in the stability of sediments (Orth 1975, 1977).

Seagrass beds can also play a critical role in enhancing the structure,
dynamics, and diversity of soft-sediment environments. The aboveground
biomass of seagrass beds increases the structural complexity of the sea floor,
increasing the surface area available for sessile epibenthic organisms and

providing a refuge from fishes and crabs (Vince et al. 1976; Kneib 1982; Summerson and Peterson 1984). Seagrass blades harbor a rich epiphytic community, including microalgae (Harlin 1975, 1980), sessile invertebrates (such as hydroids, bryozoans, sponges, and barnacles: Nagle 1968; Zieman 1981), and grazers (including amphipods and small gastropods: Nelson 1980, 1981). Few of these are obligatory seagrass epiphytes, but for many, seagrasses are their most common natural habitat (Nagle 1968; Thayer et al. 1979). The complex structure of seagrass blades also serves as a refuge from larger mobile predators. Predation on seagrass-associated prey such as grass shrimp and amphipods is much higher outside than inside seagrass beds, since prey is hidden and predator foraging is inhibited by grass cover (Heck and Thoman 1981).

Seagrasses also have enormous effects on infaunal communities. In addition to its indirect effects on infauna by stabilizing sediments and shifting sediment sizes to finer, muddier particles, seagrass cover also acts as a refuge for infaunal organisms and enhances the growth of filter feeders and larval settlement. The diversity and abundance of infaunal organisms is greater in seagrass beds than in adjacent unvegetated habitats (Orth 1977; Thayer et al. 1979; Peterson 1979). Summerson and Peterson (1984) have shown that predation by whelks is reduced by the dense root mats of seagrass beds. Bivalve growth has also been shown to be greater in seagrass beds than in unvegetated habitats, due to the increased deposition of particulate food caused by the baffling effects of seagrass blades (Peterson et al. 1984). Submerged aquatic vegetation can also enhance the recruitment of benthic organisms by increasing the passive deposition of larvae (Eckman 1979, 1983). In practice, however, it has been difficult to factor out the effects of larval supply and post-settlement mortality on the recruitment of infaunal organisms in seagrass beds (for a discussion see Olafsson et al. 1994). Without seagrasses, soft-substrate communities would clearly be much simpler, less heterogeneous, and less diverse.[🐚51]

Sand Dunes

While sand dunes, found on the terrestrial margins of shorelines, are easily overlooked by marine ecologists, they are a conspicuous component of many shorelines that deserves our attention. On sandy shorelines of the western Atlantic, sand dune plant communities are just as characteristic a feature above the high tide line as seagrass beds are below the low tide line. Like seagrass communities, sand dune communities are strongly influenced by the presence of vascular plants . The plants that initially colonize developing sand dunes are responsible for stabilizing sediments and facilitating further plant and animal colonization. Thus, like seagrass and salt marsh plant communities, sand dune communities are largely the product of plant bioengineering of soft-sediment habitats.

Sand dune plant communities are common on temperate, wave-exposed shores around the world wherever geological processes have left large sediment supplies. On the Atlantic coast, sand dune plant communities are a

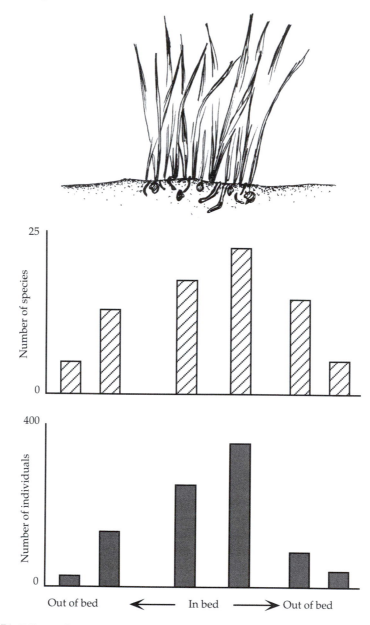

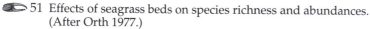

51 Effects of seagrass beds on species richness and abundances. (After Orth 1977.)

prominent coastal feature, particularly south of New England, where over 150 million years of coastal erosion have left a wide continental shelf of soft sediments.

Sand dunes are the result of onshore winds moving sediments shore-ward, forming spatially and temporally predictable sand structures. Sand is moved primarily by **saltation**, the bouncing of small sand particles picked up by the wind, and the creeping or rolling of larger sand particles over the surface. Saltation is the most important of these processes and is responsible for 75 percent of the sand movement that leads to dune formation (Pethick 1984). It begins when wind friction and drag lift a grain of sand from the surface. Once airborne, these particles move into faster winds, which further accelerate their velocity and altitude. When the particles fall and hit the sand surface, they knock other small particles into the air, continuing the process. The end result is a cloud of fine sand particles, up to a meter in height, moving landward, with increasing magnitude as wind speed picks up. Larger sand particles are moved by the impact of small particles hitting the surface during the saltation process, which causes them to creep or roll along the surface in the direction of the wind.[52]

Dunes begin to form when this movement of sand particles by wind is obstructed, leading to their deposition. This usually happens at the high tide line, where floating debris (wrack) is deposited by tides. This wrack contains plant debris, plant seeds, and seaweeds, all of which stabilize the substrate by trapping moisture and enhance deposition by slowing wind velocity. This leads to the formation of small (up to 1 meter high), transient dune structures on the seaward border of the dune community, termed **embryo dunes**. Embryo dunes without vegetation grow in height until they encounter wind velocities that limit further sand accumulation.

Without colonization by plants, embryo dune growth is limited. Plants recruit to embryo dunes in the debris deposited by tides or wind. The mois-

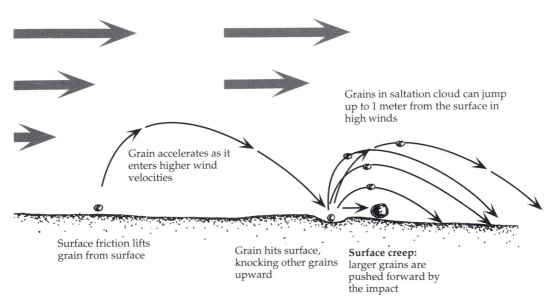

Grains in saltation cloud can jump up to 1 meter from the surface in high winds

Grain accelerates as it enters higher wind velocities

Surface friction lifts grain from surface

Grain hits surface, knocking other grains upward

Surface creep: larger grains are pushed forward by the impact

52 The process of saltation and surface creep of wind-driven sand. (After Pethick 1984.)

ture retained and nutrients released by tidal wrack foster germination of the plants characteristic of embryo dunes. These plants root in the sand, binding or fixing sand against erosion, and their aboveground mass stimulates further sediment deposition. The initial plant colonizers of embryo dunes include both solitary and clonal plants. Solitary plants, such as the sea rocket (*Cakile*), germinate in moist embryo dune sand. Their roots bind sand, and their aboveground structure obstructs wind movement and saltation clouds, enhancing sediment deposition and dune growth. Clonally propagating grasses and vines are even more effective binders of dune sediments. The beach grass *Ammophila* and sea oats (*Uniola*), for example, contribute to dune growth and stability through their vegetatively spreading rhizomes, which move horizontally and limit the mobility of dune sands. As both beach grass and sea oats stimulate sediment deposition, they continue growing out of the deposited sand, generating a strong positive feedback between plant growth and dune growth.[🐚53]

Embryo dunes form just above the high tide line, migrate landward under the influence of saltation, and are stabilized and grow vertically and horizontally as they are colonized by plants. As embryo dunes migrate landward and grow, they typically run into and merge with other embryo dunes. This precipitates sand deposition between embryo dunes and coalesces them into a single seaward dune ridge paralleling the shoreline. This process of landward migration and stabilization leads to a predictable spatial and temporal sequence of dune structures on shorelines. Closest to shore, unstable solitary embryo dunes occur. Shoreward of these dunes, a primary dune ridge, or **foredune**, of coalesced embryo dunes parallels the shore. Further shoreward, older, larger dunes occur, with the age of successive dune ridges increasing by 50–200 years as one moves from the water's edge shoreward.[🐚54]

The morphology of sand dunes is rarely as simple as I have described here, however, and is particularly influenced by variation in the strength of the winds. The simple coalescing of embryo dunes occurs in areas of low to moderate winds, where aggressive colonizing plants are capable of stabilizing dunes. As dunes grow in height, they become more and more vulnerable to erosion, and **blowouts** occur. Blowouts are common in habitats with higher wind speeds and rapid dune growth, and are often triggered by loss of binding vegetation or low vegetation cover. Thus blowouts occur when dune growth exceeds the growth of the plants binding the dunes together or when dune vegetation is disturbed. This makes dune structures particularly sensitive to human disturbance, since the simple destruction of vegetative cover can lead to dramatic shifts in dune morphology. Whereas blowouts or breaks in a foredune structure can be a part of typical foredune landscapes, **parabolic dunes** can be formed on coastlines with high sediment supplies and strong winds. In parabolic dunes, frequent blowouts and the building of dune ridges on the downwind extensions of blowouts lead to vegetated dune ridges that meander along the shore, in contrast to the neat parallel dune ridges that occur under less severe wind conditions. On shorelines with even higher onshore winds and massive sediment supplies, the ability of plants to stabilize dune sands breaks down, the development of dune

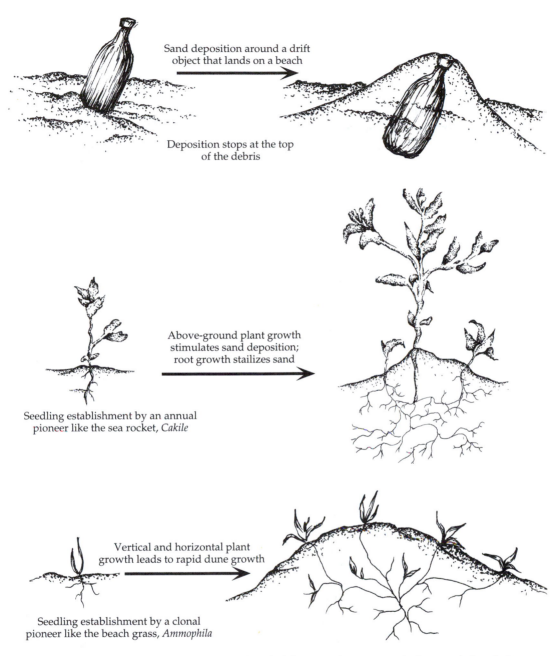

Sand deposition around a drift
object that lands on a beach

Deposition stops at the top
of the debris

Above-ground plant growth
stimulates sand deposition;
root growth stailizes sand

Seedling establishment by an annual
pioneer like the sea rocket, *Cakile*

Vertical and horizontal plant
growth leads to rapid dune growth

Seedling establishment by a clonal
pioneer like the beach grass, *Ammophila*

🐚 53 Embryo dune formation by beach debris, a solitary annual plant, and clonal plants.

structures parallel to the shore is entirely precluded by erosional processes, and dunes develop with their axis perpendicular to the direction of onshore winds. These formations are called **transgressive dune sheets**. They are highly mobile and almost entirely lack vegetation.[🐚55]

Reduced wind
speeds on lee side of
dune increases
deposition

High wind speeds at
the seaward edge
increase sand
transport leading to
erosion

High wind speeds at
end of trough between
dunes maintains
trough

54 Wind flow patterns over dune surfaces maintains their shape and spacing. Wind flow roughly follows established contours, eroding seaward edges, building up leeward edges, and eroding troughs between dune ridges.

The physical conditions on sand dunes are harsh and inhospitable to most organisms. Thus, dune development is dependent on a limited number of plant species capable of dealing with such conditions. For the vascular plants that initiate dune formation and growth, physical conditions for growth generally improve with distance from the sea. Close to the shoreline, sediment mobility can limit the establishment of most plants and can also lead to sediment size sorting, which strongly influences sediment water content. Small sediments are moved shoreward by saltation, leaving larger, more porous sands near the shoreline. As a consequence, sediment mobility and grain size decrease, whereas sediment water content increases, away from the shore. Organic matter and nitrogen levels also increase with distance from the shoreline. At the seaward border of the dunes, the sediments are composed primarily of abraded inorganic mineral sands, and usually have a high calcium carbonate content (from the animal skeletons in marine sediments). Over time, organic material and nitrogen accumulate in dune sediments through the activity of nitrogen- and carbon-fixing microorganisms, fungi, lichens, and vascular plants.[56]

The strong gradient in physical conditions across dune habitats and the limited ability of plants to deal with these conditions leads to pronounced plant zonation.[57] On the seaward border, only plants capable of dealing with extreme physical conditions are found. This zone, referred to as the **pioneer zone** or **primary dune**, is characterized by rapidly recruiting species with high growth rates that are extremely tolerant of dry conditions, high salinities, and shifting sediments. Like the plants of salt marshes (which will be discussed in the next chapter), pioneer dune plants often have mechanisms to conserve water and limit water loss. Reducing leaf area or number to reduce transpiration water loss, rolling leaves under dry conditions to

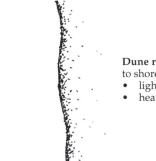

Dune ridges running parallel
to shoreline
- light onshore wind
- heavy vegetation binding sand

Parabolic dunes with blowouts
- medium onshore winds
- sparse vegetation leading to
  dune ridge failure

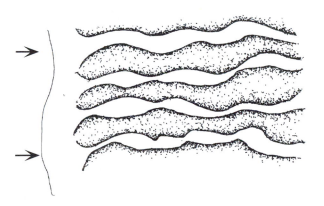

Transgressive dune sheets
running perpendicular to shore
- heavy onshore winds
- little vegetation
- high sand mobility

55 Major categories of dunes found under different wind regimes.

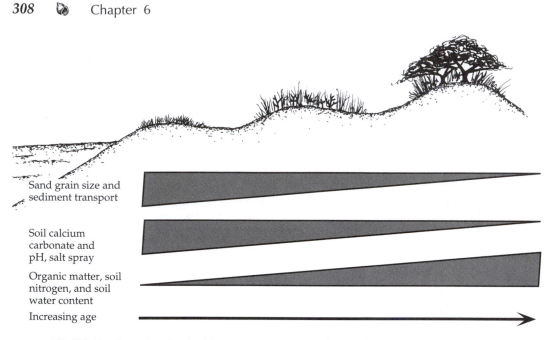

56 Gradients in physical factors across a coastal sand dune.

limit exposed leaf area, and storing water in body tissues (succulence) are all strategies that plants use to cope with dry dune habitats. Salt glands are a common mechanism for getting rid of excess salt, and heavy investment in structural support is associated with the exposure of these plants to heavy sand abrasion. Most pioneer dune plants have a **prostrate** (low to the ground) morphology, deep roots, and clonal architecture. Being low to the

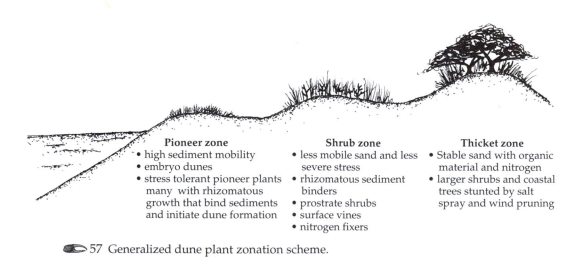

Pioneer zone
- high sediment mobility
- embryo dunes
- stress tolerant pioneer plants many with rhizomatous growth that bind sediments and initiate dune formation

Shrub zone
- less mobile sand and less severe stress
- rhizomatous sediment binders
- prostrate shrubs
- surface vines
- nitrogen fixers

Thicket zone
- Stable sand with organic material and nitrogen
- larger shrubs and coastal trees stunted by salt spray and wind pruning

57 Generalized dune plant zonation scheme.

ground minimizes exposure to wind stress, while deep roots provide access to cool, moist soil. Clonal morphology, and the ability to grow both vertically and laterally across shifting dunes, is a critical feature of pioneer dune plants that allows them to survive in shifting dune sands. Their presence facilitates the colonization of secondary dune plants.[🐚58]

At higher elevations, the pioneer zone gives way to the **dune shrub** or **secondary dune** community. The dune shrub zone is characterized by lower wind stress, more stable sediments, and increased sediment moisture. This community is dominated by plants that are unable to cope with the harsh physical conditions in the pioneer zone. The shrub community includes a mixture of short, woody shrubs such as bayberry (*Myrica pensylvanica*), wax myrtle (*Myrica cerifera*), and beach rose (*Rosa rugosa*), annual forbs such as the seaside goldenrod (*Solidago sempervirens*), and prostrate vines such as the beach pea (*Lathyrus japonicus*) and butterfly pea (*Centrosema virginianum*). Plants in the shrub community are relatively short because of wind stresses in the habitat and include many rapidly growing and nitrogen-fixing plants.[🐚59]

At even higher dune elevations, the dune shrub community gives way to the **dune thicket** community. This zone is characterized by relatively high sediment nitrogen and water levels, and features trees stunted by wind pruning and salt spray, with an understory of many of the plants found in the dune shrub.

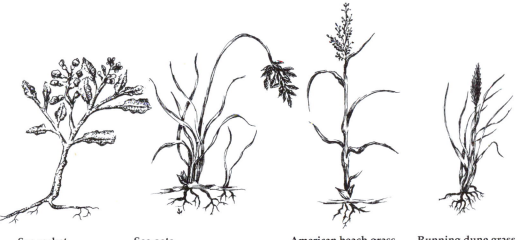

Sea rocket
Cakile edentula
Prostrate,
succulent annual

Sea oats
Uniola paniculata
Large, clonal dune stabilizer
of southern dunes; tolerant
of high winds, sand
abrasion, and salt spray

American beach grass
Ammophila breviligulata
Large clonal dune
stabilizer, but found
in northern dunes

Running dune grass
Panicum amarum
Smaller, less effective
dune stabilizer than
Uniola or *Ammophila*,
but also important

🐚58 Common dune pioneer plants responsible for colonizing and stabilizing primary dunes.

Wax myrtle
Myrica cerifera

Beach pea
Lathyrus japonicus

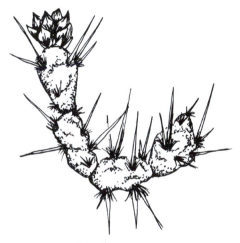

Prickly pear cactus
Opuntia compressa

🌰 59 Common shrub zone dune plants.

Sand dune succession is a powerful example of habitat modification or bioengineering by vascular plants leading to facilitated succession. The pioneer plants of dunes are clearly necessary to ameliorate the harsh physical conditions, which facilitates the colonization of the plants that dominate higher elevations. It is also evident that, over time, dune pioneers are competitively displaced by dune shrub plants in the milder habitats that the pioneers have created. In the next chapter, we will find that this classic facilitated succession is also common in salt marsh plant communities. Moreover, we will see that facilitated succession is common in marsh habitats for the same reason that it is common in sand dunes. In both of these shoreline habitats, physical and biotic stresses that potentially limit community development are ameliorated by plant colonization.

SUMMARY

Intertidal soft-sediment habitats, or tidal flats, occur wherever geological processes have left large amounts of sediments. Soft-substrate communities are energetically subsidized systems. Filter feeders and deposit feeders are at the base of the food webs, and both of these trophic groups rely primarily on external input for their food supply. Filter-feeding clams and worms sieve the water column for food, linking soft-substrate food webs to water column processes. Deposit feeders process shoreline sediments for food, and are ultimately dependent on the bacterial breakdown of organic debris for their nutrition. Thus, soft-substrate systems are tightly linked to water movement processes, not only because they dictate sediment supply and erosion patterns, but also because they determine food supply patterns.

Marine soft-substrate communities are profoundly influenced by the fact that the sediments they are built on are easily disturbed and are smaller than many of the organisms that dominate these habitats. Soft-sediment habitats are best envisioned as mosaics of patches in various stages of recovery. Storms can entirely rearrange the sediments on muddy and sandy shores. Moreover, the organisms that burrow and feed in soft sediments play a major role as biological disturbance agents by constantly reworking and processing sediments. This substrate reworking, through its effects on larval settlement and substrate productivity, is one of the most important ways in which organisms affect one another in soft-sediment communities.

Zonation in soft-sediment communities is very different from that on rocky shores. Since soft-sediment organisms can retreat into the substrate at low tides to limit heat and desiccation stresses, the physical stress gradients across these shores are less pronounced than on rocky shores. Biotic stresses on soft-substrate organisms are also different. Since most soft-substrate organisms use their habitat in three dimensions, competition for space is usually not severe. In addition, the bird, fish, and crab predators of soft-sediment organisms are very mobile, resulting in more homogeneous consumer pressure across these habitats relative to rocky shores.

Because soft-substrate habitats are so fundamentally influenced by substrate instability, they are strongly affected by organism bioengineering. Bioengineers of soft substrates stabilize sediments and prevent erosion, limit deposit feeding activity and predator effectiveness, and influence the recruitment and survivorship of virtually all soft-substrate organisms. Shallow-water seagrass beds, mussel and oyster beds, and marsh plant and sand dune communities are all dramatic examples of bioengineers in soft substrates.

SUGGESTIONS FOR FURTHER READING

Brown, A. C. and A. McLachlan. 1990. *Ecology of Sandy Shores*. Elsevier, New York. A comprehensive treatment of sandy beach ecology.

Gray, J. S. 1974. Animal-sediment relationships. *Oceanography and Marine Biology Annual Review* 12: 223–261. An old, but still pertinent, discussion of the close relationship between sediments and soft-sediment organisms.

Peterson, C. H. 1992. Intertidal zonation of marine invertebrates in sand and mud. *American Scientist* 70: 236–249.

Pethick, J. S. 1984. *An Introduction to Coastal Geomorphology*. Arnold Press, London.

Ranwell, D. S. 1971. *Ecology of Salt Marshes and Sand Dunes*. Chapman and Hall, London. A classic discussion of sand dune formation and ecology.

Reise, K. 1985. *Tide Flat Ecology: An Experimental Approach to Species Interactions*. Springer-Verlag, Berlin.

Wilson, W. H. 1991. Competition and predation in marine soft-sediment communities. *Annual Review of Ecology and Systematics* 21: 221–241.

Key Papers on Atlantic Coast Soft-Substrate Communities

Lipcius, R. N. and T. Hines. 1986. Variable functional responses of a marine predator in dissimilar homogeneous microhabitats. *Ecology* 67(5): 1361–1371.

Lopez, G. R. and J. S. Levinton. 1987. Ecology of deposit-feeding animals in marine sediments. *Quarterly Review of Biology* 2: 235–260.

Olafsson, E. B., C. W. Peterson and W. G. Ambrose. 1994. Does recruitment limitation structure populations and communities of macro-invertebrates in marine soft sediments: The relative significance of pre- and post-settlement processes. *Oceanography and Marine Biology Annual Review* 32: 65–109.

Orth, R. J. and K. A. Moore. 1983. Chesapeake Bay: An unprecedented decline in submerged aquatic vegetation. *Science* 222: 51–53.

Peterson, C. H. 1979. Predation, competitive exclusion, and diversity in the soft-sediment benthic communities of estuaries and lagoons. In *Ecological Processes in Coastal and Marine Systems*, R. J. Livingston (ed.), 233–263. Plenum Press, New York.

Peterson, C. H. 1982. Clam predation by whelks (*Busycon* spp.): Experimental tests of the importance of prey size, prey density, and seagrass cover. *Marine Biology* 66: 159–170.

Woodin, S. A. 1976. Adult-larval interactions in dense infaunal assemblages: Patterns of abundance. *Journal of Marine Research* 34: 25–41.

Woodin, S. A. 1981. Disturbance and community structure in a shallow water sand flat. *Ecology* 62: 1052–1066.

Chapter *7*

Salt Marsh Communities

*S*alt marshes are one of the most conspicuous shoreline features of the east coast of North America. They are characterized by striking zonation of highly productive, **halophytic** (salt-loving) plants that are capable of dealing with harsh physical conditions. Long considered ecologically important habitats, salt marshes are nursery grounds for fishes, shrimps, and crabs, feeding and nesting areas for birds and mammals, and buffers that protect coastal environments from erosion and nutrient loading. Despite their ecological importance, marshes are one of the shoreline habitats that have been most heavily disturbed by humans. Marshes were used for livestock grazing and heavily harvested for hay in the eighteenth and nineteenth centuries. Moreover, because marshes occur on protected shorelines, close to marine resources, they have attracted human settlement since colonial times, and have been filled and drained to promote development. The relative simplicity of marshes and their importance in the economy of coastal habitats has also attracted considerable scientific research. Their simplicity makes them amenable to experimental exploration of the forces that generate spatial patterns and influence the dynamics of nutrients and energy in natural communities.

In this chapter I examine the ecology of Atlantic coast salt marshes. I begin by introducing the plants that build marshes and give them their characteristic spatial structure. I then discuss the animal inhabitants of marsh plant habitats and their interactions and relationships with marsh vegetation.

MARSH DEVELOPMENT

Like coral reefs, salt marshes are autogenic structures built by the organisms that live in them. Salt marshes are common all along the Atlantic coast of North America, but are particularly abundant on the coasts of the Carolinas and Georgia, where sediment from continental erosion has accumulated for over 150 million years. Farther north, lack of suitable habitats and heavy winter ice scour limit salt marsh development. Not enough time has passed since Pleistocene ice sheets covered northern New England, scouring hard substrates of sediments, for soft sediments to accumulate in most northern habitats (Johnson 1920). In the Canadian Maritime provinces, Maine, and New Hampshire, winter ice continues to erode away the seaward borders of developing marshes annually, leaving them in a constant state of recovery. [🐚1]

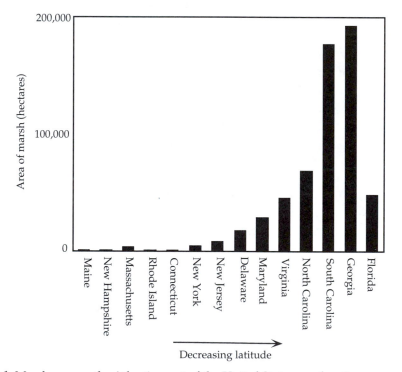

🐚1 Marsh area on the Atlantic coast of the United States as a function of latitude. (After Reimold 1977.)

Marshes develop in coastal habitats where slow-moving water leads to the accumulation of fine sediments that can be colonized by marsh plants. The mouths of estuaries and protected shores of barrier islands are common sites of sediment accumulation conducive to marsh development. On the Atlantic coast of North America, the smooth cordgrass, *Spartina alterniflora*, is the plant that colonizes the intertidal zone and initiates the marsh development process. Cordgrass is extremely tolerant of flooded soil conditions, and once established, it spreads vegetatively by sending rhizomes into surrounding sediments.[🐚2] The establishment of cordgrass dampens waves, slows water flow, and accelerates sedimentation. Its establishment on tidal flats also limits erosion as it binds sediments into a dense root mat. At the same time, algal and diatom films trap fine sediments on the marsh surface, and mussel

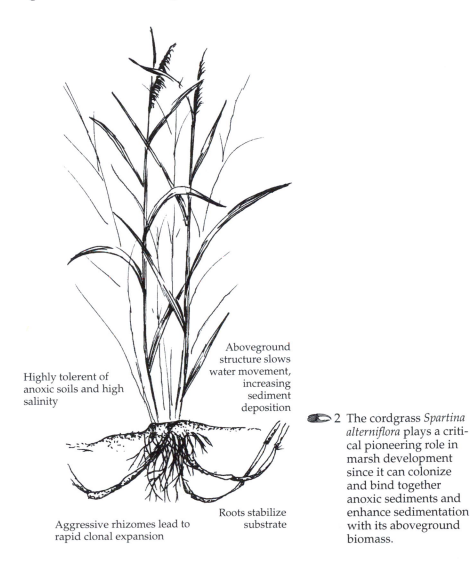

Highly tolerent of anoxic soils and high salinity

Aboveground structure slows water movement, increasing sediment deposition

Aggressive rhizomes lead to rapid clonal expansion

Roots stabilize substrate

🐚2 The cordgrass *Spartina alterniflora* plays a critical pioneering role in marsh development since it can colonize and bind together anoxic sediments and enhance sedimentation with its aboveground biomass.

and oyster beds enhance sediment deposition, in addition to binding sediments to the surface. Together, these processes lead to the seaward and vertical accretion of marshes.

Typical salt marsh development begins with cordgrass invading shallow-water sediments, so that young marshes consist almost entirely of intertidal cordgrass stands (Frey and Basan 1985). Over time, however, sediment accretion and binding lead to the vertical growth of marshes, and eventually to the development of **high marsh habitats** that are not flooded daily by tides. Marshes of intermediate age are characterized by complex drainage channels that drain tidal water from developing high marsh habitats. Continued sedimentation leads to mature marshes with few channels. These mature marshes are flooded daily by tides only on their seaward borders and on stream banks and the borders of drainage channels.[🐚3]

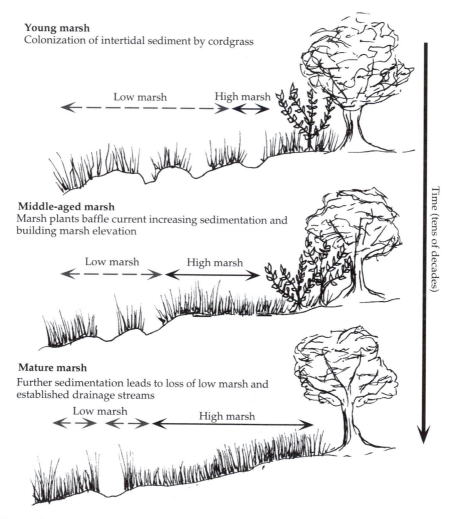

🐚3 Marsh maturation. (After Frey and Basan 1985.)

Surprisingly, core samples from northern marshes usually show evidence of the vertical accretion of belowground marsh plant debris, or **peat**, over terrestrial vegetation. This observation confused early naturalists, who hypothesized that marsh development resulted in shorelines sinking under their own weight, leading to marshes overgrowing submerged terrestrial vegetation. Recognition early in the twentieth century that global climate variation has led to changes in sea level completed our understanding of marsh development.[🐚4] During the most recent Ice Age glaciation, much of the earth's water was frozen in ice sheets, and sea level was dramatically lower. Marsh peat deposits found under 40 meters of water off the Atlantic coast are a dramatic illustration of how much sea level has changed in the past few thousand years (Rampino and Sanders 1981). Global warming over the last 4000 years has led to vertical marsh accretion that has kept pace with sea level changes, resulting in the growth of marshes over submerged terrestrial communities. A. C. Redfield's classic work (1965, 1971) on the history and development of Barnstable Marsh on Cape Cod is still one of the best understood examples of marsh development. By extensive coring, Redfield reconstructed the history of the marsh, and recognized the roles played by plant accretion, sedimentation, and fluctuating sea levels in the vertical and landward movement of marshes.[🐚5]

The development of southern marshes differs from that of northern marshes in that peat does not accumulate. The lack of peat deposition in southern marshes has never been carefully examined, but is probably a result of higher temperatures, which could enhance the decomposition of belowground debris, or the lack of strong seasonal accumulation of plant biomass

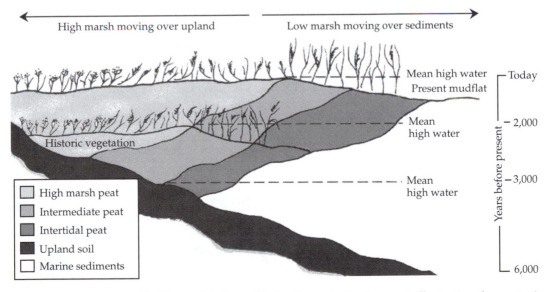

🐚4 A. C. Redfield's model of New England marsh development, illustrating the vertical growth of a marsh with increasing sea level. (After Redfield 1972.)

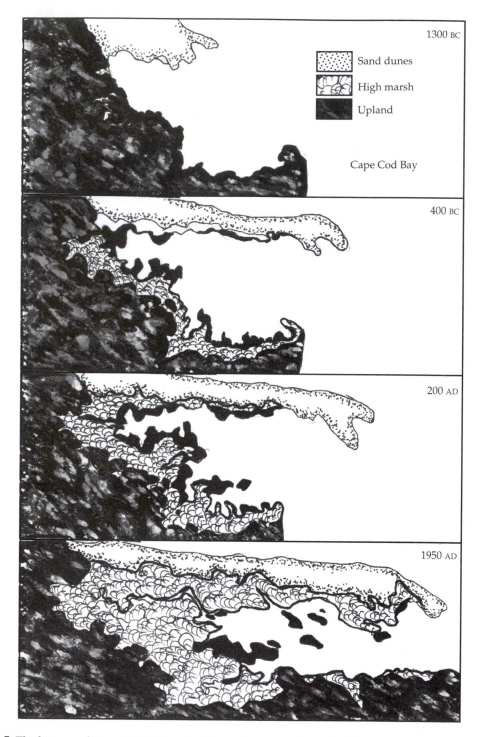

5 The history of Barnstable Marsh in Massachusetts, as described in
Redfield's classic coring study. Over time, the marsh has grown sig-
nificantly into Cape Cod Bay. (After Redfield 1972.)

belowground, which could leave less material belowground to decompose, or both. Without peat deposition, southern marshes accrete horizontally and vertically exclusively by sedimentation.

MARSH PLANT ZONATION

Plant zonation in salt marshes is usually conspicuous, with discrete, single-species bands of plants dominating specific elevations. Unvegetated free space is rare, as plants typically cover most available space as a result of vegetative growth. Along the Atlantic coast of North America, the general zonation of marsh plants is remarkably similar from Maine to Georgia, even though the plant species composition of marshes changes with latitude.

In New England, habitats that are flooded daily by tides (**low marsh habitats**) are dominated by the smooth cordgrass, *Spartina alterniflora*.[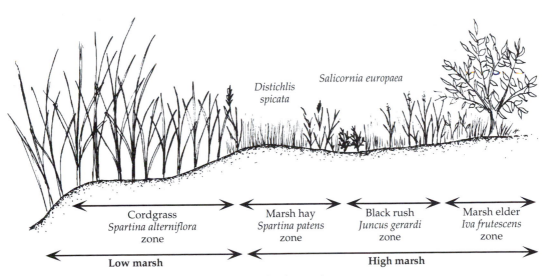6] On the seaward borders of these marshes, soils are well drained and support lush stands of cordgrass 1–2 meters tall. Immediately landward, at higher low marsh elevations, soil drainage is reduced, leading to anoxic soils and the accumulation of belowground plant debris (peat). At these elevations *Spartina alterniflora* is often less than 25 cm tall, and is referred to as short-form cordgrass.

The seaward borders of New England high marshes—those areas not flooded daily by tides—are generally dominated by marsh hay, *Spartina patens*, a perennial turf grass that grows to a height of 30 centimeters. Marsh hay is replaced at higher elevations by the black needle rush *Juncus gerardi*. Black needle rush, also a perennial with a dense turf morphology, grows to heights of 30–40 centimeters. It is less tolerant of flooding and salty soils than

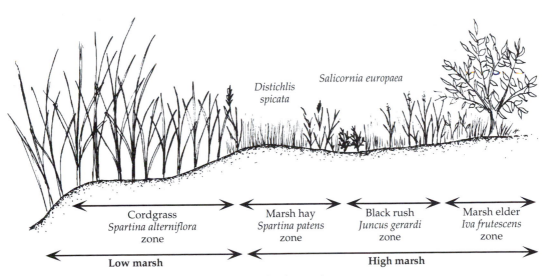

6 Plant zonation in a New England salt marsh.

marsh hay and is usually not found in habitats that are flooded more than 10 days a month (Bertness and Ellison 1987). At the terrestrial borders of New England marshes, *Juncus gerardi* is often replaced by the marsh elder, *Iva frutescens*, a woody composite. Marsh elders are extremely sensitive to tidal flooding and can survive only in well-drained soils (Bertness et al. 1992a). Consequently, they are found on the terrestrial borders of marshes, on the banks of well-drained tidal creeks, and on distinct mounds in the marsh up to 10 centimeters high, where they are flooded only by monthly tidal extremes. *Iva* forms hummock mounds with its roots, just as many wetland trees do to keep their roots out of water. The reed *Phragmites australis* is also common at the terrestrial borders of southern New England marshes, particularly when they have a history of disturbance (Dreyer and Niering 1995).

In addition to the zonal dominants in New England marsh plant communities, a number of less abundant plants are characteristic of disturbed habitats where the dense cover of perennial turf plants is reduced. The spikegrass, *Distichlis spicata*, is a common inhabitant of recently disturbed marsh habitats, which it invades vegetatively with belowground runners. Spikegrass is found at low densities among stands of marsh hay and black needle rush, and can also dominate high marsh habitats that are routinely covered with floating plant debris or have waterlogged soil.[⬤7]

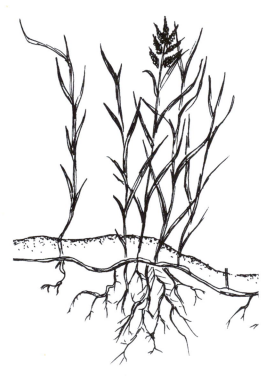

⬤7 The spikegrass, *Distichlis spicata*, is common in disturbed or physically harsh high marsh habitats. It is a rapid clonal colonizer and is tolerant of both high salinity and anoxic soils.

A number of solitary and annual plants are also common in New England high marsh habitats, usually in association with low plant cover. The glasswort *Salicornia europaea* is an annual that depends on disturbance-generated bare space in high marshes for its persistence. *Salicornia* has a distinctive morphology. It has no leaves, and is built of succulent photosynthetic stem segments. *Salicornia* seeds germinate in dense high marsh vegetation, but are usually outcompeted for light and die (Ellison 1987a). In unvegetated space, however, *Salicornia* seedlings thrive.[⬤8] A second annual glasswort, *Salicornia biglovii*, and a perennial glasswort, *Salicornia virginica*, are also found in southern New England marshes and become increasingly common at lower

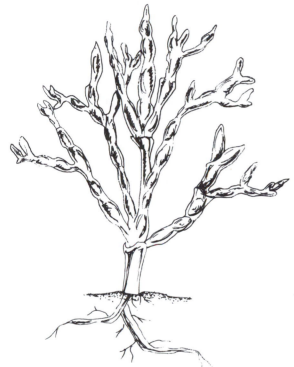

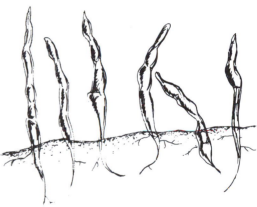

Low-density plants
- Highly branched
- Large
- High seed set and seed shadows
- Woody stems

High-density plants
- Unbranched
- Small
- Dependent on neighbors for support
- Not woody

⬤ 8 The morphology of the annual glasswort *Salicornia europaea* is highly neighbor-dependent. At low densities, it is tall, highly branched, and bushy. At high densities, however, it grows as unbranched cylinders, dependent on their neighbors for physical support. (See Ellison 1987b; Harley and Bertness 1996.)

latitudes. *Salicornia* species, in general, are extremely tolerant of high soil salinities, actually growing better in saline soils than in low-salinity soils, and are characteristic of high-salinity habitats. Other solitary marsh plants that are characteristic of disturbed patches in New England high marshes include the arrow plant (*Atriplex patula*), sea lavender (*Limonium nashii*), and marsh goldenrod (*Solidago sempervirens*). Like *Salicornia*, these species often appear to be dependent on bare space for seedling establishment. Unlike *Salicornia*, however, they often persist at low densities within dense stands of perennial marsh turfs (Miller and Egler 1950). Populations of these plants appear to be limited by low seed supply as well as low survivorship in dense clonal turf vegetation (Rand, unpublished data).[🐚 9]

Bill Niering and Scott Warren of Connecticut College (Niering and Warren 1980; Warren and Niering 1993) have shown that southern New England marsh vegetation patterns have changed over the past few centuries, particularly over the past 50 years. By resurveying marshes originally studied nearly 50 years ago (Miller and Egler 1950), they found that cordgrass has moved to higher elevations and that forbs, including *Trigloclin maritima* and *Gerardia maritima*, have invaded the black needle rush zone. They attribute these changes to contemporary increases in sea level, leading to increased waterlogging of marsh soils and thus to plants that can tolerate waterlogging. This hypothesis deserves more attention.

Marshes south of New England are also characterized by strong elevational zonation, but have a somewhat different suite of halophytic plants. Because of more intense year-round solar radiation and higher rates of evaporation, southern marshes are subjected to higher salinities. As a consequence, they are characterized by a more salt-tolerant plant assemblage (Pennings and Bertness 1998). High soil salinities in southern marshes also lead to the development of permanent salt pans (see below). Moreover, southern marshes are not subject to the strong seasonality of northern marshes. Plants in New England marshes grow from May to September, but in the late fall, after setting seeds, perennials reallocate their aboveground resources to belowground organs. This leaves only dead skeletons aboveground, which remain attached over the winter or collapse to be rafted away by tides. This seasonal production of plant debris results in large entangled mats of wrack, which can smother vegetation the following growing season or be exported from marshes in seasonal pulses. South of the Carolinas, the production of marsh grass may slow during the winter months, but it does not stop. In these southern marshes, aboveground plant debris is slowly produced year-round, without a dramatic seasonal pulse of dead plant material or investment in belowground plant parts.[🐚 10,11]

As in New England, the seaward border of most southern salt marshes is dominated by the cordgrass *Spartina alterniflora*. From Chesapeake Bay to Georgia, tall-form *Spartina alterniflora* dominates daily flooded low marsh habitats. A strikingly different feature of southern marshes is the extensive stands of short-form cordgrass dominating seaward borders of the high marsh, displacing the marsh hay (*Spartina patens*). In southern marshes,

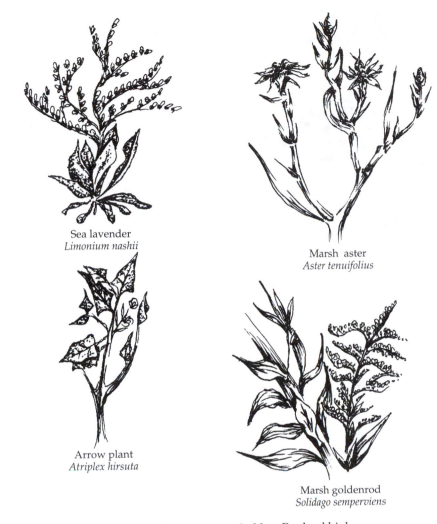

Sea lavender
Limonium nashii

Marsh aster
Aster tenuifolius

Arrow plant
Atriplex hirsuta

Marsh goldenrod
Solidago sempervirens

9 Some weedy, or fugitive, plants common in New England high marsh habitats. Their populations appear to be limited by either lack of safe sites (competition or stress-free space) or low seed supply. (Rand, unpublished.)

marsh hay is not a zonal dominant, but is occasionally found in high marsh patches, or in sand dune habitats on wave-swept beaches. Another major difference in southern marshes is that the black needle rush *Juncus romerianus* becomes the conspicuous dominant in high marsh habitats. Black needle rush-dominated salt marshes are particularly common south of Georgia on the Atlantic coast, as well as in much of the Gulf of Mexico, and are associated with lower tidal regimes and low salinities. In these areas dense, single-

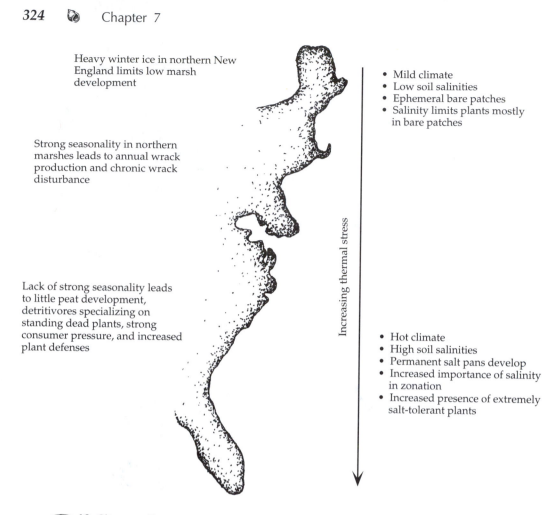

Heavy winter ice in northern New England limits low marsh development

Strong seasonality in northern marshes leads to annual wrack production and chronic wrack disturbance

Lack of strong seasonality leads to little peat development, detritivores specializing on standing dead plants, strong consumer pressure, and increased plant defenses

- Mild climate
- Low soil salinities
- Ephemeral bare patches
- Salinity limits plants mostly in bare patches

Increasing thermal stress

- Hot climate
- High soil salinities
- Permanent salt pans develop
- Increased importance of salinity in zonation
- Increased presence of extremely salt-tolerant plants

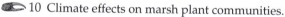 10 Climate effects on marsh plant communities.

species stands of *Juncus romerianus* displace both *Spartina patens* and *Juncus gerardi* from high marsh habitats.[🐚 12]

Large, permanent bare areas with sparse plant cover, called **salt pans,** are characteristic of southern high marshes, forming a distinct zone at intermediate elevations. Salt pans typically have extremely high soil salinities, usually in excess of 100 parts per thousand (more than three times higher than full-strength seawater) and often appear white due to the buildup of salt. Unlike bare areas in New England marshes, southern salt pans are permanent rather than transitional, successional habitats. Salt pans are usually inhabited only by extreme halophytes, if at all. They are believed to result from the excessive accumulation of soil salts where even the most salt-tolerant marsh plants cannot survive and become established to ameliorate these soil conditions. (Pethick 1974). Their elevational position in marsh habitats is a consequence of water movement patterns across the tidal gradient. Tidal flooding limits the

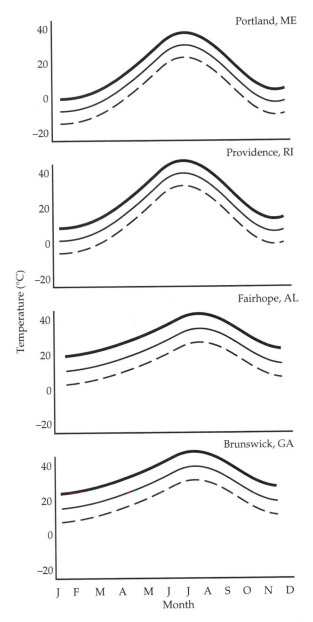

11 Mean daily high, low, and average air temperatures for four locations along the east coast of the United States. Data are from the National Weather Service and are 100-year daily average temperatures.

accumulation of soil salts at low marsh elevations, whereas runoff and rains do so at high elevations. At intermediate elevations, however, where salt pans occur, however, soil salts build up due to evaporation and lead to potentially lethal accumulations.[13] Northern marshes are not exposed to the high-

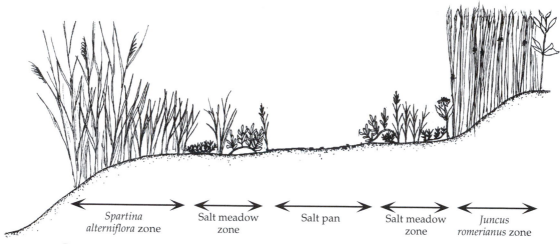

12 Generalized vertical zonation of southern salt marshes.

intensity solar radiation of southern marshes, and during northern winters, rain and cool weather ameliorate high soil salinities. The presence of salt pans in southern marshes is thus largely a consequence of climate.

    Salt pans on southern marshes have a characteristic suite of associated plants. Short black needle rush, probably stunted by the high salinities, is

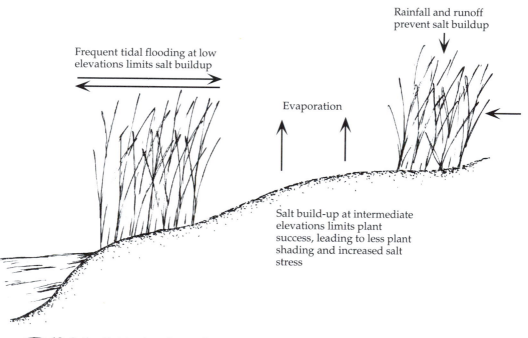

13 Soil salinities in salt marshes are typically highest at intermediate elevations, where the effects of evaporative salt accumulation are the strongest.

usually found on the edges of salt pans. Invading ramets of the spikegrass *Distichlis spicata* are also common on the edges of salt pans, as are a number of highly salt-tolerant succulents. The northern annual glasswort, *Salicornia europaea*, the perennial glasswort, *Salicornia virginica*, and the annual glasswort, *Salicornia bigelovii*, are common in southern salt pans, as is *Batis maritima*, a salt-tolerant succulent that creeps into pans on clonal runners. Most of these plants are restricted to the edges of pans. In the centers of southern salt pans, high soil salinities can entirely preclude even the hardiest plants. *Salicornia bigelovii* seems to be the plant that is the best able to cope with high soil salinities, and is often found growing in the center of large bare patches. *Distichlis spicata* and *Batis maritima* also often penetrate into the centers of salt pans, but cannot survive there without being physiologically connected to clone mates beyond the patch perimeter (Pennings and Callaway, unpublished data).[14]

Batis maritima

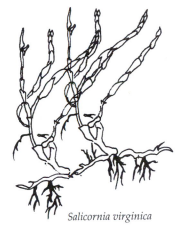

Salicornia virginica

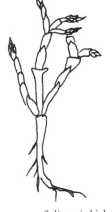

Salicornia biglovii

Distichlis spicata

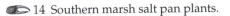14 Southern marsh salt pan plants.

Around salt pans in southern marshes, dense monocultures of *Juncus romerianus* are common. At the terrestrial borders of these marshes, flooded only by monthly tidal extremes, the salt shrub *Borrichia frutescens* often occurs. *Borrichia* is often mixed with the marsh elder, *Iva frutescens*, which dominates this habitat at higher latitudes.

The marsh plant communities that dominate soft-sediment shoreline habitats on the mid-Atlantic coast of North America are replaced in central Florida by mangroves (Chapman 1974). The northern distributional limit of mangroves in Florida appears to be set by periodic winter freezing. The floating seeds of red mangroves regularly colonize north of central Florida, but the mangroves die back every few years during harsh winters. The southern limit of marsh grasses is likely set by competitive exclusion by mangroves.

Elevational Size Gradients

Conspicuous intraspecific elevational gradients in plant stature are a common feature of marsh plant zonation in western Atlantic salt marshes. In general, these size gradients reflect the steep physical gradients of marsh environments. Size variation in the cordgrass *Spartina alterniflora* has received the most attention, even though equally dramatic elevational size gradients are found in the marsh elder (*Iva frutescens*) in New England and in the black needle rush (*Juncus romerianus*) and the salt shrub (*Borrichia frutescens*) in southern marshes.

Elevational size gradients are a particularly striking feature of cordgrass populations from New England to Florida (Valiela et al. 1978; Mendelssohn 1979b; Howes et al. 1981). At low marsh elevations, *Spartina alterniflora* is typically tall, often reaching heights of 2 meters and producing large amounts of biomass. At higher marsh elevations, cordgrass is often less than 30 centimeters tall, much denser, and often less than a quarter as productive as at lower elevations. The transition between tall- and short-form cordgrass zones can be abrupt, with tall- and short-form plants growing adjacent to each other, though they can also gradually grade into each other.

Most evidence suggests that cordgrass growth forms are largely a phenotypic response to local variation in abiotic soil (**edaphic**) conditions (see Anderson and Treshow 1980 for discussion). On the seaward borders of low marsh habitats, the soil is well drained and nutrients are plentiful. At higher elevations, soil drainage is limited, leading to anoxic soils composed of nutrient-poor peat, which may stunt plant growth.[15] Experimental additions of nutrients have shown that short-form plants are typically nitrogen-limited, but that tall-form plants are not (Valiela and Teal 1974; Mendelssohn 1979b). Adding nitrogen to the soil typically increases the stature and production of short-form cordgrass, but not tall-form. Artificially draining short-form cordgrass stands increases cordgrass height (King et al. 1982; Wiegert et al. 1983), while decreasing the drainage of tall-form cordgrass stands decreases it (Bertness 1984b).

Evidence for a genetic component in the two forms of cordgrass is mixed. Transplant studies in New England, in which tall- and short-form plants

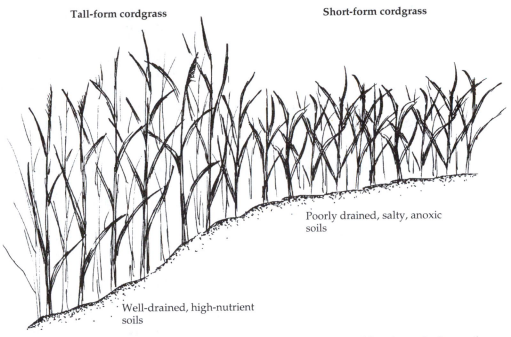

Tall-form cordgrass

Short-form cordgrass

Poorly drained, salty, anoxic soils

Well-drained, high-nutrient soils

15 The cordgrass *Spartina alterniflora* dominates the seaward borders of salt marshes from Canada to Florida, and typically occurs in two height forms. Tall-form cordgrass occurs nearest the water's edge, and short-form cordgrass in higher, less well drained habitats.

were grown under identical conditions, found no evidence that cordgrass growth forms have a genetic basis (Shea et al. 1975). However, the most thorough examination of a possible genetic basis for cordgrass height forms, done in Delaware, found a strong genetic component (Gallagher et al. 1988). In this study, rhizomes of tall- and short-form plants were transplanted to a common garden, and tall- and short-form plants remained distinct after over 5 years. This evidence strongly suggests that the height forms of *Spartina alterniflora* have a genetic basis. It may be that the cordgrass forms in the north, exposed to routine ice disturbance, are largely phenotypic responses to physical conditions. In southern marshes, however, where short-form plants have invaded high marsh habitats, the short-statured morphology may have become genetically specialized. In the only other study to examine a potential genetic basis for salt marsh plant height forms, Antlfinger and Dunn (1979) examined tall and short forms of *Borrichia frutescens* in Georgia, and found no evidence of genetic differentiation.

PHYSICAL STRESSES ON MARSH PLANTS

Although salt marshes are highly productive and superficially lush, they are extremely stressful physical habitats for vascular plants. The plants that dominate salt marshes are specialized to deal with the problems associated with waterlogged, saline soils. Most vascular plants would die under typical salt marsh soil conditions.

Tidal flooding of salt marshes creates waterlogged soils, particularly at low, frequently flooded elevations and in poorly drained sediments. The physiological problem with waterlogged soils is that they become depleted of oxygen. Marsh plants produce oxygen as a by-product of photosynthesis aboveground. Their belowground tissues, however, consume oxygen through metabolic activities faster than the soil oxygen levels can be replenished, because of the lower diffusion rates into waterlogged as compared with dry soil. Without oxygen, plant roots are unable to provide nutrients to aboveground plant structures.

Wetland plants have both architectural and metabolic adaptations to cope with waterlogged soils. Architecturally, plants that live in waterlogged soils usually have roots near the soil surface to facilitate root oxygenation. [🖝16] Smooth cordgrass, which dominates frequently flooded salt marsh habitats on the Atlantic coast of North America, has surface roots in the top 2–3 centimeters of the marsh surface to oxygenate deeper root structures (Anderson 1974). Mangroves, which dominate tropical wetlands, have aboveground prop roots, or "pneumatophores," that oxygenate belowground roots. Plants that dominate waterlogged soils also typically have well-developed **aerenchyma** tissue, which transports oxygen to belowground structures. The aerenchyma tissue forms a system of air passages (conduits) extending from the leaves to belowground plant parts that allows oxygen to passively diffuse to belowground roots (Armstrong 1979). [🖝17] The marsh plants that are common in waterlogged low marsh soils, including *Spartina alterniflora*, *Distichlis spicata*, and *Juncus roemerianus*, all have well-developed aerenchyma tissue, and many marsh plants respond to waterlogged soils by increasing the size of this tissue (Seliskar 1983). Marsh plants that live in drier high marsh habitats have less developed aerenchyma tissue than those found in low marsh habitats. In addition to these structural modifications, many marsh plants also have well-developed anaerobic metabolic pathways, which reduce their dependence on root oxygenation (Mendelssohn et al. 1981; King et al. 1982).

Salt stress is the second serious physical problem for salt marsh plants. High soil salinities osmotically draw water from roots, making salt marshes physiologically dry habitats. Consequently, marsh plants are similar to desert plants in terms of the mechanisms they use to conserve water.

Salt marsh plants have several strategies to deal with high soil salinities. To minimize osmotic water loss through the roots, most marsh plant tissues contain high concentrations of solutes. This ensures that their vascular systems will continue to draw fluid from the soil and distribute nutrients to aboveground body parts. Since high salt concentrations can be detrimental to plant metabolism, however, plants often use organic solutes, not salt, to maintain these high osmotic vascular pressures (Flowers et al. 1977, 1986). [🖝18]

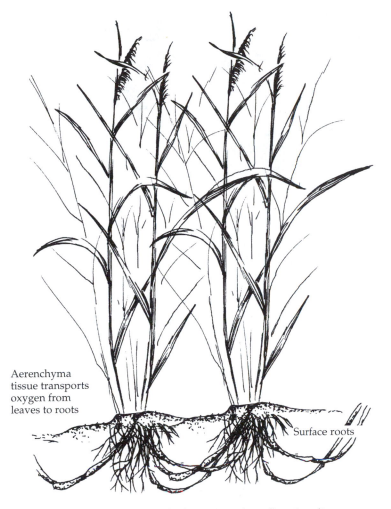

Aerenchyma
tissue transports
oxygen from
leaves to roots

Surface roots

16 *Spartina alterniflora* has several adaptations that allow it to live
in waterlogged, anoxic habitats, including surface roots in oxygenated soil, anaerobic metabolic pathways, and aerenchyma
tissue that delivers oxygen to belowground parts.

To manage tissue salt concentrations, many salt marsh plants have **salt
glands,** specialized leaf cells that concentrate and excrete salt. Examples of
marsh plants with well-developed salt glands include the cordgrass (*Spartina
alterniflora*), spikegrass (*Distichlis spicata*), and arrow plant (*Atriplex patula*).
These plants deal with high soil salinities by passively taking up high-salinity
fluid from the rhizosphere (root zone), removing excess salts with their salt
glands, and balancing their systems osmotically with organic solutes.
Examination of the leaves of these plants in the field reveals salt crystals
demarcating the locations of the salt glands.

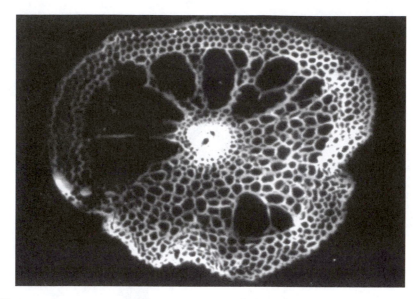

17 Aerenchyma tissue of marsh plants found in waterlogged soils is a system of air passages allowing movement of oxygen from the leaves to roots. (Photograph by Irv Mendelssohn.)

Another common strategy for dealing with saline soils is the maintenance of high tissue water concentrations as a buffer against osmotic water loss. This **succulence** strategy is similar to that of desert plants, but is complicated in salt marsh plants by the need to maintain high internal osmotic concentrations. Marsh succulents, such as *Salicornia*, *Batis*, and *Suaeda*, accomplish this by isolating stored water in enlarged storage cells, while maintaining high osmotic concentrations in their vascular systems (Flowers et al. 1977). Marsh succulents have other conspicuous morphological modifications to reduce water loss, including reduced leaf surface area and photosynthetic stems. *Salicornia* has entirely lost its leaves and is essentially a low surface area photosynthetic stem. *Batis* and *Suaeda* have needlelike leaves with low surface areas and photosynthetic stems. To minimize water loss, succulents have only a very few, recessed stomata, which limit transpiration (Flowers et al. 1977).

COMPETITION AMONG MARSH PLANTS

Marsh plant communities worldwide are characterized by strong intertidal height zonation of extensive single-species stands of halophytic plants (Chapman 1974). While waterlogged and saline soils clearly play an important role in this striking pattern, plant competition is equally important in generating marsh plant zonation. That harsh physical conditions at low marsh elevations preclude plants that are not physiologically equipped to deal with them is widely appreciated. However, the physical stresses at low

Salicornia, Suaeda, Batis
- Reduced leaf surface area
- Photosynthetic stems
- Recessed stomata and reduced transpiration
- Water storage in tissues
- High organic solvent concentrations to maintain osmotic balance

Salicornia   *Suaeda*   *Batis*

Succulence: Store water, reduce water loss, maintain high osmotic pressure to draw water from soil

Spartina alterniflora, Spartina patens, Distichlis spicata
- Salt excretion
- High solvent or salt concentrations in tissues to draw fluid from soil
- High transpiration rates

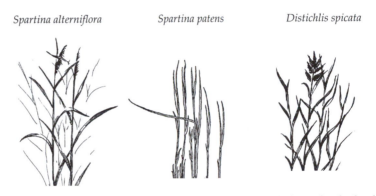

Spartina alterniflora   *Spartina patens*   *Distichlis spicata*

Salt glands: Maintain high osmotic pressure, excrete salt through salt glands

18 Mechanisms of salt tolerance in common marsh plants.

marsh elevations do not explain why marsh plants are usually restricted to marsh habitats, nor do they explain the distinct upper elevational limits of most marsh plants. What limits the upper elevational boundaries of marsh plants where physical conditions are less severe? In general, the answer to this question is plant competition.

The role played by competition in setting marsh plant species boundaries is nicely illustrated in the sharp zonation between the cordgrass *Spartina alterniflora* and salt marsh hay *Spartina patens* on the Atlantic coast of North America. [19] Throughout its range, *Spartina alterniflora* dominates low marsh habitats. Since other marsh plants are not equipped to deal with low marsh physical conditions, cordgrass flourishes at low marsh elevations without competitors. Throughout much of its geographic range, low marsh monocultures of cordgrass are bordered on their terrestrial margins by dense, single-species stands of marsh hay. Marsh hay is not well equipped to oxygenate its roots in anoxic soils (Gleason 1980; Mendelssohn 1979b), and dies if transplanted to low marsh habitats (Bertness 1991b). Thus, marsh hay is precluded from living at low eleva-

Cordgrass is capable of living in the high
marsh, but is precluded by the competitive
dominance of marsh hay.

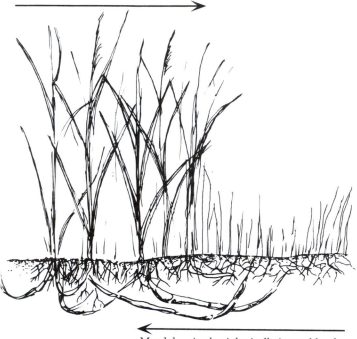

Marsh hay is physiologically incapable of
surviving in the low marsh due to physical
stress

19 The interface between the low marsh cordgrass zone and the high
marsh marsh hay zone. This boundary is maintained by the competitive dominance of the dense marsh hay turf in the high marsh
and the inability of marsh hay to tolerate low marsh physical conditions. (After Bertness 1991b.)

tions by physical stress alone. In contrast, cordgrass is capable of living under
the less stressful conditions of the high marsh. Cordgrass commonly invades
high marsh habitats after disturbances (Miller and Egler 1950; Bertness and
Ellison 1987), and when transplanted to the high marsh without neighbors, does
as well or better than it does at low marsh elevations (Bertness 1991b). Yet cordgrass does not invade vegetated high marsh habitats, and is outcompeted and
dies if transplanted to high marsh habitats with neighbors (Bertness 1991b).
Clearly, the terrestrial border of cordgrass is set not by physical limitations, but
by the competitive dominance of high marsh plants.

    In general, the lower intertidal limits of marsh plant species are set by
physical stresses, whereas the upper intertidal limits are set by plant competition (Bertness 1991a,b, 1992; Snow and Vince 1984). The generality of this
pattern is best seen in New England, where most major marsh plant interac-

tions have been quantified. In New England marshes, *Iva* outcompetes black needle rush at the terrestrial borders of marshes, displacing black needle rush to lower high marsh elevations. In turn, black needle rush competitively displaces marsh hay to lower high marsh elevations, and marsh hay displaces cordgrass to low marsh habitats. The ability to handle harsh edaphic conditions is most often seen in plants that dominate lower elevations, whereas the plants that dominate higher elevations do so because they are superior competitors.[20] In southern marshes, however, elevated soil salinities resulting from increased solar radiation may limit plants at high elevations, and competitive dominants may displace subordinates to higher, saltier habitats, rather than lower, wetter ones (Pennings and Callaway 1992). The processes dictating the zonation of northern and southern marshes seem to vary predictably with latitude and may be tightly linked to climate. (See Keddy (1989) for a general discussion of the common trade-off between competitive ability and tolerance of physical stress in plants.)

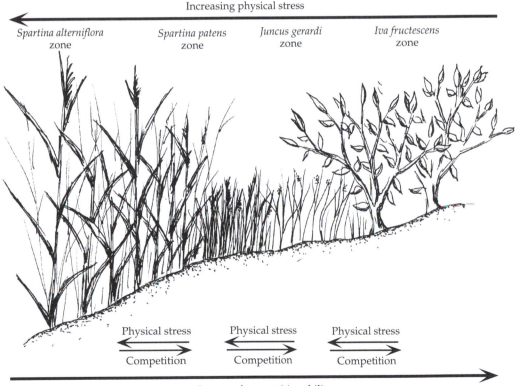

20 The elevational zonation of plants in southern New England is the product of competitively dominant plants displacing competitive subordinates to lower elevations and the inability of those competitive dominants to live at more stressful lower elevations.

In addition to influencing the elevational distributions of marsh plant species, competition also limits seedling establishment in the dense vegetation of many marshes. In northern marshes, where seasonal cordgrass production generates floating wrack that kills underlying vegetation, seedling establishment is often limited to transient, disturbance-generated bare space (Bertness and Ellison 1987). Seedlings of many marsh species routinely germinate in the dense canopies of zonal dominants, but die in competition with the much larger turfs (Ellison 1987a; Shumway and Bertness 1992).

Competitive Dominance

Competitive dominance among marsh plant species is largely the product of variation in growth morphology and size. Most plants that dominate salt marsh landscapes are perennial grasses and rushes that form extensive monocultures through vegetative growth. Among clonal perennial species, those that grow as dense turfs are generally competitively dominant to species that grow with more diffuse morphologies. Harper (1977) has termed dense turf morphologies **phalanx** growth forms and more diffuse runner morphologies **guerrilla** growth forms. Phalanx clones invade available space both above and below the ground as a dense mat of roots, rhizomes, and aboveground tillers (stems).[🐚21] In general, they are successful competitors for space and belowground resources. In contrast, plants with guerrilla morphologies invest heavily in belowground rhizomes and are capable of rapid expansion, but are poor competitors for space and belowground resources.

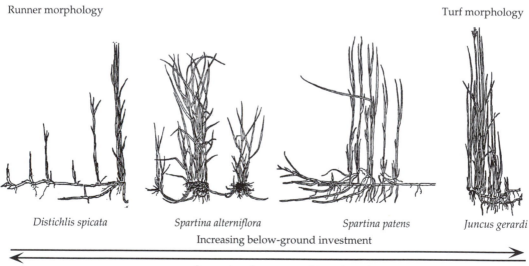

Runner morphology

Turf morphology

Distichlis spicata *Spartina alterniflora* *Spartina patens* *Juncus gerardi*

Increasing below-ground investment

Increasing above-ground investment

🐚 21 Plant clonal growth forms.

The effect of clonal morphology on marsh plant interactions is illustrated by the dominance of marsh hay and rushes in the high marsh over the spikegrass *Distichlis spicata*. Whereas marsh hay (*Spartina patens*) and rushes (*Juncus gerardi* and *Juncus roemarianus*) have dense phalanx morphologies, spikegrass has a guerrilla morphology and a weedy or fugitive distribution across Atlantic coast marshes. *Distichlis* invests heavily in belowground runners and sparingly in roots, spreads rapidly through high marsh habitats by means of clonal growth, and is common in recently disturbed habitats and persistent salt pans with low vegetation cover. However, it is restricted to ephemeral or marginal habitats by the dense turfs of other plants that dominate high marsh habitats: *Spartina patens* and *Juncus gerardi* in northern marshes, and *Juncus roemerianus* in southern marshes.

In a similar fashion, the competitive dominance of *Spartina patens* over *Spartina alterniflora*, described earlier, results from morphological differences. Cordgrass grows as diffuse culms (clumps) of tillers separated by belowground runners. In contrast, marsh hay grows as a dense turf of roots, rhizomes, and tillers. The dense turf morphology of marsh hay prevents cordgrass invasion, and chokes out less dense plants in mixtures.

Clonal Organization

One of the conspicuous features of most marsh plant communities is the dominance of clonal plants. Clonal, or modular, plants are built of a series of plant modules that are produced by the vegetative growth of rhizomes. All of the genetically identical units (termed **ramets**) produced by a clonal organism are referred to collectively as a **genet**. In most clonal plants, individual ramets remain physiologically connected to their clone mates. In contrast, solitary plants are produced sexually through seeds, and neighbors are both genetically distinct and physiologically independent.

Clonal plants have two major advantages over solitary plants in marsh environments. Physiological integration among the ramets of a clonal plant allows them to share resources and thus buffer one another from physical and biological stresses. In marshes, clone mates can share water and fixed carbon, which allows them to buffer one another from salt and nutrient stresses (Shumway 1995). Clonal plants can also spread through marsh habitats vegetatively without the need for sexual reproduction. This allows them to colonize and monopolize habitats where successful seedling establishment is difficult (Pennings and Callaway, unpublished data).[🫘22]

POSITIVE FEEDBACK IN MARSH PLANT COMMUNITIES

Whereas physical stresses and competition play dominant roles in generating the spatial structure and dynamics of salt marsh plant communities, positive interactions among plants are also a pervasive feature of these communities. Positive interactions are particularly common under harsh edaphic conditions because marsh plant neighbors can ameliorate the effects of both anoxic and salty soils.

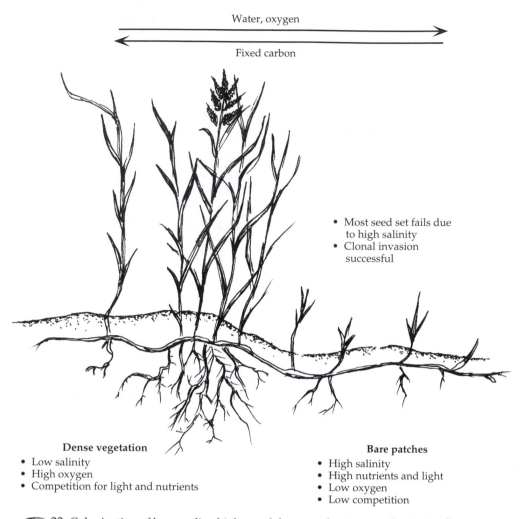

Water, oxygen

Fixed carbon

- Most seed set fails due to high salinity
- Clonal invasion successful

Dense vegetation
- Low salinity
- High oxygen
- Competition for light and nutrients

Bare patches
- High salinity
- High nutrients and light
- Low oxygen
- Low competition

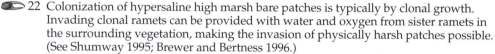 22 Colonization of hypersaline high marsh bare patches is typically by clonal growth. Invading clonal ramets can be provided with water and oxygen from sister ramets in the surrounding vegetation, making the invasion of physically harsh patches possible. (See Shumway 1995; Brewer and Bertness 1996.)

Positive interactions among marsh plant neighbors are common at low marsh elevations as the product of neighbor amelioration of low soil oxygen levels (Howes et al. 1981, 1986; Bertness 1991b). As already discussed, most marsh plants capable of living in anoxic soils transport oxygen to their roots through aerenchyma tissue. As a consequence of the leakage of oxygen from roots, the rhizosphere surrounding roots is oxygenated (Teal and Kanwisher 1966; Howes et al. 1981; Armstrong 1979; but see Howes and Teal 1994). Rhizosphere oxygenation has been shown in the field (Howes et al. 1986; Bertness 1991a) and greenhouse (Shat 1984) to lead to positive plant associations in anoxic habitats.

The success of cordgrass in muddy, anoxic soils results largely from group benefits. Cordgrass seedlings and small plants have higher growth and survival rates in the high marsh in the absence of competitors than they do in the low marsh (Bertness 1991b). However, large cordgrass stands are more productive in the low marsh than in the high marsh. This size-dependent success in the low marsh comes from the group benefits of rhizosphere oxygenation. Cordgrass seedlings and small plants are unable to oxygenate low marsh soils successfully, but large cordgrass stands have a communal effect on soil oxygen levels. This allows them to take advantage of the high nitrogen levels found at low marsh elevations, resulting in enhanced plant growth. This growth leads to further rhizosphere oxygenation, which in turn enhances growth, until nutrients limit plant growth. Thus, the interaction between rhizosphere oxygenation and nitrogen availability in low marsh habitats leads to a positive feedback loop resulting in highly productive cordgrass stands (Howes et al. 1981, 1986).[🫘23]

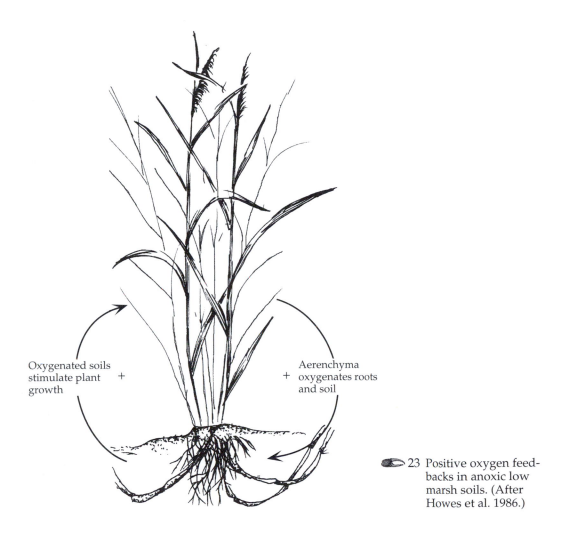

Oxygenated soils stimulate plant growth   +

+   Aerenchyma oxygenates roots and soil

🫘 23 Positive oxygen feedbacks in anoxic low marsh soils. (After Howes et al. 1986.)

In contrast, positive associations among plants at higher elevations are more commonly due to the reduction of salt stress. Soil salinities generally decrease with marsh elevation in dense vegetation, but are highest at intermediate elevations in the absence of vegetation because of evaporative water loss and salt accumulation. Frequent tidal flooding limits salt accumulation at low elevations, whereas at the highest elevations, rain and runoff limit salt buildup. At intermediate elevations, plant cover plays a major role in limiting soil salinities. At these elevations, plant cover shades the soil, limiting the evaporation of surface water and the resultant accumulation of salt. A major consequence of this feedback between plant cover and soil conditions is that plant neighbors often buffer one another from salinity stress.[⬅24]

Marsh elders are relatively intolerant of high soil salinities, and as a consequence, both seedling recruitment and the persistence of adult *Iva* in salty habitats are dependent on neighbor amelioration of soil conditions. Most *Iva* seeds fall and germinate within a meter of the parent plant. Shading by dense perennial turfs or adult *Iva*, however, usually limits seedling success in undisturbed vegetation; the seedlings receive little light and usually die within a month. Tidally transported wrack, however, often kills the turf vegetation underlying adult *Iva*. This releases *Iva* seedlings from competition with the surrounding turfs, and seedling survivorship increases dramatically. High *Iva* seedling survivorship in these bare patches, however, is often dependent on positive neighbor associations. *Iva* seedlings that germinate in large bare

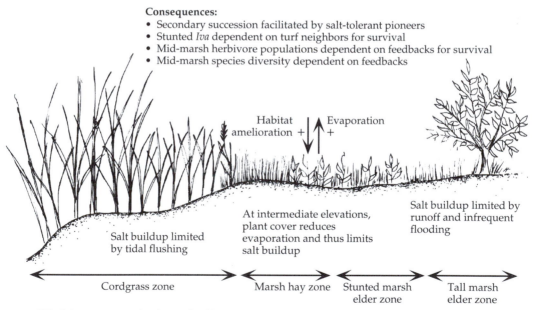

Consequences:
- Secondary succession facilitated by salt-tolerant pioneers
- Stunted *Iva* dependent on turf neighbors for survival
- Mid-marsh herbivore populations dependent on feedbacks for survival
- Mid-marsh species diversity dependent on feedbacks

Habitat amelioration + ↓ ↑ + Evaporation

Salt buildup limited by tidal flushing

At intermediate elevations, plant cover reduces evaporation and thus limits salt buildup

Salt buildup limited by runoff and infrequent flooding

Cordgrass zone   Marsh hay zone   Stunted marsh elder zone   Tall marsh elder zone

⬅24 Positive soil salinity feedbacks in New England salt marshes. After Bertness and Shumway 1993, Bertness and Hacker 1995, Hacker and Bertness 1996, 1997, Hacker and Gaines 1997.

spaces in the lower *Iva* zone without plant neighbors die due to the accumulation of salt in unshaded marsh soils. The presence of adult nurse plants or dense concentrations of seedling neighbors can shade the marsh surface, limit soil salinity increases, and enhance seedling survival. In this way, the seed recruitment of *Iva* and other marsh plants can be controlled by positive interactions within and among plant species.[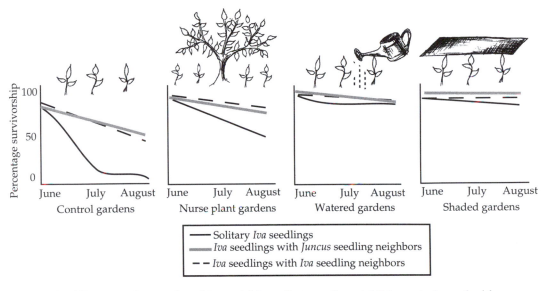25]

Positive interactions among adults also influence marsh elder zonation. On the seaward border of the *Iva* zone, short, stunted adult *Iva* occur at low densities surrounded by dense perennial turfs. Removal of the turf neighbors of *Iva* in this zone leads to decreased soil oxygen availability, increased soil salinities, and the death of established adult marsh elders. Adult *Iva* cannot survive without neighbors on the seaward border of the *Iva* zone (Bertness and Hacker 1994). However, neighbors compete rather than cooperate on the uppermost terrestrial border of the *Iva* zone. At higher elevations, soils are well oxygenated and have low salinities with or without plants present, and removal of neighbors leads to competitive release and increased *Iva* growth. Thus, the nature of the interaction between *Iva* and its neighbors changes with environmental stress levels. In stressful habitats of the *Iva* zone, neighbors buffer both seedlings and adults from high soil salinities and low oxygen levels and increase *Iva* growth and survivorship. In contrast, at higher elevations, where edaphic conditions are not potentially limiting, neighbors compete for light and nutrients.[26]

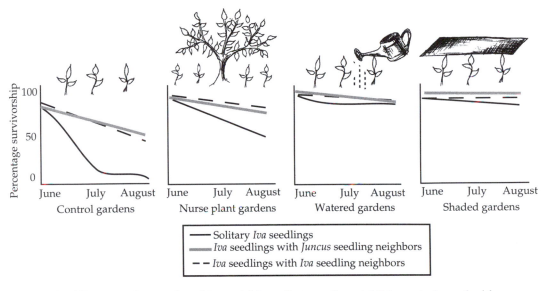

25 Nurse plant and positive neighbor effects on the establishment of marsh elder seedlings due to neighbor amelioration of high soil salinities. These group benefits disappear when plants are watered or shaded to reduce soil salinities. (After Bertness and Yeh 1994.)

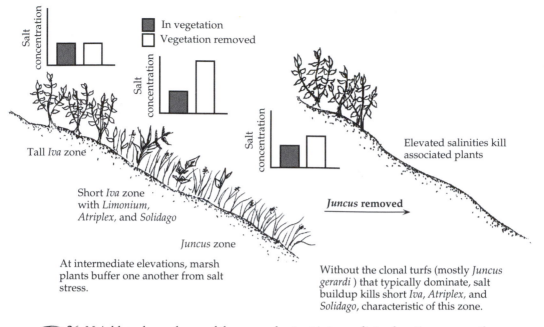

Salt concentration

■ In vegetation
□ Vegetation removed

Salt concentration

Salt concentration

Salt concentration

Tall *Iva* zone

Short *Iva* zone
with *Limonium,*
Atriplex, and *Solidago*

Juncus zone

Juncus **removed**

Elevated salinities kill
associated plants

At intermediate elevations, marsh
plants buffer one another from salt
stress.

Without the clonal turfs (mostly *Juncus
gerardi*) that typically dominate, salt
buildup kills short *Iva, Atriplex,* and
Solidago, characteristic of this zone.

🫘 26 Neighbor dependence of the rarer plants at intermediate elevations on southern
New England marshes. Bar graphs show the effect of vegetation cover on soil salin-
ity as a function of marsh elevation. (After Bertness and Hacker 1994; Hacker and
Bertness 1998.)

The occurrence of permanent salt pans in low-latitude salt marshes may
be a result of salinity buildup to the point at which plants can no longer provide sufficient habitat amelioration to lower salinities and permit plant survival (Pennings and Bertness 1998). Salt pans at high intertidal elevations are
also characteristic of low-latitude Pacific coast marshes and tropical mangrove forests worldwide.

DISTURBANCE IN MARSH PLANT COMMUNITIES

Natural disturbances play an important role in the dynamics of marsh plant
communities, particularly at northern latitudes, where they are seasonally
predictable and intense. In Canadian and New England marshes, harsh winters can cause ice damage, and large quantities of seasonally produced plant
debris can smother and kill marsh plants each year. Southern marshes, in
contrast, are not exposed to either of these predictable annual disturbances,
and are not as conspicuously affected by disturbance.

Winter ice in New England can severely damage marshes by destroying
large areas of established vegetation and rafting intact marsh turfs to new
locations. In southern New England, ice sheets up to 30 centimenters thick
can cover intertidal marsh habitats during severe winters (Redfield 1972).

These ice sheets can include waterlogged low marsh banks, which may be lifted free in large, intact portions during high tides (Redfield 1971). If rafted to the high marsh, they can kill underlying vegetation. If rafted to low elevations, they can nucleate new marsh growth, or kill the rafted plants if they are deposited in deep water.

Ice damage generates strong biogeographic patterns in salt marsh structure and development. In northern New England, ice damage is so severe that it limits low marsh development (Johnson 1920). Low marsh habitats in Maine and New Hampshire are continually in a state of recovery, with cordgrass colonizing lower elevations that have been scoured. In southern New England, ice damage is less severe, and while often important in low marsh habitats, not always conspicuous. South of New England, ice disturbance is not an important force in marshes. As a consequence, sediment deposition on the seaward edge of the marsh often leads to the development of characteristic **levies**—also known as **berms**—or elevated mounds. This is particularly the case in the marshes of Georgia, which have a higher tidal range (about 3 meters) than other southern marshes (Wiegert and Freeman 1990).[27]

A second chronic disturbance in northern marshes is the burial of vegetation by rafts of floating dead plant material, or **wrack** (Reidenbaugh and Banta 1982; Hartman et al. 1983; Bertness and Ellison 1987). In northern marshes, wrack mats can often be 100–1000 square meters in area and 10–30

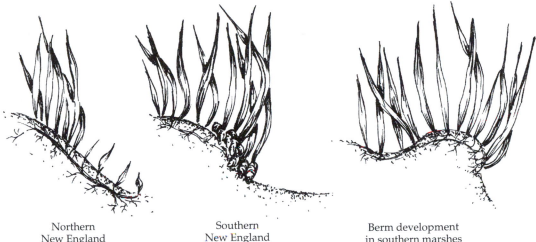

Northern
New England
(Constant recovery
from ice disturbance)

Southern
New England

Berm development
in southern marshes

27 Latitudinal variation in the seaward edges of marshes. In northern New England, winter ice regularly destroys low marsh plants, and cordgrass is continually reinvading. In southern New England, ice damage is reduced, and low marsh borders are stabilized by ribbed mussels. In southern marshes without ice disturbance, sediment buildup leads to the development of elevated mounds, or berms, on the seaward edge of marsh vegetation.

centimeters thick. In the early spring, extreme high tides can move wrack to the high marsh, where it can be stranded long enough to kill underlying vegetation. Wrack that remains stationary for most of a growing season generally kills the underlying vegetation, leaving an unvegetated bare patch that triggers secondary succession. Unlike the effects of ice disturbance, wrack disturbance is most severe in the high marsh. At low marsh elevations, wrack is constantly moved by tides, and is less likely to stay in place long enough to kill vegetation.

Wrack disturbance does occur in southern marshes, but is much more subtle. Without the seasonal production of dead aboveground material, southern marsh wrack is produced slowly, as individual plants die back and break off. In the winter in Georgia, this can lead to light wrack mats only a few centimeters thick and usually less than 10 square meters in size. These wrack mats can lead to some mortality of underlying vegetation, but rarely kill large areas, and their effects on the plant community are minimal. In fact, Steve Pennings of the University of Georgia has recently found that the light wrack common in southern marshes may often benefit vegetation by shading the soil and limiting surface evaporation and the buildup of surface salts (Pennings and Richards, unpublished).

SECONDARY SUCCESSION IN MARSH PLANT COMMUNITIES

Disturbance-generated bare patches in salt marsh plant communities can trigger plant recolonization and secondary succession. In New England, secondary succession follows a relatively predictable time course. In low marsh habitats, where physical stresses limit the invasion or persistence of most marsh plants, secondary succession usually involves the clonal reinvasion of cordgrass, the only plant capable of thriving in these habitats (Hartman et al. 1983). In high marsh habitats, however, secondary succession is a more complicated process involving the more numerous plant species that live there.

Initial invasion of high marsh bare patches is by fugitive plants, which are uncommon in undisturbed vegetation. These fugitive plants invade bare patches either by seed dispersal or belowground rhizomes. The most common seed-dispersed fugitive in New England bare patches is the annual glasswort *Salicornia europaea*. *Salicornia* is competitively subordinate to the clonal turfs that dominate New England marsh landscapes, and depends on disturbance-generated bare space for its persistence (Ellison 1987b). *Salicornia* has an extremely plastic morphology with both high and low density forms. It grows up to 45 centimeters tall as a highly branched shrub with woody stems at low densities, and as slender, etiolated, unbranched individuals resembling grass blades at high densities. At high densities, *Salicornia* grow to less than one-tenth the size of uncrowded individuals (Ellison 1987a), and the top-heavy seedlings invest so little in structural support that they become entirely dependent on their neighbors for support (Harley and Bertness 1996).

Salicornia skeletons with seeds still attached are common in the floating wrack that forms marsh bare patches. Moreover, *Salicornia* seeds are covered with hairs that make them stick to wrack. Consequently, *Salicornia* seeds are commonly transported with wrack to new patches. Seeds of other marsh fugitives such as *Atriplex, Suaeda*, and *Limonium* are also commonly wrack-dispersed, and their seedlings are commonly found in bare patches. Whether or not these seedlings survive, however, is dependent on their ability to deal with the salt and anoxia stresses common in marsh bare patches (Hacker and Bertness 1999).

Other common patch-dependent fugitives rely entirely on vegetative colonization. The spikegrass, *Distichlis*, is a characteristic clonally invading fugitive plant. It commonly occurs at low densities among the numerically dominant perennial turfs in high marsh habitats, but dramatically increases in abundance following wrack disturbance. Spikegrass dominates many recently disturbed habitats because it is able to rapidly colonize bare patches and is resistant to disturbance mortality. Spikegrass stays alive when buried under wrack because its sharp tillers can penetrate wrack cover and because aboveground ramets may sustain ramets buried under wrack. The runner morphology and clonal growth form of spikegrass also allows rapid growth and spreading to suitable habitats. Recently disturbed bare patches may contain high nutrient supplies, since plant cover and competition for nutrients is reduced, and decomposing wrack may "fertilize" bare patches. Spikegrass is known to increase in abundance in habitats with high nutrient supplies (Valiela et al. 1985; Levine et al. 1998).

In New England high-marsh habitats, disturbance patches dominated by fugitive plants are generally short-lived. Over time, the numerically dominant turfs vegetatively colonize these patches and competitively displace the fugitives. The competitively subordinate fugitive plants in many marshes appear to depend on frequent disturbances to avoid local extinction.[🐚28]

Examination of the mechanisms involved in bare patch colonization has revealed that positive interactions often drive marsh plant secondary succession. These positive interactions are driven by the ameliorating effects of plant cover on marsh soil salinities. Without plant cover, the evaporation of soil moisture predictably leads to high soil salinities, which can strongly influence both sexual and vegetative colonization of bare patches.

High soil salinities in bare patches can preclude most seedling establishment (Shumway and Bertness 1992). Many high marsh plant seeds simply do not germinate at the high soil salinities found in many bare patches, and those that do often cannot cope with the osmotic stress, and die.[🐚29] As a result, only seeds and seedlings that are extremely salt-tolerant, such as *Salicornia*, successfully germinate and grow in salty bare patches. High soil salinities can also limit the vegetative colonization of clonal perennial plants that are not salt-tolerant. In southern New England, neither marsh hay nor black needle rush, both of which dominate high marsh habitats, is capable of rapidly invading patches with high soil salinities. Spikegrass, however, can invade even extremely salty patches. After a disturbance, *Distichlis* rapidly invades bare patches with long rhizomes. Ramets in bare patches are released from competition with the surrounding turfs, but must rely on water trans-

Wrack cover kills vegetation

Disturbance-generated bare patch

Invasion of fugitives such as *Salicornia europacea* and *Distichlis spicata*

Invasion of dominants such as *Spartina patens* and *Juncus gerardi*

🌱 28 Annual production of dead plant debris, or wrack, in northern marshes often results in wrack smothering vegetation in the spring, creating unvegetated bare patches that trigger plant secondary succession.

ported to them by their clone mates outside the patches, which are not under osmotic stress. Conversely, clone mates in the dense background vegetation receive the products of photosynthesis in exchange.

The initial invaders of hypersaline bare patches facilitate further invasion of the patches because they shade the soil, ameliorate hypersaline soil condi-

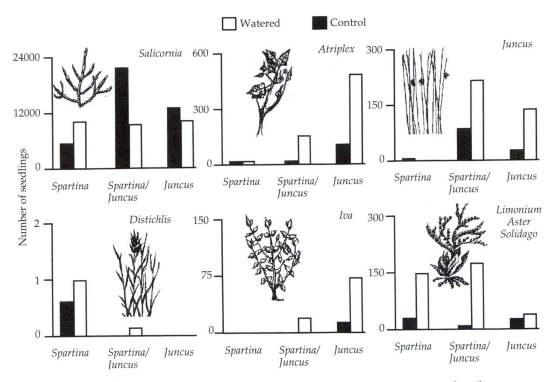

29 Soil salinity inhibition of marsh seedling germination. Watering marsh soil across marsh zones with fresh water to reduce soil salinity increases the germination of all common marsh plants except for the extremely salt-tolerant *Salicornia*. (After Shumway and Bertness 1993.)

tions, and thus create conditions conducive for the invasion of less salt-tolerant, but competitively dominant, plants. Because this pattern of facilitated secondary succession in marsh bare patches depends on soil conditions, positive interactions are predictably important only in the colonization of osmotically stressful patches. Consequently, factors that influence salt accumulation in bare patches have a predictable effect on the mechanisms of patch colonization. Small patches that are relatively shaded and thus not prone to salt buildup are colonized without facilitated succession, whereas succession in larger salty patches is facilitated. Moreover, patches in well-drained and frequently flooded habitats generally do not become hypersaline and probably do not close by facilitated succession, whereas colonization of patches in poorly drained and infrequently flooded habitats is often characterized by positive interactions. Facilitated colonization of bare space is probably more prevalent in southern than northern Atlantic coast marshes, since they receive more intense solar radiation without a seasonal break. In northern marshes, soil salinities are highly seasonal, climbing over the summer months and returning to low levels during the winter, when evaporation from the soil surface is low.[30]

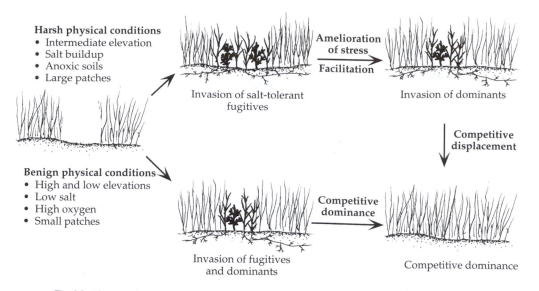

Harsh physical conditions
- Intermediate elevation
- Salt buildup
- Anoxic soils
- Large patches

Invasion of salt-tolerant fugitives

Amelioration of stress

Facilitation

Invasion of dominants

Competitive displacement

Benign physical conditions
- High and low elevations
- Low salt
- High oxygen
- Small patches

Invasion of fugitives and dominants

Competitive dominance

Competitive dominance

30 The mechanisms of high marsh bare patch closure vary as a function of physical stress. In stressful habitats, succession is facilitated by initial stress-tolerant invaders, whereas in more benign habitats, interactions among invaders are entirely competitive. (After Bertness and Shumway 1992.)

MARSH PRODUCTION AND ITS EFFECT ON ESTUARINE HABITATS

Salt marshes are generally acknowledged to be one of the most productive natural vascular plant communities on earth (Whittaker 1975; Odum 1970), and they clearly play a major role in the economy of estuaries as exporters of carbon (Teal 1962; Woodwell et al. 1977; Wiegert and Pomeroy 1981). Aboveground primary production of tall-form cordgrass *Spartina alterniflora* can be as high as 3000 grams per square meter per year, exceeding the productivity of most agricultural crops. Furthermore, belowground cordgrass production often equals or exceeds aboveground production (Valiela et al. 1976). Cordgrass production varies latitudinally, increasing with decreasing latitude from New England to the Gulf Coast.[31] Stronger solar radiation, higher average annual temperatures, and a longer growing season at lower latitudes are probably responsible for this variation (Turner 1976).

The high productivity of marshes has long been thought to play a major role in providing nutrients to adjacent marine habitats. John Teal, in his seminal study of energy flow in a Georgia salt marsh (Teal 1962), estimated that 40 percent of cordgrass production was exported from the marsh, supplying fixed carbon and nutrients to other marine habitats. Subsequent studies (Woodwell et al. 1977; Wiegert and Pomeroy 1981) also found that marshes were major nutrient exporters. This led to the view of marshes as important habitats in estuarine food chains (Odum and De la Cruz 1967; Odum 1978). This result, however, has not proved to apply generally (Nixon and Oviatt

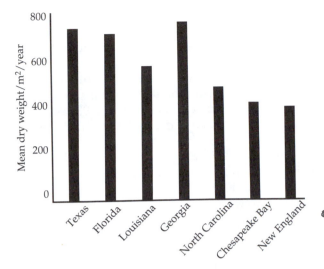

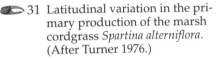 31 Latitudinal variation in the primary production of the marsh cordgrass *Spartina alterniflora*. (After Turner 1976.)

1973; Haines 1977). Marshes vary in their export/import characteristics as a function of a wide variety of factors, including level of primary production, geomorphology, oceanographic conditions, and marsh age. As a result, some marshes are exporters, or sources, and others may be primarily importers, or sinks (for reviews of this literature see Nixon 1980; Odum 1978).

Marsh plant primary production, like the primary production of other marine plants, is generally nitrogen-limited. Nitrogen fertilization experiments in a wide variety of North American marsh systems (Valiela and Teal 1974; Jefferies and Perkins 1977; Sullivan and Daiber 1974; Mendelssohn 1979b; Gallagher 1975) have shown that added nitrogen leads to increased plant growth, while other nutrients (such as phosphorus and potassium) do not influence marsh production alone.

An important consequence of the nitrogen limitation of marsh plant communities is that they are likely to be very responsive to the anthropogenic eutrophication that is so pervasive in nearshore waters. Evidence suggests that eutrophication leaves strong signatures on marsh plant communities. In southern New England, cordgrass production across a eutrophication gradient shows a strong positive correlation with nutrient inputs (Oviatt et al. 1977; Nixon and Oviatt 1978). In Narragansett Bay, cordgrass near nutrient sources was as much as five times more productive than cordgrass stands on offshore islands without high nutrient inputs. Long-term (10-year) nitrogen fertilization experiments at Sippiwissett Marsh on Cape Cod make a strong case for the role of bottom-up forces in dictating plant and animal populations in marshes (see Chapter 2). In the Sippiwissett study, nitrogen fertilization increased aboveground and belowground cordgrass production, resulting in plants with a higher nitrogen content, which ultimately led to dramatic changes in populations of marsh plant consumers (Vince et al. 1981). The insect herbivores of cordgrass increased along with primary production, showing that nitrogen supply also influenced herbivore densities.[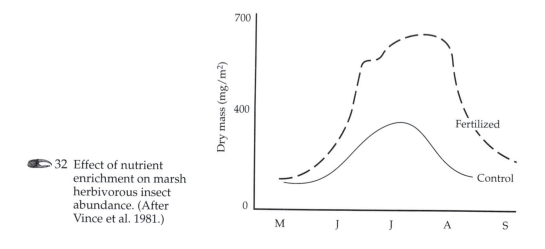32] Carnivorous insects also increased in response to fertilization, demonstrating that nitrogen supply to marshes can have dramatic bottom-up consequences on marsh food webs.

Nitrogen supply, however, may play an even more powerful bottom-up role in marsh plant communities by altering the distributions and abundances of plant species across marsh landscapes (Levine et al. 1998). Nitrogen fertilization can entirely reverse the outcome of competition among marsh plants, demostrating that the distributions of these plants may be largely a function of nitrogen levels. *Distichlis spicata*, for example, is typically a competitive subordinate that inhabits recently disturbed and physically stressful high marsh habitats. However, at high experimental nitrogen levels, spikegrass outcompetes both *Juncus gerardi* and *Spartina patens*. As a consequence, *Distichlis* invades high marsh habitats at high nutrient levels. Similarly, *Spartina alterniflora* outcompetes high marsh perennials and invades high marsh habitats at high nitrogen levels.[33] These dramatic results appear to be the consequence of a luxurious nitrogen supply shifting the arena of competition from belowground (competition for nutrients) to aboveground (competition for light). Nutrient supplies thus appear to play a major bottom-up role in marsh plant communities. Nitrogen regulates primary production and modulates the outcome of marsh plant competitive encounters, thereby regulating the very fabric of marsh communities.

These results imply that the typical zonation of plants across New England marshes is largely the result of low nutrient levels, and that nutrient resources are depleted by marsh plants during the course of marsh development. Furthermore, they suggest that the current eutrophication of North American estuaries may be having marked, yet largely unappreciated, effects on marsh plant community structure by shifting the competitive balance among plants and potentially altering the plant species composition of marsh plant communities.

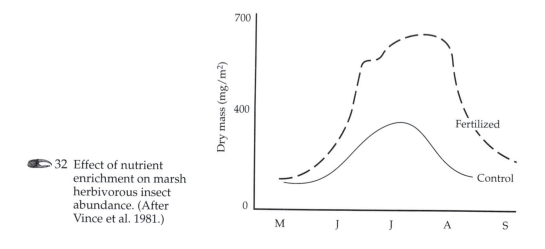

32 Effect of nutrient enrichment on marsh herbivorous insect abundance. (After Vince et al. 1981.)

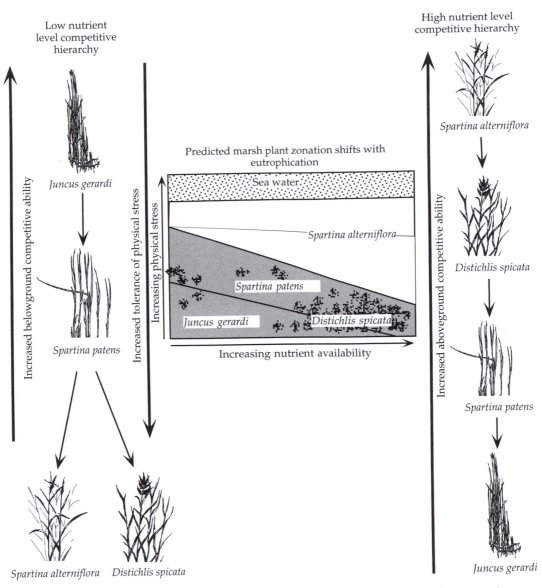

🐚 33 Nitrogen effects on the competitive relations of New England marsh grasses. At low nutrient levels, plants that exploit belowground resources best are competitively dominant. At high nutrient levels, competitive dominance shifts to plants that are the best aboveground competitors. This competitive switch in response to nitrogen availability predicts the movement of typically subordinate plants into the high marsh with eutrophication. (After Levine et al. 1998.)

ANIMALS OF SALT MARSHES

Salt marsh food webs are thought to be detritus-based, with most primary production reaching consumers via the decomposition of plant material.

Direct consumption of living plant material by herbivores accounts for only a small portion of the primary production of marsh vascular plants.[🖾34] Typically, only about 10 percent of the carbon fixed by marsh halophytes is eaten by herbivores (Smalley 1960; Teal 1962; Nixon and Oviatt 1973; Valiela and Teal 1979; Wiegert and Pomeroy 1979).

Why aren't the dominant vascular plants of marshes more readily utilized by herbivores? In contrast to many shoreline algae, which are heavily consumed by herbivores, most of the vascular plants in marshes (particularly the dominant clonal turfs) are structurally defended from many herbivores by heavy fortification with tough cellulose and silica. In one of the few studies to examine herbivory on marsh plants, Pennings and his colleagues (1998) found that plant toughness was the most important determinant of food choice for the wharf crab *Armases*, a generalist herbivore. Marsh plants may also be covered with salt excretions or have salty vascular fluids. In either case, these high salt concentrations may cause water balance problems for their consumers. Marsh plants may also contain secondary chemicals that discourage herbivores, but little work has been done on the defensive chemistry of marsh plants. In New England, *Spartina alterniflora* contains high enough phenolic concentrations to limit direct herbivory and initial decomposition (Buchsbaum et al. 1981; Barlocher and Newell 1994b). Recent evidence has also suggested that marsh plant chemical defenses are stronger at lower latitudes (Pennings, unpublished data). Structural defenses, low nutritive quality, and defensive secondary compounds all probably contribute to the low levels of herbivory on many numerically dominant marsh plants.

While the somatic tissues (tissues not involved in gamete production) of dominant marsh plants are not heavily consumed, the flowers and seeds of these plants, as well as the less numerous annual and solitary plants that live

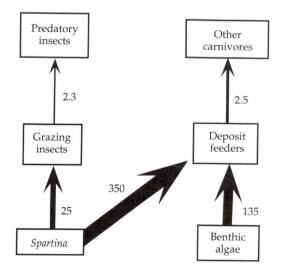

🖾 34 Energy flow (in grams of carbon per square meter per year) in a Georgia salt marsh. (After Teal 1962.)

inconspicuously in dense vegetation cover, are often heavily consumed. In New England, chewing insects often destroy more than 90 percent of marsh plant flowers and can limit seed production (Bertness et al. 1987). Marsh plant flowers and seeds do not appear to be as well defended as somatic tissues, and probably represent a high-quality food source. Heavily eaten marsh annuals such as *Salicornia* and *Atriplex* also appear to have fewer structural or chemical defenses than the clonal turfs that dominate most marsh landscapes. These plants probably escape heavy herbivore damage by living hidden in the dense matrix of relatively unpalatable clonal turfs (Ellison 1987b), relying on low apparency and associational defenses for their persistence.

Detritivores

While marshes are among the most productive environments known, little of their annual primary production is directly consumed as living tissue (Teal 1962). Most of the vascular plant biomass of marshes dies, is attacked by bacteria and fungi, and only then reaches invertebrate detritivores, deposit feeders, and filter feeders (Lopez and Levinton 1987; Haines 1977; Haines and Montague 1979; Cammen 1980; Newell 1993). Whether or not the production of salt marshes ultimately contributes to the trophic support of mobile consumers in nearby estuarine and nearshore waters is an important, but largely unresolved, question (Odum 1978; Nixon 1980; Dame 1996). Kneib (1997) has recently suggested that **trophic relays** may be critical in transforming marsh primary production into secondary production in adjacent, energetically subsidized habitats. Trophic relays can be envisioned as juvenile animals that tap the detritivore food chain in marshes before moving into adjacent habitats, where they transfer marsh-produced organic matter to larger consumers and water column food chains.

A wide variety of marsh animals can be broadly defined as detritivores. Many snails, crabs, and amphipods are direct detritivores that feed on decomposing marsh plant tissue, the microbes responsible for its decay, or both. The fiddler crabs, snails, worms, and shrimps that make their living as deposit feeders in marsh habitats by removing organic material from sediments are also detritivores. In contrast, marsh filter feeds such as clams, mussels, and oysters depend primarily on phytoplankton as their food source (Haines and Montague 1979).

The detrital link in marsh plant food webs has been most thoroughly examined in the southern marshes of Georgia (see Newell 1993 for a nice review). In these marshes, plants grow year-round and routinely produce senescent leaves, which remain attached to the living plants as they undergo initial decay. The decay of standing dead plant matter involves bacteria and fungi, and is enhanced by shredder invertebrates, which break down the cell walls of the decaying plants. Decomposing standing cordgrass stems may enhance the growth of live neighboring plants by continuing to supply oxygen to potentially anoxic soils via aerenchyma tissue as well as recycling nutrients to live neighbors through vascular connections (Newell et al. 1991, Newell 1996).

The marsh periwinkle *Littoraria irrorata* is one of the most common and conspicuous detritivores on southern marshes. It can reach densities as high as 600 individuals per square meter (Barlocher and Newell 1994a), but is excluded from creek banks and low marsh elevations by blue crab predation (Hamilton 1976; Schindler et al. 1994). It exhibits the curious behavior of migrating up grass stems out of the water at high tide and migrating to lower stem elevations at low tide. Hamilton (1976) showed experimentally that this migration pattern limited the snails' contact with blue crabs, their most common predators. While this finding may partially explain their unusual migration pattern, which keeps these gill-breathing snails out of water, more recent evidence suggests that *Littoraria* migrate to maximize their food intake. *Littoraria* is widely thought to be a detritivore shredder specialist (Barlocher and Newell 1993a). Most *Littoraria* are found on free-standing dead plants, and their diets appear to consist of both the fungi that dominate the initial decay process and partially decomposed plant cells. *Littoraria* may migrate up and down with the tide to forage on recently wetted dead plant material, which is the most nutritious and easily processed state for grass stems (Newell 1996).[🦪35]

Snail exclusion studies in Virginia marshes in recent years have begun to challenge the characterization of *Littoraria* as a detritivore specialist and question the current dogma that herbivores play a relatively minor role in marsh systems. When *Littoraria* were removed from penned cordgrass stands, Silliman (1998) found that cordgrass production increased by as much as 25 percent, while adding *Littoraria* to penned areas decreased cordgrass production by up to 40 percent. These intriguing results suggest that these snails may exert top-down control of the cordgrass standing crop, and may ultimately change our understanding of these systems. Silliman has suggested that *Littoraria* are capable of controlling cordgrass production, but are themselves kept in check by crab predators.

The amphipod *Uhlorchestia spartinophilia* is another southern marsh animal that appears to be a detritivore specialist on decaying cordgrass stems (Covi and Kneib 1995). *Uhlorchestia* is a small amphipod, only 5–8 millimeters in length, that is found commonly at densities of 150–350 individuals per square meter in Georgia marshes. It lives primarily on decomposing cordgrass, particularly on free-standing dead cordgrass between stems and decomposing leaves (Kneib et al. 1997). This microhabitat gives *Uhlorchestia* protection from the fish and crab predators that forage in these habitats during high tide, and also provides it with a fungus-rich detrital food sources. Since these amphipods are common food items for transient marsh consumers, they may be an important trophic relay between decaying cordgrass and the secondary production of adjacent habitats.[🦪36]

Many other common marsh invertebrates are also important detritivores. The lung-breathing snail *Melampus*, for example, lives at high marsh elevations and is probably a detritivore specialist. These snails can be found at high densities (over 100 per square meter), and often migrate up plants during high tides to avoid submersion (Fell et al. 1991). Because they are slow-

35 The southern marsh periwinkle, *Littoraria irrorata*, migrates up and down cordgrass stems, staying just above the waterline. This behavior minimizes losses to blue crab predators and maximizes access to bacterial and fungal food on decaying cordgrass stems. (See Hamilton 1976; Barlocher and Newell 1994a.)

moving and sensitive to desiccation, *Melampus* at high elevations are found under plant debris or in moist cryptic habitats under vegetation. These thin-shelled gastropods are easy prey for ducks and fishes, and predation on small *Melampus* by killifish restricts them to high elevations and dense vegetation. Larger *Melampus* reach a size refuge from killifish predation and occur at lower elevations and in sparser vegetation, where their food is more abundant (Vince et al. 1976).

Other marsh consumers opportunistically feed on detritus or go through detritivore phases during their development. Wharf crabs, *Armases*, and marsh crabs, *Sesarma*, while primarily herbivorous, will readily consume large detritus particles. Moreover, many juvenile fishes (such as the killifish, *Fundulus*) are largely detritivores, switching to a carnivorous diet as they grow older.

An interesting question about detritivores in Atlantic coast salt marshes is whether their role and importance in these systems changes with latitude. As I have just outlined, southern marshes are characterized by relatively continuous production of free-standing dead leaves, which are attacked by fungi and are shredded and eaten by snails and amphipods that appear to be specialized for this role. In northern marshes, dead plants are produced season-

36 Amphipods are common detritivores and scavengers in salt marsh communities. (Fell et al. 1982.)

ally en masse, and there are no apparent detritivore specialists that capitalize on this temporally and spatially predictable resource.

Herbivores

Herbivores in marsh plant communities include insects, snails, crabs, and vertebrates. Insects are the most abundant and conspicuous herbivores in most cases. Marsh insect herbivores can usually be characterized as either chewing or sucking feeders. Chewing insects include grasshoppers and beetles. Marsh grasshoppers, *Conocephalus spartinae* in northern marshes and *Orchelium fidicinum* in southern marshes, are particularly common. In Georgia, grasshoppers consume over 10 percent of *Juncus roemerianus* production, eating primarily younger portions of plants (Smalley 1960). In New England marshes, grasshopper consumption of the leaves and stems of vascular plants is minimal, but their consumption of flowers and seeds is intense. However, no insect herbivore seems to be capable of limiting the distribution and abundance of the clonal turfs that typically dominate marsh landscapes. Herbivorous beetles can be common at high marsh elevations, where tidal flooding is uncommon, and appear to be capable of limiting the distributions of their preferred food plants. The leaf beetle *Erynephala maritima*, for example, is capable of limiting the distribution of *Salicornia* in New England marshes (Hartman and Engler 1982). Leaf beetle herbivory may also exclude the seaside goldenrod, *Solidago sempervirens*, from bare patches where it would be released from competition (Hacker, unpublished data). Both the arrow plant (*Atriplex*) and seaside goldenrod survive in high marsh habitats where they are surrounded by vegetation, ostensibly since they are not detected by beetle herbivores in dense vegetation.[37]

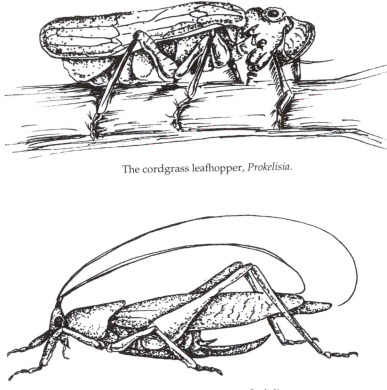

The cordgrass leafhopper, *Prokelisia*.

The marsh grasshopper, *Orchelium*.

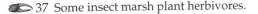 37 Some insect marsh plant herbivores.

In contrast to grasshoppers and most beetles, which use masticating mouthparts to chew plant tissue, sap-sucking insects have mouthparts modified to suck vascular fluids from plants. Leafhoppers and aphids are the most common sap-sucking insects in salt marshes. Leafhoppers are usually abundant on low marsh cordgrass stands, but do not have strong negative effects on cordgrass distributions or productivity (Denno 1980). Leafhoppers are highly mobile, which allows them to live in the low marsh but avoid submergence during high tides. The leafhopper *Prokelisia marginata*, which lives on cordgrass along the Atlantic coast of North America, has been extensively studied. It has two morphs: flightless, short-winged brachypters with high reproductive rates dominate stable habitats, and flying macropters with fully developed wings, but lower reproductive rates, dominate ephemeral habitats. The development of these morphs is triggered by environmental conditions. Under stable conditions with good food supplies, flightless brachypters with high reproductive output are favored, whereas under harsh physical conditions and deteriorating food supplies, winged macropters capable of migration are favored (Denno and Grissell 1979; Denno et al. 1985).[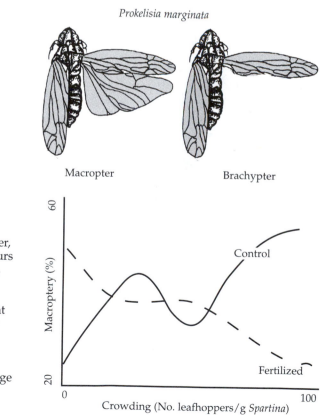38]

Aphids are more sedentary than leafhoppers, and are found on plants near the infrequently flooded terrestrial borders of marshes. The life history of aphids usually gives them a highly clumped distribution. Winged aphids called alates colonize host plants and produce sedentary, asexually reproducing daughters (fundatrices). These females produce a number of genetically identical feeding and asexually reproducing daughters. Once host resources are depleted, however, they produce sexual male and female offspring, which in turn produce flying alates that colonize new hosts or leave the marsh to overwinter.[39]

In southern New England, aphids' asexual life cycles and their predators lead to a characteristic patchy distribution. The salt marsh aphid, *Uroleucon ambroisae*, is usually found only on *Iva*. As might be expected, these aphids have their highest growth potential on the tall, productive marsh elders farthest from the water. Nonetheless, aphids are rare on tall *Iva*. In fact, they are found almost exclusively on the short, stunted marsh elders closest to the water, where aphid growth rates are lowest. This interesting pattern is driven

Prokelisia marginata

Macropter   Brachypter

38 The cordgrass leafhopper, *Prokelisia marginata*, occurs in two morphs: a flying, colonizing macropter, and a short-winged, flightless brachypter that has a high reproductive output. These morphs develop in response to food supply conditions during the nymphal stage of development. (After Denno et al. 1985.)

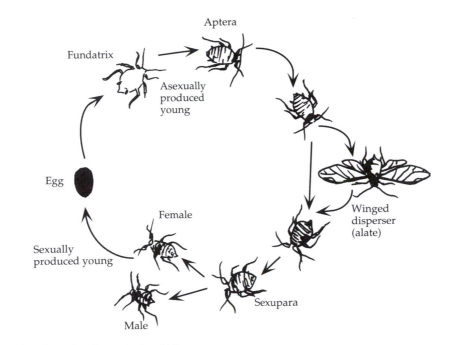

🐚 39 Life cycle of a typical aphid.

by ladybird beetle (*Hippodamia convergens*) predation on aphids. Ladybird beetles preferentially forage on tall, conspicuous vegetation, and predation pressure thus restricts aphids to stunted, poor-quality hosts. On these stunted host plants, aphids have a very patchy distribution due to their low mobility and asexual reproduction (Hacker and Bertness 1995a).[🐚40]

Snails that use a tonguelike radula to rasp diatoms and algal films from hard surfaces may be more important marsh herbivores than currently believed. In New England marshes, the rocky shore periwinkle, *Littorina littorea*, can often be found in marshes with coarse sandy substrates, but is rare in marshes with muddy sediments. South of Chesapeake Bay, the marsh periwinkle *Littoraria irrorata* is a conspicuous component of marshes, where it may graze on diatom epiphytes on cordgrass stems (see Silliman 1998).

Herbivorous crabs that consume the vascular plants of salt marshes are also abundant in many Atlantic Coast marshes. The wharf crab, *Armases* (formerly *Sesarma*) *cinereum*, and the marsh crab, *Sesarma reticulatum*, are the most common of these. Wharf crabs are small, semi-terrestrial crabs that burrow in high intertidal marsh habitats with high salinities (over 30 parts per thousand) and sandy soils. They occur from Chesapeake Bay to the Caribbean and often wander up to 100 meters inland, where they can be garden pests. Marsh crabs, in contrast, are found at lower marsh elevations, in muddy soils and brackish (15–20 parts per thousand) water. Marsh crabs

Ladybird beetles forage almost exclusively in tall *Iva*, excluding aphids from this area

Sap-sucking aphids are restricted to short *Iva*, where they have low growth rates and are dependent on *Juncus* to buffer *Iva* from physical stresses

Tall *Iva* zone   Short *Iva* zone   *Juncus* zone

Well-drained, low-salinity soils — High *Iva* growth results in *Iva* completely displacing *Juncus*

Juncus facilitates *Iva* growth and survival by buffering from salt and oxygen stresses

Waterlogged, saline soils *Iva* precluded due to severe conditions

40 Relationship among ladybird beetles, aphids, and the marsh elder *Iva frutescens* on southern New England marshes. (After Hacker and Bertness 1995a.)

usually live in large communal burrows at intermediate marsh elevations. Marsh crabs are usually slightly larger than wharf crabs (up to 3 centimeters carapace width) and have a more northern distribution, ranging from Cape Cod in the north to the Gulf Coast of Texas.[41]

Whereas wharf and marsh crabs are best known as consumers of marsh plants, they are actually omnivorous, opportunistic feeders. Both will scavenge, and marsh crabs are known to prey on fiddler crabs. Very little is known, however, about the field diets of these crabs (but see Pennings et al. 1998) or the effects they have on marsh vegetation. In Delaware Bay, marsh crab herbivory may be responsible for denuding large (over 5 square meters) areas of cordgrass (Crichton 1960), though this claim has not been tested. Throughout the range of marsh crabs, high densities of burrows are found associated with large cleared areas. It is uncertain, however, whether these areas are kept cleared of vegetation by crab herbivory or whether the crabs preferentially burrow in habitats lacking vegetation. Both marsh and wharf crabs are also common prey of birds, which, when abundant, may limit these

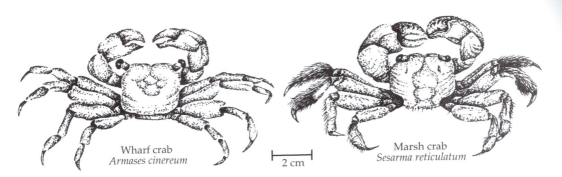

Wharf crab
Armases cinereum

2 cm

Marsh crab
Sesarma reticulatum

41 Wharf and marsh crabs are common in Atlantic coast marshes and can be important consumers of marsh grasses.

crabs to living in their burrows and in dense vegetation cover. In Georgia marshes, these crabs, along with fiddler crabs, have been found to make up the bulk of the diets of the marsh clapper rails *Rallus longirostris* and *Rallus waynei* (Oney 1954).

Vertebrate grazers are not generally important herbivores in most North American marshes. Grazing by large herbivorous mammals and geese, however, can be locally important, and may have been important in the past. Historically, North American marshes were used for livestock grazing and harvested for hay, which most certainly affected marsh plant community species distributions and abundances. Sheep grazing on European marsh plants is known to play an important role in regulating production and species composition (Ranwell 1961). The only large mammals grazing on Atlantic coast marshes today are the wild horse populations on a few of the barrier islands off Maryland and the Carolinas. These populations are remnants of horses that were introduced to the islands by European explorers in the sixteenth and seventeenth centuries to breed. Their grazing on cordgrass and marsh hay reduces the standing stock of grazed grasses and influences marsh plant species composition by reducing plant cover and selectively removing competitively dominant plants (Furbish and Albano 1994).

Geese are probably the most common vertebrate herbivores on Atlantic coast marshes. Migrating Canada geese (*Branta canadensis*) often stop on marshes to feed on belowground rhizomes. Large flocks of geese can entirely denude large (tens of square meters) areas of marsh plants (Smith and Odum 1981; Cargill 1984; Cargill and Jefferies 1984; Bazely and Jefferies 1985) and may be extremely important consumers in some marshes. In Canadian subarctic marshes, snow goose (*Chen caerulescens*) herbivory is predictable and intense. In these habitats, goose grazing has been shown to increase marsh grass production by thinning marsh grass stands, reducing plant competition, and adding nitrogen in the form of goose feces (Cargill and Jefferies 1984).[42]

Increased marsh grass production increases food supply for geese

+

Goose feeding thins vegetation and root mat. Goose feces increase nitrogen supply, increasing marsh grass production +

🝔 42 Positive feedbacks between goose foraging and marsh grass production. (After Cargill and Jefferies 1984.)

Filter Feeders

Animals that actively or passively filter organic food from the water column are often a conspicuous and important trophic group in marsh systems. When abundant, filter feeders can clear the water column, increasing light transmission. Moreover, filter feeders can transfer nutrients from the water column to the marsh surface, where they are available to rooted plants. Variation in water column primary production is probably directly reflected in the growth, biomass accumulation, and sediment deposition of filter feeders, which tightly couple water column and marsh food webs.

Most filter feeders in marshes live at low elevations, where they are frequently covered by tides. Clams are often abundant in marsh sediments, and hard surfaces in marshes are covered by barnacles. The effects of clams and barnacles on marsh communities, however, probably do not match those of mussels and oysters. Both ribbed mussels (*Geukensia demissa*) and oysters (*Crassostraea virginica*) commonly form dense aggregations, or **beds,** of hundreds to thousands of individuals per square meter in low marsh habitats. These beds are the product of gregarious settlement of larvae and the survivorship advantages of living in groups. Both mussels and oysters are active filter feeders that pump water over their gills with cilia, filtering planktonic food particles from the water. When dense beds of these bivalves cover the seaward borders of marshes, they are often capable of processing all of the water entering the marsh system every tide cycle (Jordan and Valiela 1982; Dame 1996). Thus, their potential role in coupling water column and benthic marsh processes is substantial.

The group dynamics of ribbed mussels have received the most attention. Dense beds of ribbed mussels are characteristic of marshes from southern New England to Chesapeake Bay. Ribbed mussels, like most bivalves, are external fertilizers that broadcast gametes seasonally into the water column. After a feeding planktonic larval phase lasting a number of weeks, ribbed mussel recruits settle with other mussels. They are rarely found outside of marshes or not associated with adult mussels, suggesting that *Geukensia* larval settlement is very specific. Within marshes, *Geukensia* settle in association with cordgrass stems, and gregariously with adults and other mussel recruits (Bertness and Grosholz 1985). In New England, this leads to the development of dense beds on the seaward borders of marshes. Like the blue mussel beds of rocky shores, ribbed mussel beds have both costs and benefits to group members (Bertness and Grosholz 1985). The major cost is competition for food and space. However, the fact that mussel and oyster beds are a common feature of shorelines is strong evidence that the benefits of beds typically outweigh their costs.[43]

While intertidal oyster beds were common in colonial times as far north as southern New England, they are now found only in the southern marshes. Overharvesting, disease, and eutrophication have probably all contributed to the decline of oysters. Harvesting has selectively removed oysters from shallow-water habitats and reduced the height of oyster beds. These effects of oyster harvesting have increased the vulnerability of remaining oysters to the habitat-deteriorating effects of eutrophication. Eutrophication in estuaries commonly leads to elevated microbial respiration and the depletion of oxygen in bottom water. In shallow estuaries like Chesapeake Bay, anoxia events are common in the summer because of high temperatures and slow water circulation. Overharvesting has restricted oysters to deep-water habitats, where they are vulnerable to mortality in anoxic bottom water (Lenihan and Peterson 1998). As previously discussed (see Chapter 6), the loss of oyster beds has had large repercussions for east coast estuaries, since bed structures enhance larval settlement, protect juveniles from predators, increase water clarity, and accelerate nutrient cycling (Newell 1988).

Positive associations between vascular plants and sessile filter feeders appear to be widespread; specific cases include marsh grasses (Bertness 1984b; Dame 1996), seagrasses (Peterson et al. 1984), and mangroves (Ellison et al. 1996). Wetland plants are generally nitrogen-limited (see Valiela 1984) and are vulnerable to erosion and other disturbances (Redfield 1971). Sessile filter-feeding organisms attached to their roots can provide plants with nitrogen by transferring particulate material from the water column to the soil surface as feces, as well as stabilizing plants by their attachment to the substrate. Conversely, sessile organisms in wetland habitats are often limited by

Group costs
- Competition for food
- Competition for space
- Reduced growth rates

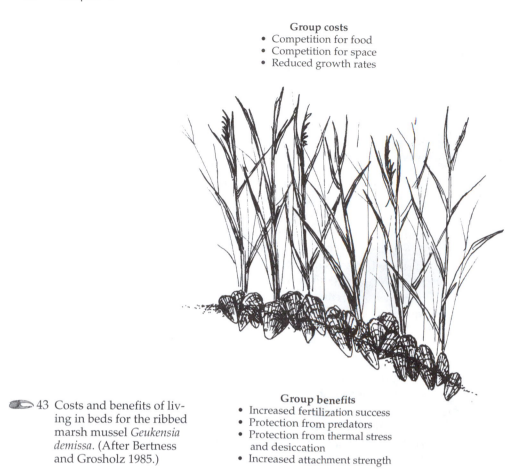

43 Costs and benefits of living in beds for the ribbed marsh mussel *Geukensia demissa*. (After Bertness and Grosholz 1985.)

Group benefits
- Increased fertilization success
- Protection from predators
- Protection from thermal stress and desiccation
- Increased attachment strength

the availability of attachment sites and detritus-based food sources, and thus can benefit from close association with plants. [44]

The relationship between the ribbed mussel, *Geukensia demissa*, and the cordgrass, *Spartina alterniflora*, in southern New England marshes is a good example of a marsh grass/filter feeder mutualism. In southern New England, ribbed mussels are often found in extremely dense (300–1500 individuals per square meter) beds on the seaward edges of marshes, attached to cordgrass roots and one another. The dense ribbed mussel beds characteristic of southern New England are not found in northern New England, since winter ice scour in the low marsh is too severe. They are also not found south of Chesapeake Bay due to predation by blue crabs (Stiven and Gardner 1992). Experimental studies of the effects of ribbed mussels on cordgrass in southern New England have demonstrated that mussels increase the growth of cordgrass by providing nutrients to the otherwise nutrient-limited plants (Bertness 1984b). Ribbed mussels also stabilize the seaward edges of marshes

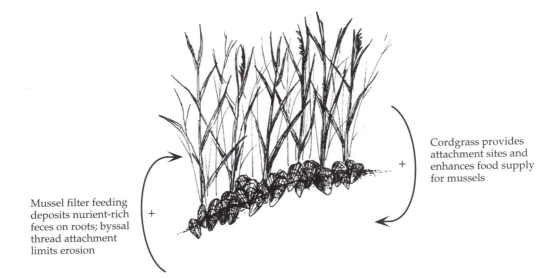

Cordgrass provides attachment sites and enhances food supply for mussels

Mussel filter feeding deposits nurient-rich feces on roots; byssal thread attachment limits erosion

44 Facultative mutualism between marsh mussels and cordgrass. (See Jordan and Valiela 1982 and Bertness 1984b.)

through the strength of their communal attachment to the substrate and their effect on cordgrass root production. Mussels attach themselves to the substrate by proteinaceous byssal threads, and in dense beds, the communal strength of these byssal thread attachments is extremely high. Cordgrass also responds to mussel feces by increased production of nutrient-absorbing roots, further stabilizing marsh sediments. In southern marshes, where ribbed mussels are restricted to living in tight clumps at high elevations, almost entirely buried below the surface, mussel clumps are often conspicuously demarcated by localized areas of increased marsh grass height.

Deposit Feeders

Deposit feeders that ingest sediments or sort through them for particulate organic material or small organisms (meiofauna) are also common in marsh communities. They are critical links in marsh food chains as consumers of detritus and the small organisms responsible for detritus decomposition. They also play a crucial role in marsh sediment metabolism, since bioturbation of sediments is a major by-product of their feeding activity. They are essentially the earthworms of marsh systems, constantly plowing and tilling marsh soils. There is a wide variety of deposit feeders in marshes, including worms, grass shrimp, and even some fishes, but mud snails and fiddler crabs are the most conspicuous and abundant.

The mud snail, *Ilyanassa obsoletus*, is an extremely common member of salt marsh communities from Cape Cod to Georgia. It is often found at densities of thousands of individuals per square meter and can play an important role in regulating sediment metabolism. At high densities, mud snails can limit marsh sediment productivity, but when they are rare, sediment productivity is limited by nutrients and the accumulation of wastes (Pace 1979). Intermediate snail densities, however, can stimulate the productivity of marsh sediments through bioturbation and nutrient cycling (Connor et al. 1982). *Ilyanassa* populations are currently being displaced by the introduced European periwinkle *Littorina littorea* in New England marshes. Although *Littorina* and *Ilyanassa* do not compete for food, *Ilyanassa* is dependent on hard substrates, to which it must attach its eggs, and hard surfaces are in short supply in marsh habitats. *Littorina* is a hard-substrate scraping herbivore that grazes hard surfaces in marshes and dislodges *Ilyanassa* egg capsules. *Ilyanassa* actively avoids contact with periwinkles, leading to *Littorina* displacing mud snails from marsh habitats (Brenchley and Carlton 1983).

The fiddler crab is one of the other characteristic animals of salt marsh communities. Fiddler crabs are the most common deposit-feeding crabs on western Atlantic coast shorelines. They feed using their claws to scoop sediments into their mouths, selectively ingesting small organic particles and microbes and processing the organic material (Miller 1960; Robertson and Newell 1982). A common sight on the marsh is the piles of consolidated balls of processed sediment left by crabs on the marsh surface.[⬤45] The crabs feed during low tides when sediments are exposed to air, and hide from predators in burrows during high tides.

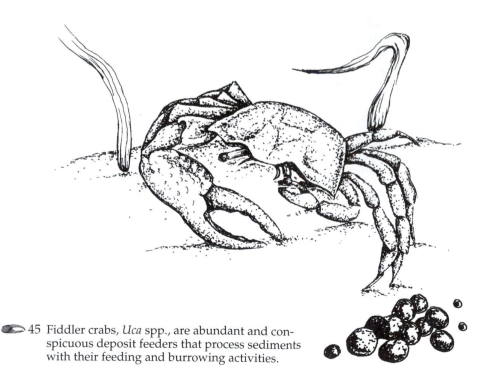

⬤45 Fiddler crabs, *Uca* spp., are abundant and conspicuous deposit feeders that process sediments with their feeding and burrowing activities.

Three fiddler crab species are common in Atlantic coast marshes. The mud fiddler crab, *Uca pugnax*, is characteristic of muddy marsh habitats, whereas the sand fiddler, *Uca pugilator*, is found on coarser, sandy soils. Both of these crabs are usually found at low marsh elevations. The larger orange-clawed fiddler crab (*Uca minax*) burrows in the high marsh but forages in lower marsh sediments. All of these fiddler crabs are deposit feeders and are characterized by the sexual dimorphism in claw size discussed earlier (see Chapter 3).[⟋46]

Fiddler crabs are year-round marsh residents. In New England, they hibernate in burrows during the winter. In southern marshes, they are active year-round. At all latitudes, fiddler crab activity is conspicuously temperature-dependent. Crabs are relatively inactive at cooler temperatures, and increase their activity as the temperature rises.

Fiddler crab burrowing can have an enormous effect on marsh sediments. Fiddler crab burrows can be very abundant (more than 200 per square meter), up to 30 cm centimeters deep, and can have a high turnover (Bertness and Miller 1984). Particularly in soft sediments, where burrows collapse without regular maintenance, high burrow turnover can force crabs to burrow constantly, reworking sediments down to 10–25 centimeters below the surface. Conversely, harder sediments (particularly dense peat) are difficult to burrow in and can limit crab activity (Ringold 1982; Bertness and Miller 1984). Turnover of burrows in dense peat is very slow. While they are more difficult to excavate, most crabs prefer burrows in hard sediments, as these homes are relatively maintenance-free and can last for years.

Positive associations between marsh plants and deposit-feeding crabs appear to be widespread. Wetland plants are often limited by anoxic, poorly

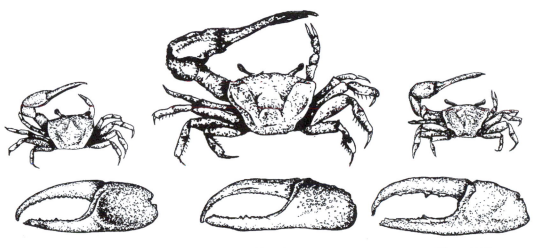

Uca pugilator   *Uca minax*   *Uca pugnax*

⟋46 Fiddler crabs of the Atlantic coast.

Fiddler crabs
- Oxidize soils
- Increase drainage
- Enhance decomposition
- Enhance cordgrass production

Cordgrass
- Roots and plant debris support burrows
- Production enhances food supply

🐚 47 Fiddler crab-cordgrass facultative mutualism. See Montague 1982, and Bertness 1984.

drained soils, and the burrowing and feeding activities of crabs can potentially alleviate these soil conditions and increase plant production. Through their feeding and burrowing activities, fiddler crabs increase soil drainage, extend the depth of oxygen-rich soils and associated microfauna, and increase the decomposition of belowground plant debris, as well as enhancing soil nitrogen levels with their waste products (Bertness 1985; Montague 1982).[🐚47] Fiddler crabs have been shown to enhance cordgrass production in New England (Bertness 1985) and Georgia (Montague 1982). They may also enhance the growth of mangroves in tropical wetlands (Smith et al. 1992).

Like the relationship between sessile filter feeders and marsh plants, the relationship between fiddler crabs and marsh plants is best characterized as a facultative mutualism. In soft sediments, cordgrass stems and roots structurally support fiddler crab burrows, reducing burrow maintenance costs (Bertness and Miller 1984). Cordgrass detritus is probably also an important food source for fiddler crabs (Robertson and Newell 1982), and shading by marsh plants may both protect the crabs from predators and keep the soft sediments that they feed on free of seaweeds.

Predators

Marsh habitats are characterized by a variety of predators that are important in dictating the abundances and distributions of marsh organisms, ranging from interstitial meiofauna (e.g., nematodes and harpacticoid copepods) to large molluscs and fishes. By doing so, these predators can have important direct and indirect effects on marsh invertebrates and sediment metabolism. While less studied than on rocky shores and in unvegetated soft-sediment habitats, predation plays an equally important role in marshes.

Predation on small nematodes, copepods, and larvae that inhabit the top 2–5 centimeters of marsh sediments is intense. Grass shrimp and killifish that forage on sediments during high tide can limit the size, distribution, and abundance of meiofaunal populations across marsh habitats, as well as limiting their prey to spatial and temporal refuges (Bell and Coull 1978; Bell 1980). As a result, meiofauna are found in abundance only at high marsh elevations and among dense vegetation, where predation pressure is reduced (Bell 1980).

Predatory crabs and fishes also play important roles in most marsh communities by limiting the distributions and abundances of their prey (Vince et al. 1976; Micheli 1997). Crab predation south of Cape Cod, particularly by the blue crab, *Callinectes sapidus*, limits the distributions of many important marsh organisms. Blue crabs are large, highly mobile, voracious swimming predators that forage during high tide in lower marsh habitats. Dense vegetation and reduced foraging time appear to limit blue crab predation on the high marsh. As a consequence, the prey of blue crabs are often found at high elevations, where they have relatively low growth rates and reduced reproductive success (Vince et al. 1976). For example, intense blue crab predation in southern marshes restricts ribbed mussels to high marsh elevations, where they have reduced feeding time (Stiven and Gardner 1992). Blue crabs also exclude marsh periwinkles from habitats where they cannot migrate out of the water during high tides to avoid being eaten (Hamilton 1976; Schindler et al. 1994), and probably force many slow-moving infaunal organisms to live in association with marsh grass roots or in other refuges. Since blue crabs themselves are vulnerable to being eaten by birds such as herring and ring-billed gulls, blue crab foraging may be the most intense on the seaward edges of marshes (Micheli 1997), which are both accessible to crabs at high tide and provide them with cover from their enemies.

Large, mobile fishes and birds are also important predators of macroinvertebrates in marshes. Of the fishes that forage on marsh invertebrates, adult killifish are generally the most common and have received the most study. In New England marshes, they are often the only common pelagic predator, and have been shown to limit the distributions of grazing snails (*Melampus*) and amphipods to the high marsh (Vince et al. 1976). In Chesapeake Bay, amphipods are limited to living in dense cordgrass stands by killifish predation (Van Dolah 1978), and in Georgia and North Carolina marshes, Kneib (1997) has shown that a wide variety of juvenile fishes and invertebrates are generally restricted to high marsh elevations by killifish predation. Wading birds such as egrets and herons can also be important predators of fishes and shrimps in marshes.[🐚48]

MARSHES AS NURSERY GROUNDS

Marshes have long been known to play a critical role in marine environments as nursery grounds for fishes, shrimps, and crabs (Odum 1978). This knowledge has been important in justifying conservation efforts, since many recre-

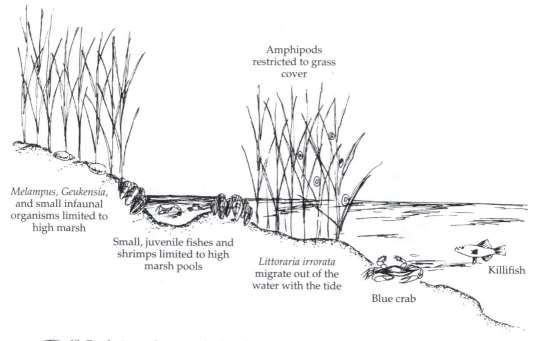

Amphipods
restricted to grass
cover

Melampus, Geukensia,
and small infaunal
organisms limited to
high marsh

Small, juvenile fishes and
shrimps limited to high
marsh pools

Littoraria irrorata
migrate out of the
water with the tide

Blue crab

Killifish

48 Predation at low marsh elevations restricts many organisms to high marsh and refuge habitats.

ationally and commercially important marine organisms utilize marsh habitats as juveniles. The list of commercially and ecologically important species that utilize marshes as nursery grounds is impressive, and includes blue crabs (*Callinectes sapidus*), grass shrimp (*Palaemonetes* spp.), the Atlantic croaker (*Micropogonias undulatus*), the southern flounder (*Paralichthys lethostigma*), the hogchoker (*Trinectes maculatus*), and the spot (*Leiostomus xanthurus*). Many of these species enter the shallow portions of marshes and estuaries in the winter or spring at high densities as juvenile recruits, and leave shallow marsh areas for deeper water as they grow. Considerable evidence supports the argument that marshes are important nursery grounds (Turner 1977; Kneib 1997). It is less clear, however, why marshes harbor so many juveniles.

Early arguments for the importance of marshes as nurseries were based on the idea that marshes provide abundant food for juveniles (Odum 1978). Many fishes and crustaceans that utilize marshes as juveniles rely heavily on detritus or the microbes associated with decaying detritus when young, but shift their diets with size and age. The common marsh killifish, *Fundulus heteroclitus*, for example, exhibits this sort of dietary shift with age (Fell et al. 1991). As juveniles, *Fundulus* live in shallow high marsh pools and make their living as deposit feeders and scavengers (Kneib 1986). Adult *Fundulus* live at low marsh elevations on the seaward edge of marshes and are primarily predators.[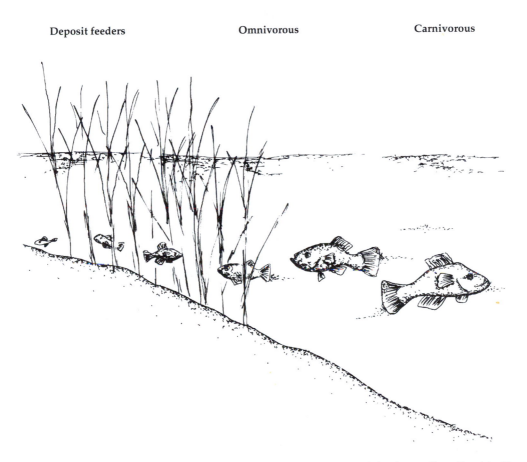49]

Recent work, however, has suggested that the nursery function of marshes may be most strongly driven by predation (Boesch and Turner 1984; Rozas and Odum 1987; Vermeij 1972; Carr 1994; Heck and Thoman 1981 for similar nursery ground arguments from other habitats). Juveniles utilize high marsh habitats at considerable risk. Juveniles that forage on the marsh surface during high tide or live in high marsh pools risk being stranded out of water at low tide or in shallow pools that can become hot and hypersaline. These risks suggest that there must be some corresponding benefit to living in these habitats. Predation at low marsh elevations has been shown to restrict small snails (Vince et al. 1976), crustaceans (Kneib 1985), and fishes (Kneib 1986; Halpin, unpublished data) to high marsh habitats. Halpin has

Deposit feeders   **Omnivorous**   **Carnivorous**

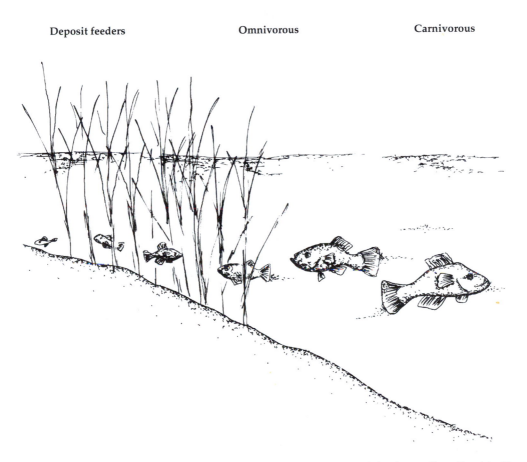

49 Developmental shifts in the diet of the killifish *Fundulus heteroclitus*. Recruits live at high marsh elevations and are deposit feeders. Adults live at the lowest elevations and are carnivores.

shown that juvenile killifish have lower growth rates at high intertidal heights than they would at lower intertidal heights, but that they are precluded from living at low marsh elevations by predation. The shallow water and high plant stem densities found in high marsh habitats probably reduce predator foraging efficiency and create prey refuges.

MARSH CONSERVATION

Salt marshes are clearly important habitats, but until recently, they have been abused and underappreciated. As just described, marshes are critical nursery grounds for numerous marine populations, including many commercially and recreationally valuable species. Turner (1977) has shown that, at large spatial scales, the yields of coastal fisheries worldwide are highly positively correlated with wetland area, suggesting a tight link between the nursery ground function of marshes and the size of adult populations.[50]

The role of marshes on shorelines is not restricted to their nursery ground function. As major elements of shoreline ecosystems around the world, they play a significant role in the global cycling of carbon, nitrogen, and sulfur. Salt marshes are potentially large sinks for the excess carbon that is being produced by the burning of fossil fuels, and chemical processes in marsh sediments can reduce or limit the accumulation of the nitrogen and sulfur by-products of human agriculture and industry.

Coastal marshes also buffer shorelines from flood and storm damage. After heavy rains, most flood damage occurs when runoff hits coastal floodplains. Marshes intercept and absorb storm runoff, slowing its discharge and minimizing its impact on coastal habitats. Coastal marshes also buffer shorelines from oceanic storms, both by absorbing the energy of storms as they hit the land and by limiting storm-induced erosion by binding shoreline sediments.

Coastal marshes are also widely believed to enhance water quality and serve as natural sewage treatment systems. The often large area of marshes along shorelines slows the movement of runoff, which increases sedimentation and the removal of particulate wastes from runoff. Moreover, both anaerobic and aerobic chemical processes, along with the action of a wide variety of decomposers, act to remove organic and inorganic wastes from water supplies. Because of the ability of salt marshes to process wastes and improve water quality, they are currently being examined for use in the processing of human sewage.[51]

In spite of the conspicuous societal value of marshes, they have a long history of abuse by humans. In the seventeenth and eighteenth centuries, particularly in New England, high marshes were used for grazing livestock and harvested for livestock feed and bedding. Colonial farmers also started the practice of ditching and digging marshes. This was initially done to increase grass production and enhance the value of marshes as farmland. Marsh plants were also harvested and burned for their ash, which was used for fertilizer and the manufacturing of glass, but this was more common in Europe than in North America. The common name for *Salicornia*, glasswort, appears to have come from the use of *Salicornia* ash for making glass in Europe.

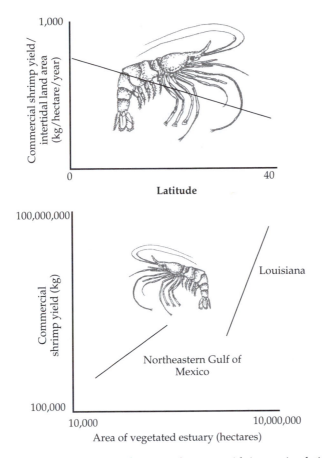

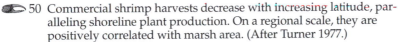

🦐 50 Commercial shrimp harvests decrease with increasing latitude, paralleling shoreline plant production. On a regional scale, they are positively correlated with marsh area. (After Turner 1977.)

The draining of marshes by ditching to control mosquitoes became popular in the late nineteenth century and continued well into the twentieth century. Massive marsh draining projects were initiated after the Civil War to stem the northern spread of yellow fever and malaria. These drainage projects were continued and escalated during the depression of the 1920s, when Civilian Conservation Corps (CCC) workers ditched large expanses of northern marshes.

The practice of filling marshes with dredge spills, which accelerated in the early twentieth century, was probably the worst single assault on our coastal marshes. As a by-product of the dredging of shallow waterways to keep them open for increasingly larger ships, large amounts of dredge material were produced. Industrious developers used this material to fill marshes near population centers, converting them into industrial parks, parking lots, airports, and shopping malls (see Dreyer and Niering 1995).

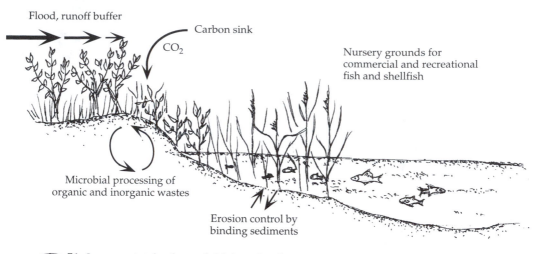

Flood, runoff buffer

Carbon sink

CO_2

Nursery grounds for commercial and recreational fish and shellfish

Microbial processing of organic and inorganic wastes

Erosion control by binding sediments

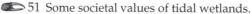 51 Some societal values of tidal wetlands.

Public transportation projects have had a particularly large impact on coastal marshes. Utilizing the relatively flat and unobstructed terrain of high marsh habitats skirting the coastline, engineers in the early twentieth century filled marshes to build highways and railroads. This led not only to widespread loss of marshes, but also to the isolation of remaining marshes from tidal flows. Without tidal flushing, salt marshes stagnated, gradually becoming freshwater marshes, and were invaded by the reed *Phragmites australis*.

The invasion of *Phragmites* in disturbed wetlands has become a serious problem in central Atlantic and southern New England marshes. The spread and dominance of *Phragmites* seems to be precipitated by several factors. First are tidal flow restrictions, which reduce the salinity of coastal marshes. Disturbances that create safe sites for *Phragmites* establishment are also a key factor. Eutrophication and nitrogen enrichment also appears to accelerate its spread at the cost of native plants, which may be better competitors when nutrients are in poor supply. Invasion of *Phragmites* has a number of undesirable consequences. *Phragmites* is a fierce competitor and can outcompete and entirely displace the native marsh plant community. This can lead to the local extinction not only of many native marsh plants, but also of the insects and birds that depend on them. Moreover, *Phragmites* forms large clones that may have considerable effects on shoreline ecosystems. *Phragmites* invasion typically results in marsh soil drying due to its ability to wick and transpire water out of its habitat. Thus, at large spatial scales, *Phragmites* invasion may lead to the loss of wetland area, wetland plant and animal diversity, and the services that marsh communities provide to shorelines.

Recognition of the societal and ecological importance of wetlands and the alarming rate at which they have been lost over the last two centuries has led to accelerated marsh conservation efforts over the last two decades. Particularly in the central Atlantic states and New England, where human population densities are the highest, marsh losses have been severe. By 1970, for example, Connecticut is thought to have lost at least 30 percent of its marshes. Rhode Island's marsh losses have been estimated to be as high as 95 percent! Current wetland protection laws, most of which date to the 1970s, now protect wetlands, and no-net-loss wetland policies have led to restoration efforts. The success of marsh restoration efforts, however, is clearly tied to how well we understand marshes. We cannot fix what we do not understand.

SUMMARY

Salt marsh plant communities are a conspicuous feature of Atlantic shorelines that develop in sheltered soft-sediment habitats. Like coral reefs, salt marshes are biogenic habitats, built by the plants and animals that live in them. Salt marshes begin when salt-tolerant plants colonize intertidal habitats, binding and accumulating sediments and leading to vertical and horizontal marsh growth. High soil salinities and low oxygen levels preclude most non-marsh plants from invading marshes.

The striking elevational plant zonation of marshes is the product of strong physical stress gradients across marsh landscapes and plant competition. At lower elevations, marsh soils are typically more waterlogged, anoxic, and salty than at higher elevations. Marsh plant habitats are also generally space-limited, leading to intense competition for space. As a general rule, competitively superior marsh plants dominate more physically benign elevations and competitively displace competitive subordinates to lower, more stressful habitats.

Whereas competition for space plays a major role in affecting the distributions and abundances of plants, positive interactions among plants, as well as between marsh plants and animals, are also important in marshes. Both within- and among-species group benefits enable many plants to live in marsh habitats. At low elevations, anoxic soils can limit individual survivorship, but dense groups of plants can thrive at low elevations through plant oxygenation of soils. Similarly, at higher marsh elevations, plant cover is necessary to limit the evaporative buildup of salt, which can kill most marsh plants. Consequently, while plant competition is an important determinant of plant spatial patterns across marsh habitats, plants could not even live in many marsh habitats without positive neighbor effects.

Marsh habitats are important nursery grounds for estuarine fishes and crustaceans. Like those of other soft-sediment habitats, marsh food chains are detritus-based, and many fishes that use marshes as nursery grounds are detritivores as juveniles. Protection from large mobile predators, however, may be the most important reason marshes are heavily utilized as nursery grounds.

Marshes provide humans with a number of important societal services. Marshes buffer coastlines from erosion, filter terrestrial inputs of nutrients and sediments, and are important habitats for fish and shellfish, including many commercially harvested species. Nonetheless, marshes have long been ditched, farmed, and filled, and are an extremely threatened and vulnerable habitat.

SUGGESTIONS FOR FURTHER READING

Adam, P. 1995. *Saltmarsh Ecology*. Cambridge University Press, Cambridge. A very well-done, current overview of salt marsh ecology.

Chapman, V. J. 1974. *Salt Marshes and Salt Deserts of the World*. Cramer Press, Lehre, England.

Dreyer, G. D. and W. A. Niering. 1995. *Tidal Marshes of Long Island Sound*. Connecticut College Arboretum Bulletin no. 34.

Nixon, S. W. 1980. Between coastal marshes and coastal waters: A review of twenty years of speculation and research on the role of salt marshes in estuarine productivity. In *Estuarine and Wetland Processes*, P. Hamilton and K. MacDonald (eds.), 437–525. Plenum Press, New York.

Nixon, S. W. 1982. *The Ecology of New England High Salt Marshes: A Community Profile*. Biological Report 81(55). U.S. Fish and Wildlife Service, Washington, D.C.

Ranwell, D. S. 1971. *Ecology of Salt Marshes and Sand Dunes*. Chapman and Hall, London.

Teal, J. 1985. *The Ecology of New England Low Marsh Habitats: A Community Profile*. U.S. Fish and Wildlife Service.

Wiegert, R. G. and B. J. Freeman. 1990. *Tidal Salt Marshes of the Southeast Atlantic Coast: A Community Profile*. Biological Report 85(7.29) U.S. Fish and Wildlife Service, Washington, D.C.

Key References for Atlantic Coast Salt Marsh Communities

Bertness, M. D. and A. M. Ellison. 1987. Determinants of pattern in a New England marsh plant community. *Ecological Monographs* 57: 129–147.

Bertness, M. D. and S. D. Hacker. 1994. Physical stress and positive associations among plants. *American Naturalist* 144:363–372.

Howes, B. L., J. W. H. Dacey and D. D. Goehringer. 1986. Factors controlling the growth form of *Spartina alterniflora:* Feedbacks between aboveground production, sediment oxidation, nitrogen and salinity. *Journal of Ecology* 74: 881–898.

Kneib, R. T. 1997. Tidal marshes offer a different perspective on estuarine nekton. *Annual Review of Oceanography and Marine Biology* 35: 1–120.

Mendelssohn, I. A. 1979. Nitrogen metabolism in the height forms of *Spartina alterniflora* in North Carolina. *Ecology* 60: 106–112.

Niering, W. A. and R. S. Warren. 1980. Vegetation patterns and processes in New England salt marshes. *BioScience* 30: 301–307.

Nixon, S. W., and C. A. Oviatt. 1973. Ecology of a New England salt marsh. *Ecological Monographs* 43: 463–498.

Redfield, A. C. 1972. Development of a New England salt marsh. *Ecological Monographs* 42: 201–237.

Valiela, I., and J. M. Teal. 1974. Nutrient limitation of salt marsh vegetation. In *Ecology of Halophytes*, R. J. Reimold and W. H. Queen (eds.), 547–563. Academic Press. New York.

Glossary

abiotic Of or relating to physical (e.g., heat, water loss) rather than biological (e.g., competition, predation) processes.

acclimation The process by which an organism adjusts physiologically to new physical conditions.

aerenchyma Tissue in vascular plants characteristic of waterlogged sediments that permits the diffusion of air to belowground parts.

Allee effect Low reproductive, recruitment, or feeding success due to low population densities; named after W. C. Allee.

amphipod Small (1–2 cm long) crustaceans with laterally compressed bodies. They are common scavengers on rocky and soft sediment beaches.

amensalism A relationship among organisms in which they negatively affect one another due to their effects on the habitat.

anaerobic Not requiring oxygen.

anoxic Without oxygen.

associational defense A form of group benefit in which a palatable organism gets protection from its consumers by living with less palatable prey.

ballast Heavy material carried on a ship to give it stability. Rocks and sand were used in wooden sailing ships, while today's ships carry water ballast..

benthic Of or relating to the sea bottom.

bioengineers Organisms that affect their habitats either by providing structure for other organisms (autogenic bioengineers) or by converting resources from one form to another (allogenic bioengineers).

biogenic Made by organisms.

biogeographic Pertaining to the large-scale distributions of organisms.

biotic Of or relating to biological (competition, predation) rather than physical (heat, water loss) processes.

bivalves The class of molluscs that are laterally compressed and covered with two shells, or valves; clams and mussels.

boundary layer A thin layer of slow-moving air or water over a surface, caused by friction.

brooding Carrying developing eggs.

bryozoans A phylum of clonal invertebrates that are often encrusting and important fouling organisms.

carbon fixer An organism that converts CO_2 to organic carbon molecules, usually through photosynthesis.

chemoreception The ability to sense chemicals.

chemotrophic Of or relating to an organism that oxidizes chemicals as an energy source.

chitons A class of molluscs that are scraping herbivores with a shell made up of seven dorsal plates of calcium carbonate.

clonal Of or relating to a modular organism made up of genetically identical, asexually produced units that are usually physiologically connected.

cnidarians The phylum of invertebrates that includes jellyfishes, anemones, and corals.

community The organisms living together in a habitat and their interrelationships.

compensation depth The deepest point at which a plant or alga can live, at which its photosynthetic energy gains are matched by metabolic costs.

competent Ready to settle.

competition Negative interactions among organisms contesting access to resources, such as food or space.

conspecifics Organisms of the same species.

continental drift The movement of continental plates on the earth's semi-liquid core.

crustaceans The class of arthropods, or organisms with jointed appendages, that are aquatic and have gills; includes crabs, shrimps, amphipods, isopods, and copepods.

cyprid The terminal settling larval form of barnacles and some other crustaceans.

density-dependent Of or relating to effects that are dependent on the density of individuals.

deposit feeders Organisms, such as fiddler crabs and many worms, that feed by processing sediments and extracting organic material.

desiccation Water loss.

detritivore An organism that utilizes detritus, or decomposing plant and animal material, as an energy source.

diatoms Single-celled algae with a silica shell, many of which are colonial.

dinoflagellates Single-celled algae with a cellulose shell, many of which are colonial or filamentous.

diploid Having two sets of genetic instructions, onefrom each parent.

direct development Development into free-living juveniles in the parental habitat.

dispersal The movement of propagules among habitats.

disturbance Any physical or biological process that kills organisms and thus provides opportunities for new colonists.

DOM Dissolved organic material, utilized as a nutrition source by many filter feeders.

drag The force on an object in a moving fluid that acts parallel to and in the same direction as the flow.

dynometer A device used to measure wave forces.

echinoderms The phylum of invertebrates that have a water vascular system; includes starfishes, sea urchinss, brittle stars, and sand dollars.

edaphic Of or relating to the physical conditions of sediments and soils.

eelgrass The temperate seagrass *Zostera*.

ephemeral Short-lived.

epibenthic Living on the substrate surface.

epiphytes Organisms that live on the surface of plants and seaweeds.

etiolated Having a tall, thin, light-seeking morphology as a result of low light availability.

eutrophication The chemical enrichment of natural habitats by the addition of nutrients, as by sewage and fertilizerrunoff.

facilitation The enhancement of the success of a species in a habitat by the presence of another species.

facultative Not necessary; optional.

filter feeders Organisms that feed by sieving small particulate material from the water column.

flume A laboratory tank with controlled water flow used to study how organisms are influenced by water movement; also called a flow tank.

flux The movement of something through a habitat (such as food, nitrogen, or larvae); the product of concentration and flow rate or delivery.

fouling organisms Organisms such as barnacles, bryozoans, tunicates, and hydroids that commonly colonize and "foul" docks and boat bottoms.

fugitive species Species that depend on rapid recruitment to live in ephemeral habitats often created by disturbances. Also known as weedy species.

gastropods The class of molluscs that includes snails, nudibranchs, and slugs.

geotactic Able to sense and respond to gravity.

group benefits Benefits gained by organisms from living together.

guerrilla strategy A clonal growth strategy with high investment in runners, allowing rapid spread and leading to a diffuse morphology.

habitat amelioration Improvement of habitat conditions.

halophyte A salt-loving plant that has the physiological machinery to deal with high salt concentrations.

haploid Having only a single set of genetic instructions.

hermaphroditic Having both male and female reproductive organs in the same individual.

heteromorphic Having two morphologically distinct life history stages.

heterospecifics Organisms of different species.

Holocene The recent or current epoch of geological time.

hummock An elevated mound of barnacles, mussels, or plants composed of crowded, morphologically modified individuals.

hydroids A class of cnidarians consisting mostly of inconspicuous fouling organisms.

hypotonic Having a salt concentration less than that of an organism, resulting in water passively moving into the organism and flooding it if it has no way of getting rid of excess water.

indeterminate growth A growth pattern in which individuals do not stop growing larger as they get older, as in lobsters and crabs.

indirect development Development that includes a larval dispersal stage.

indirect effects Effects of one species on another mediated by its effects on the habitat or on a third species.

infaunal Living in rather than on the substrate.

intermediate disturbance hypothesis A hypothesis predicting that species diversity will be greatest at intermediate levels of disturbance, at which competitive dominants and weedy species coexist; at low disturbance levels a competitive dominant takes over, whereas at high levels only weedy species persist.

internal waves Shoreward-moving waves generated when oceanic waves hit the continental shelf and become compressed.

interstitial Between sediment grains.

introduced species Species that have been inadvertently or intentionally brought into a new region by the activities of humans.

isopod Small (1–2 cm long) crustaceans with dorsoventrally compressed bodies.

keystone species Species that play a pivotal role in the abundances and distributions of other species in a natural community.

lecithotrophic Developing in the plankton, but nourished by parentally provided yolk.

lift The force on an object in a moving fluid that acts perpendicular to the direction of the flow.

limiting resource A resource, such as food or space, that limits the abundance of a species in a given habitat.

limpets Gastropods with uncoiled, cone-shaped shells.

littoral On or along the coast.

mariculture The culturing of marine organisms.

meiofauna Organisms that live between sediment grains; typically copepods, nematodes, and small annelids.

meiosis Reduction cell division, in which chromosome number is reduced from diploid to haploid.

mitosis Cell division without a change in chromosome number.

molluscs The invertebrate phylum containing gastropods, bivalves, and squids.

mutualism A relationship between two species in which both benefit.

native species A species with a long history of living in a particular region.

nauplius larva One of the planktonic larval forms of many crustaceans.

neap tides Monthly lows in the magnitude of tides that occur when the gravitational forces of the sun and moon oppose each other.

nurse plant An adult plant that provides some benefit, usually in the form of habitat amelioration, to seedlings.

nursery grounds Habitats that are used by juveniles for food and protection from predators and physical stresses.

osmotic stress The problem of living in a solution with a salt concentration different from that of one's body. In dilute (hypotonic) solutions, water floods the body, whereas in more saline (hypertonic) solutions, water is drawn out of the body.

passive deposition The deposition of sediments and/or larvae in microhabitats due to low water movement rather than active habitat choice.

pelagic In the water column; the open water above the seafloor.

phalanx strategy A clonal growth strategy resulting in dense aboveground and belowground structure and little investment in rapid expansion; contrast with guerrilla strategy.

phototactic Able to detect and respond to light.

phylogeny A pattern of evolutionary relationships among organisms.

physical stresses Abiotic (heat, water loss, wave impacts) stresses as opposed to biotic (competition, predation) stresses.

planktonic Of or relating to microorganisms living in the water column.

planktotrophic Developing in the water column while feeding on other planktonic organisms.

plasticity Expression of a trait that is not fixed and varies with the environment.

Pleistocene The epoch in geological history that ended 10,000 years ago; characterized by the appearance of modern humans.

polychaetes Marine segmented worms with lateral appendages.

positive feedback The enhancement by some interaction of similar interactions in the future.

positive interactions Interactions in which at least one species benefits, while the other is either not influenced or also benefits.

positive switches Positive feedbacks or interactions that control thresholds that dictate state changes in habitats.

primary production The fixing of inorganic carbon from an outside energy source, usually solar energy.

propagule Generic term for the seeds, spores, and larvae used by plants, algae, and animals to reproduce themselves. Usually produced sexually.

pseudofeces In clams and mussels, particulate material filtered from the water column that is disposed of without going through the digestive system.

radula The scraping tongue of gastropods, chitons, and octopods.

ramets The vegetatively produced individual units of a clonal plant or invertebrate.

recruitment The addition of new members to a population, either through sexual reproduction or the immigration of individuals.

rhizome A rootlike stem, usually horizontal, sent out from a plant.

rhizosphere The root zone of a plant.

salt stress The presence of high water or sediment salinities that draw water osmotically out of organisms.

sea cucumbers A class of echinoderms that are deposit or filter feeders.

secondary succession The recovery of a community from a disturbance event.

seed bank The supply of propagules that are present in a habitat and will germinate if an opportunity to do so occurs.

semi-diurnal tides A tidal pattern in which there are two high and two low tides a day, each of different magnitude; by far the most common tidal pattern on earth, and the only tidal pattern found on northern Atlantic shorelines.

shellfish A common term for clams, mussels, shrimps, and lobsters; commercially harvested marine organisms other than fishes.

sheltered Not exposed to heavy waves.

slicks Parcels of surface water that are moved by local oceanographic conditions, such as internal waves, winds, or tidal flows.

spring bloom The seasonal spring pulse in primary production in most temperate zone bodies of water.

spring tides Monthly extremes in the magnitude of tides that occur when the gravitational forces of the sun and moon act together on the earth's water masses.

stomata Cells on the leaves of vascular plants that regulate gas and water exchange.

succession Changes in community composition and structure over time.

supply-side ecology The study of the role played by recruit supply in the structure and dynamics of communities.

swell Waves in large bodies of water that are the product of the coalescing of wind- and tide-generated waves.

top-down forces Forces that control population and community structure through consumers.

transpiration The loss of water by vascular plants that occurs passively as a by-product of gas exchange.

trophic Of or relating to feeding or food webs.

trophic cascade An indirect consumer effect in which a consumer influences the abundance of its prey's food.

tunicates Sea squirts; a phylum of filter-feeding organisms that are typically clonal and are common fouling organisms.

turbulent flow Fluid flow dominated by eddies and chaotic water movement, as opposed to organized parallel (laminar) flow.

turf morphology A clonal plant architecture in which ramets (individual clonal units) are tightly packed, leading to a dense, competitively dominant morphology.

upwelling The movement of cold, nutrient-rich bottom water into coastal habits to replace water moved offshore by winds or ocean currents.

vegetative reproduction Asexual reproduction.

veliger The terminal planktonic larval form of most gastropod and bivalve molluscs.

vivipary Bearing live young.

water column The water mass lying above the benthic (seafloor) habitat; the open water where planktonic and pelagic organisms live.

weedy species Species equipped to colonize rapidly; also known as fugitive species.

wind waves Waves that are the result of winds.

wrack Floating dead plant material (often including seeds) that is carried by winds and currents onto shorelines.

zooids Individuals in a bryozoan or hydroid colony.

Bibliography

Ackerman, J. D. 1997. Submarine pollination in the marine angiosperm *Zostera marina*. II. Pollen transport in flow fields and capture by stigmas. *American Journal of Botany* 84: 1110–1119.

Adam, P. 1990. *Saltmarsh Ecology*. Cambridge University Press, Cambridge.

Ahn, Y., G. Lopez and R. Malouf. 1993. Effects of the gem clam *Gemma gemma* on early post-settlement emigration, growth and survival of the hard clam *Mercenaria mercenaria*. *Marine Ecology Progress Series* 99: 61–70.

Allee, W. C. 1923. Studies in marine ecology. IV. The effect of temperature in limiting the geographic range of invertebrates of the Woods Hole littoral. *Ecology* 4: 341–354.

Allen, E. A., P. E. Fell, M. A. Peck, J. A. Gieg, C. R. Guthke and M. D. Newkirk. 1994. Gut contents of common mummichogs, *Fundulus heteroclitu*, in a restored impounded marsh and in a natural reference marshes. *Estuaries* 17: 462–471.

Ambrose, W. G. 1984a. Influences of predatory polychaetes and epibenthic predators on the structure of a soft-bottom community in a Maine estuary. *Journal of Experimental Marine Biology and Ecology* 81: 115–145.

Ambrose, W. G. 1984b. Role of predatory infauna in structuring marine soft-bottom communities. *Marine Ecology Progress Series* 17: 109–115.

Ambrose, W. G. 1986. Estimate of removal rate of *Nereis virens* (Polychaeta: Nereidae) from an intertidal mudflat by gulls (*Larus* spp.). *Marine Biology* 90: 243–247.

Ambrose, W. G. 1991. Are infaunal predators important in structuring marine soft-bottom communities? *American Zoologist* 31: 849–860.

Anderson, C. E. 1974. A review of structures of several North Carolina salt marsh plants. In *Ecology of Halophytes*, R. J. Reimold and W. H. Queen (eds.), 307–344. Academic Press, New York.

Anderson, C. M. and M. Treshow. 1980. A review of environmental and genetic factors that affect *Spartina alterniflora* height forms. *Estuaries* 3: 168–176.

Anderson, D. 1993. *Barnacle Biology*. Chapman and Hall. London.

Antlfinger, A. E. and E. L. Dunn. 1979. Seasonal patterns of CO_2 and water vapor exchange of three salt marsh succulents. *Oecologia* 43: 249–260.

Appleton, R. D. and A. R. Palmer. 1988. Water-borne stimuli released by predatory crabs and damaged prey induce more predator-resistant shells in a marine gastropod. *Proceedings of the National Academy of Science USA* 85: 4387–4391.

Armstrong, W. 1979. Aeration in higher plants. *Advances in Botanical Research* 7: 225–332.

Atsatt, P. R. and D. J. O'Dowd. 1976. Plant defense guilds. *Science* 193: 24–29.

Barlocher, F. and S. Y. Newell. 1994a. Growth of the salt marsh periwinkle *Littoraria irrorata* on fungal and cordgrass diets. *Marine Biology* 118: 109–114.

Barlocher, F. and S. Y. Newell. 1994b. Phenolics and proteins affecting palatability of *Spartina* leaves to the gastropod *Littoraria irrorata*. *Marine Ecology* 15(1): 65–75.

Barnes, H. 1962. Note on variations in the release of nauplii of *Balanus balanoides* with special reference to the spring diatom outburst. *Crustaceana* 4: 118–122.

Barnes, H. and M. Barnes. 1957. Resistance to desiccation in intertidal barnacles. *Science* 126: 358.

Barnes, H. and H. T. Powell. 1950. The development, general morphology, and subsequent elimination of barnacle populations after a heavy initial settlement. *Journal of Animal Ecology* 19: 175–179.

Barnes, R. D. 1987. *Invertebrate Zoology*, 4th Ed. Saunders, Philadelphia.

Bascom, W. 1980. *Waves and Beaches*. Doubleday, New York.

Bayne, B. L. 1964. Primary and secondary settlement in *Mytilus edulis*. *Journal of Animal Ecology* 33: 516–523.

Bayne, B. L. 1969. The gregarious behavior of the larvae of *Ostrea edulis* at settlement. *Journal of the Marine Biological Association* (UK) 49: 327–356.

Bazely, D. R. and R. L. Jefferies. 1985. Goose feces: A source of nitrogen for plant growth in a grazed salt marsh. *Journal of Applied Ecology* 22: 693–703.

Bell, E. C. and M. W. Denny. 1994. Quantifying "wave exposure": A simple device for recording maximum velocity and results of its use at several field sites. *Journal of Experimental Marine Biology and Ecology* 181: 9–29.

Bell, S. S. 1980. Meiofauna-macrofauna interactions in a high salt marsh habitat. *Ecological Monographs* 50(4): 487–505.

Bell, S. S. and B. Coull. 1978. Field evidence that shrimp predation regulates meiofauna. *Oecologia* 35: 141–148.

Bergeron, P. and E. Bourget. 1986. Shore topography and spatial partitioning of crevice refuges by sessile epibenthos in an ice disturbed environment. *Marine Ecology Progress Series* 28: 129–145.

Bert, T. M. 1986. Speciation in western Atlantic stone crabs (genus *Menippe*): The role of geological processes and climatic events in the formation and distribution of species. *Marine Biology* 93: 157–170.

Bertness, M. D. 1981. Competitive dynamics of a tropical hermit crab assemblage. *Ecology* 2(3): 751–761.

Bertness, M. D. 1984a. Habitat and community modification by an introduced herbivorous snail. *Ecology* 65(2): 370–381.

Bertness, M. D. 1984b. Ribbed mussels and the productivity of *Spartina alterniflora* in a New England salt marsh. *Ecology* 65: 1794–1807.

Bertness, M. D. 1985. Fiddler crab regulation of *Spartina alterniflora* production on a New England salt marsh. *Ecology* 66: 1042–1055.

Bertness, M. D. 1988. Peat accumulation and the success of marsh plants. *Ecology* 69: 703–713.

Bertness, M. D. 1989. Intraspecific competition and facilitation in northern acorn barnacles. *Ecology* 70(1): 257–268.

Bertness, M. D. 1991a. Interspecific interactions among high marsh perennials in a New England salt marsh. *Ecology* 72(1): 125–137.

Bertness, M. D. 1991b. Zonation of *Spartina patens* and *Spartina alterniflora* in a New England salt marsh. *Ecology* 72(1): 138–148.

Bertness, M. D. 1992. The ecology of New England salt marsh plant communities. *American Scientist* 80: 260–268.

Bertness, M. D. and R. Callaway. 1994. Positive interactions in communities: A post Cold War perspective. *Trends in Ecology and Evolution* 9: 191–193.

Bertness, M. D. and C. Cunningham. 1981. Crab predation and gastropod shell architecture. *Journal of Experimental Marine Biology and Ecology* 50: 213–230.

Bertness, M. D. and A. M. Ellison. 1987. Determinants of pattern in a New England marsh plant community. *Ecological Monographs* 57: 129–147.

Bertness, M. D. and S. D. Gaines. 1992. Larval dispersal and local adaptation in acorn barnacles. *Evolution* 47: 316–320.

Bertness, M. D. and T. Grosholz. 1985. Population dynamics of the ribbed mussel *Geukensia demissa*: The costs and benefits of a clumped distribution. *Oecologia* 67: 192–204.

Bertness, M. D. and S. D. Hacker. 1994. Physical stress and positive associations among plants. *American Naturalist* 144: 363–372.

Bertness, M. D. and G. Leonard. 1997. The role of positive interactions in communities: Lessons from the intertidal. *Ecology* 78: 1976–1989.

Bertness, M. D. and T. Miller. 1984. The distribution and dynamics of *Uca pugnax* (Smith) burrows in a New England salt marsh. *Journal of Experimental Marine Biology and Ecology* 83: 211–237.

Bertness, M. D. and S. W. Shumway. 1993. Competition and facilitation in marsh plants. *American Naturalist* 142: 718–724.

Bertness, M. D. and S. M. Yeh. 1994. Cooperative and competitive interactions in the recruitment of marsh elders. *Ecology* 75: 2416–2429.

Bertness, M. D., S. D. Garrity and S. C. Levings. 1981. Predation pressure and gastropod foraging patterns: A latitudinal pattern. *Evolution* 35(5): 995–1007.

Bertness, M. D., P. O. Yund and A. F. Brown. 1983. Snail grazing and the abundance of algal crusts on a sheltered New England rocky beach. *Journal of Experimental Marine Biology and Ecology* 71: 147–164.

Bertness, M. D., C. Wise and A. M. Ellison. 1987. Consumer pressure and seed set in a salt marsh perennial plant community. *Oecologia* 71: 190–200.

Bertness, M. D., T. Chatkupt and K. Wikler. 1992a. Flood tolerances and the distribution of *Iva frutescens* across New England salt marshes. *Oecologia* 91: 171–178.

Bertness, M. D., L. Gaugh and S. W. Shumway. 1992b. Salt tolerances and the distribution of plants across a New England salt marsh. *Ecology* 72: 1842–1851.

Bertness M. D., S. D. Gaines and R. Wahle. 1996. Wind-driven spatial patterns in the settlement of benthic shoreline invertebrates. *Marine Ecology Progress Series* 137: 103–110.

Bertness M. D., S. D. Gaines and S. M. Yeh. 1998. Making mountains out of barnacles: The causes and consequences of barnacle hummocks. *Ecology* 79: 1382–1394.

Bingham, B. L. and L. J. Walters. 1989. Solitary ascidians as predators of invertebrate larvae: Evidence from gut analyses and plankton samples. *Journal of Experimental Marine Biology and Ecology* 131: 147–159.

Blinks, L. R. 1955. Photosynthesis and productivity of littoral marine algae. *Journal of Marine Research* 14: 363–373.

Blundon, J. A. and V. S. Kennedy. 1982. Refuges for infaunal bivalves from blue crab, *Callinectes sapidus* (Rathbun), predation in Chesapeake Bay. *Journal of Experimental Marine Biology and Ecology* 65: 67–81.

Boesch, D. F. and R. E. Turner. 1984. Dependence of fishery species on salt marshes: The role of food and refuges. *Estuaries* 7: 460–468.

Botton, M. L. 1984a. Effects of laughing gull and shorebird predation on the intertidal fauna at Cape May, New Jersey. *Estuarine, Coastal, and Shelf Science* 18: 209–220.

Botton, M. L. 1984b. The importance of predation by horseshoe crabs, *Limulus polyphemus*, to an intertidal sand flat community. *Journal of Marine Research* 42: 139–161.

Boulding, E. G. 1984. Crab-resistant features of the shells of burrowing bivalves: Decreasing vulnerability by increasing handling time. *Journal of Experimental Marine Biology and Ecology* 76: 201–223.

Bourget, E. and E. Holm. 1994. Large-scale population genetics of the barnacle *Semibalanus balanoides* in the Gulf of St. Lawrence and the NW Atlantic. In *Changes in Fluxes in Estuaries*, K. Dyer and R. Orth (eds.), 295–300. Olsen and Olsen, Fredensborg, Denmark.

Breitburg, D. L. 1984. Residual effects of grazing: Inhibition of competitor recruitment by encrusting coralline algae. *Ecology* 65(4): 1136–1143.

Brenchley, G. A. and J. T. Carlton. 1983. Competitive displacement of native mud snails by introduced periwinkles in the New England intertidal zone. *Biological Bulletin* 165: 543–558.

Brewer, J. S. and M. D. Bertness. 1996. Disturbance and intraspecific variation in the clonal morphology of salt marsh perennials. *Oikos* 77: 107–116.

Brown, A. C. and A. McLachlan. 1990. *Ecology of Sandy Shores*. Elsevier, Amsterdam.

Buchsbaum, R., I. Valiela and J. M. Teal. 1981. Grazing by Canada geese and related aspects of the chemistry of salt marsh grass. *Colonial Waterbirds* 4: 126–131.

Burrell, V. G. and W. A. Engel. 1976. Predation by and distribution of a ctenophore *Mnemiopsis leidyi* Agassiz, in the York River estuary. *Estuarine and Coastal Marine Science* 4: 235–242.

Burrows, E. M. and S. M. Lodge. 1950. Note on the inter-relationships of *Patella, Balanus* and *Fucus* on a semi-exposed coast. *Reports of the Biological Station at Port Erin* 62: 30–34.

Buss, L. W. 1979a. Bryozoan overgrowth interactions: The interdependency of competition for space and food. *Nature* 281: 475–477.

Buss, L. W. 1979b. Habitat selection, directional growth, and spatial refuges: Why colonial animals have more hiding places. In *Biology and Systematics of Colonial Organisms*, B. Rose and G. Larwood (eds.), 459–499. Academic Press, London.

Buss, L. W. 1980. Competitive intransitivity and the size-frequency distributions of interacting populations. *Proceedings of the National Academy of Sciences USA* 77: 5355–5359.

Buss, L. W. 1986. Competition and community organization on hard surfaces in the sea. In *Community Ecology*, J. Diamond and T. J. Case (eds.), 517–536. Harper & Row. New York.

Buss, L. W. and J. B. C. Jackson. 1979. Competitive networks: Nontransitive competitive relationships in cryptic coral reef environments. *American Naturalist* 113(2): 223–234.

Butman, C. A. 1987. Larval settlement of soft-sediment invertebrates: The spatial scales of pattern explained by active habitat selection and the emerging role of hydrodynamical processes. *Oceanography and Marine Biology Annual Review* 25: 113–165.

Butman, C. A. 1989. Sediment-trap experiments on the importance of hydrodynamical processes in distributing settling invertebrate larvae in near-bottom waters. *Journal of Experimental Marine Biology and Ecology* 134: 37–88.

Butman, C. A., J. P. Grassle and C. M. Webb. 1988. Substrate choices made by marine larvae in still water and in a flume flow. *Nature* 333: 771–773.

Caffey, H. M. 1985. Spatial and temporal variation in settlement and recruitment of intertidal barnacles. *Ecological Monographs* 55: 312–332.

Callaway, R. M. 1994. Facultative and interfering effects of *Arthrocnemum subterminale* on winter annuals in a California salt marsh. *Ecology* 75: 681–686.

Cammen, L. M. 1980. The significance of microbial carbon in the nutrition of the deposit feeding polychaete, *Nereis succinea*. *Marine Biology* 61: 9–20.

Carefoot, T. 1977. *Pacific Seashores*. J. J. Douglas, Vancouver, Canada.

Cargill, S. M. 1984. The effects of grazing by lesser snow geese on the vegetation of a sub-arctic salt marsh. *Journal of Applied Ecology* 21: 669–686.

Cargill, S. M. and R. L. Jefferies. 1984. Nutrient limitation of primary production in a sub-arctic salt marsh. *Journal of Applied Ecology* 21: 657–668.

Carlton, J. T. 1982. The historical biogeography of *Littorina littorea* on the Atlantic coast of North America and implications for the interpretation of the structure of New England intertidal communities. *Malacological Review* 15: 146.

Carlton, J. T. 1985. Transoceanic and interoceanic dispersal of coastal marine organisms: The biology of ballast water. *Oceanography and Marine Biology Annual Review* 23: 313–371.

Carlton, J. T. 1987. Patterns of transoceanic marine biological invasions in the Pacific Ocean. *Bulletin of Marine Science* 41(2): 452–465.

Carlton, J. T. 1992. Blue immigrants: The marine biology of maritime history. Mystic Seaport Museum 44(2): 31–36.

Carlton, J. T. 1993. Neoextinctions of marine invertebrates. *American Zoologist* 33: 499–509.

Carlton, J. T. and J. A. Scanlon. 1985. Progression and dispersal of an introduced Alga: *Codium fragile* (Chlorophyta) on the Atlantic coast of North America. *Botanica Marina* 27: 155–165.

Carlton, J. T., G. J. Vermeij, D. R. Lindberg, D. A. Carlton and E. C. Dudley. 1991. The first historical extinction of a marine invertebrate in an ocean basin: The demise of the eelgrass limpet *Lottia alveus*. *Biological Bulletin* 180: 72–80.

Carpenter, S. R.and J. F. Kitchell. 1985. Cascading trophic interactions and lake productivity. *BioScience* 35: 634–639.

Carpenter, S. R. and J. F. Kitchell. 1993. *Trophic Cascades in Lakes*. Cambridge University Press, Cambridge.

Carr, M. H. 1994. Effects of macroalgal dynamics on the recruitment of a temperate reef fish. *Ecology* 75: 1320–1333.

Carriker, M. 1955. *The Biology of the Oyster Drill, Urosalpinx cinera*. Special Report. U.S. Fish and Wildlife Service, Washington, D.C.

Carrington, E. 1990. Drag and dislodgment of an intertidal alga: Consequences of morphological variation in *Mastocarpus*. *Journal of Experimental Marine Biology and Ecology* 139: 185–200.

Castilla, J. C. and L. R. Duran. 1985. Human exclusion from the rocky intertidal zone of central Chile: The effects on *Concholepas concholepas* (Gastropoda). *Oikos* 45: 391–399.

Caswell, H. 1978. Predator mediated coexistence: A nonequilibrium model. *American Naturalist* 112: 127–154.

Chapman, A. R. O. 1986. Population and community ecology of seaweeds. *Advances in Marine Biology* 23: 1–161.

Chapman, A. R. O. and J. S. Craigie. 1977. Seasonal growth of *Laminaria longicruris*: Relations with dissolved inorganic nutrients and internal reserves of nitrogen. *Marine Biology* 40: 197–205.

Chapman, V. J. 1970. *Seaweeds and Their Uses*. Methuen and Company, London.

Chapman, V. J. 1974. *Salt Marshes and Salt Deserts of the World*. Interscience Press, New York.

Charnov, E. L. 1976. Optimal foraging: The marginal value theorem. *Theoretical Population Biology* 9: 129–136.

Clements, F. E. 1916. *Plant Succession*. Publication 242. Carnegie Institute, Washington, D.C.

Coleman, J. S. 1933. The nature of intertidal zonation of plants and animals. *Journal of the Marine Biology Association* (UK) 61: 71–93.

Commito, J. A. 1982. Importance of predation by infaunal polychaetes in controlling the structure of a soft-bottom community in Maine. *Marine Biology* 68: 77–81.

Commito, J. A. and E. M. Boncavage. 1989. Suspension-feeders and coexisting infauna: An enhancement counterexample. *Journal of Experimental Marine Biology and Ecology* 125: 33–42.

Commito, J. A. and P. B. Shrader. 1985. Benthic community response to experimental additions of the polychaete *Nereis virens*. *Marine Biology* 86: 101–107.

Connell, J. H. 1961. The influence of interspecific competition and other factors on the distribution of the barnacle *Chthamalus stellatus*. *Ecology* 42: 710–723.

Connell, J. H. 1972. Community interactions on marine rocky shores. *Annual Review of Ecology and Systematics* 3: 169–192.

Connell, J. H. 1978. Diversity in tropical rain forests and coral reefs. *Science* 199: 1302–1310.

Connell, J. H. 1983. On the prevalence and relative importance of interspecific competition: Evidence from field experiments. *American Naturalist* 122(5): 661–696.

Connell, J. H. 1985. Consequences of variation in settlement vs. recruitment in rocky intertidal communities. *Journal of Experimental Marine Biology and Ecology* 95: 11–45.

Connell, J. H. and R. O. Slatyer. 1977. Mechanisms of succession in natural communities and their role in community stability and organization. *American Naturalist* 111(982): 1119–1144.

Connor, M. S., J. M. Teal and I. Valiela. 1982. The effect of feeding by mud snails, *Ilyanassa obsoleta* (Say), on the structure and metabolism of a laboratory benthic algal community. *Journal of Experimental Marine Biology and Ecology* 65: 29–45.

Cooper, A. 1982. The effects of salinity and waterlogging on the growth and cation uptake of salt marsh plant. *New Phytologist* 90: 263–275.

Costa, J. E. 1988. Distribution, production and historical changes in the abundance of eelgrass in southeastern Massachusetts. Ph.D. dissertation, Boston University.

Cottam, C. and D. A. Munro. 1954. Eelgrass status and environmental relations. *Journal of Wildlife Management* 18: 449–460.

Coull, B. C. 1985. Long-term variability of estuarine meiobenthos: An 11 year study. *Marine Ecology Progress Series* 24: 205–218.

Covi, M. P. and R. T. Kneib. 1995. Intertidal distribution, population dynamics and production of the amphipod *Uhlorchestia spartinophila* in a Georgia, USA, salt marsh. *Marine Biology* 121: 447–455.

Crane, J. 1975. *Fiddler Crabs of the World*. Princeton University Press, Princeton, NJ.

Crichton, O. W. 1960. Marsh crab: Intertidal tunnel-maker and grass-eater. *Estuarine Bulletin* 5: 3–10.

Crisp, D. J. 1955. The behavior of barnacle cyprids in relation to water movement over a surface. *Journal of Experimental Biology* 32: 569–590.

Crisp, D. J. and P. S. Meadows. 1962. The chemical basis of gregariousness in cirripedes. *Proceedings of the Royal Society of London* B 156: 500–520.

Croker, A. R. 1967. Niche diversity in five sympatric species of intertidal amphipods (Crustacea: Haustoriidae). *Ecological Monographs* 37: 173–200.

Dahl, E. 1952. Some aspects of the ecology and zonation of the fauna of sandy beaches. *Oikos* 4: 1–27.

Dame, R. F. 1976. Energy flow in an intertidal oyster population. *Estuarine and Coastal Marine Science* 4: 243–253.

Dame, R. F. 1993. The role of bivalve filter feeding in material fluxes in estuarine ecosystems. In *Bivalve Filter Feeders in Estuarine and Coastal Ecosystem Processes*, R. F. Dame (ed.), 245–269. Springer-Verlag, New York.

Dame, R. F. 1996. *Ecology of Marine Mussels: An Ecosystem Approach*. CRC Press, Boca Raton, Florida.

Davis, R. A. 1994. *The Evolving Coast*. Scientific American Library. W. H. Freeman, New York.

Dayton, P. K. 1971. Competition, disturbance and community organization: The provision and subsequent utilization of space in a rocky intertidal community. *Ecological Monographs* 41: 351–389.

Dayton, P. K. 1973. Dispersion, dispersal, and persistence of the annual intertidal alga *Postelsia palmaeformis*. *Ecology* 54: 433–438.

Dayton, P. K. 1975. Experimental evaluation of ecological dominance in a rocky intertidal algal community. *Ecological Monographs* 54: 253–289.

Dean, T. A. and L. E. Hurd. 1980. Development in an estuarine fouling community: The influence of early colonists on later animals. *Oecologia* 46: 295–301.

Den Hartog, C. 1970. *The Seagrasses of the World*. North-Holland, Amsterdam.

Denno, R. F. 1980. Ecotype differentiation in a guild of sap-feeding insects on the salt marsh grass, *Spartina patens*. *Ecology* 61(3): 702–714.

Denno, R. F. and E. E. Grissell. 1979. The adaptiveness of wing-dimorphism in the salt marsh-inhabiting planthopper, *Prokelisia marginata* (Homoptera: Delphacidae). *Ecology* 60(1): 221–236.

Denno, R. F., L. W. Douglass and D. Jacobs. 1985. Crowding and host plant nutrition: Environmental determinants of wing-form in *Prokelisia marginata*. *Ecology* 66(5): 1588–1596.

Denny, M. W. 1987. Life in the Maelstrom: The biomechanics of wave-swept rocky shores. *Trends in Ecology and Evolution* 2(3): 61–66.

Denny, M. W. 1988. *The Biology and Mechanics of the Wave-Swept Environment*. Princeton University Press. Princeton, NJ.

Denny, M. W. and M. Shibata. 1989. Consequences of surf-zone turbulence for settlement and external fertilization. *American Naturalist* 134(6): 859–889.

Denny, M. W., T. Daniel and M. A. R. Koehl. 1985. Mechanical limits to size in wave swept organisms. *Ecological Monographs* 55: 69–102.

Dethier, M. N. 1981. Heteromorphic algal life histories: The seasonal pattern and response to herbivory of the brown crust, *Ralphsia californica*. *Oecologia* 49: 333–339.

Dethier, M. N. and D. O. Duggins. 1984. An "indirect commensalism" between marine herbivores and the importance of competitive hierarchies. *American Naturalist* 124(2): 205–219.

Devries, A. L. 1971. Glycoproteins as biological antifreeze agents in antarctic fishes. *Science* 172: 1152–1155.

Dinsmore, R. P. 1972. Ice and its drift into the North Atlantic ocean. *International Commission of Northwestern Atlantic Fisheries (ICNAF) Special Publication* 8: 89–128.

Doering, P. H., C. A. Oviatt and J. R. Kelly. 1986. The effects of the filter-feeding clam *Mercenaira mercenaria* on carbon cycling in experimental marine mesocosms. *Journal of Marine Research* 44: 839–861.

Doering, P. H., J. R. Kelly, C. A. Oviatt and T. Sowers. 1987. Effect of the hard clam *Mercenaira mercenaria* on benthic fluxes of inorganic nutrients and gases. *Marine Biology* 94: 377–383.

Doty, M. S. 1946. Critical tide factors that are correlated with the vertical distribution of marine algae and other organisms along the Pacific coast. *Ecology* 27: 315–328.

Dreyer, G. D. and W. A. Niering. 1995. *Tidal marshes of Long Island Sound*. Connecticut College Arboretum Bulletin 34.

Dring, M. J. 1981. Chromatic adaptation of photosynthesis in benthic marine algae: An examination of its ecological significance using a theoretical model. *Limnology and Oceanography* 26: 271–284.

Dring, M. J. and F. A. Brown. 1982. Photosynthesis of intertidal brown algae during and after periods of emersion: A renewed search for physiological causes of zonation. *Marine Ecology Progress Series* 8: 301–308.

Druehl, L. D. 1967. Vertical distribution of some benthic marine algae in British Columbia as related to some environmental factors. *Journal of the Canadian Fisheries Research Board* 24: 33–46.

Dudgeon, S. R. and A. S. Johnson. 1992. Thick vs. thin: Thallus morphology and tissue mechanics influence differential drag and dislodgement of two co-dominant seaweeds. *Journal of Experimental Marine Biology and Ecology* 165: 23–43.

Dudgeon, S. R., I. R. Davison and R. L. Vadas. 1989. Effect of freezing on photosynthesis of intertidal macroalgae: Relative tolerance of *Chondrus cris-*

pus and *Mastocarpus stellatus* (Rhodophyta). *Marine Biology* 101: 107–114.

Duffy, J. E. 1990. Amphipods on seaweeds: Partners or pests? *Oecologia* (Berlin) 83: 267–276.

Duffy, J. E. and M. E. Hay. 1990. Seaweed adaptations to herbivory. *BioScience* 40(5): 368–375.

Duggins, D. O., J. E. Eckman and A. T. Sewell. 1990. Ecology of understory kelp environments. II. Effects of kelps on recruitment of benthic invertebrates. *Journal of Experimental Marine Biology and Ecology* 143: 27–45.

Dumas, J. V. and J. D. Witman. 1993. Predation by gulls on two rocky intertidal crab species. *Journal of Experimental Marine Biology and Ecology* 169: 89–101.

Dungan, M. L. 1986. 3–way interactions in a Sonoran desert rocky intertidal zone. *American Naturalist* 127: 292–316.

Ebling, F. J., J. A. Kitching, L. Muntz and M. Taylor. 1964. Experimental observations on the destruction of *Mytilus edulis* and *Nucella lapillus* by crabs. *Journal of Animal Ecology* 33: 73–82.

Eckman, J. E. 1979. Small-scale patterns and processes in a soft-substratum intertidal community. *Journal of Marine Research* 37: 437–457.

Eckman, J. E. 1983. Hydrodynamic processes affecting benthic recruitment. *Limnology and Oceanography* 18: 241–257.

Eckman, J. E. 1985. Flow disruption by an animal-type mimic affects sediment bacterial colonization. *Journal of Marine Research* 43: 419–435.

Eckman, J. E., D. O. Duggins and A. T. Sewell. 1989. Ecology of understory kelp environments. I. Effects of kelps on flow and particle transport near the bottom. *Journal of Experimental Marine Biology and Ecology* 129: 173–187.

Edwards, D. C. and J. D. Huebner. 1977. Feeding and growth rates of *Polonices duplicatus* preying on *Mya arenaria* at Barnstable Harbor, Massachusetts. *Ecology* 58: 1218–1236.

Ellis, M. J. and B. C. Coull. 1989. Fish predation on meiobenthos: Field experiments with juvenile spot *Leiostromus xanthurus* Lacepede. *Journal of Experimental Marine Biology and Ecology* 130: 19–32.

Ellison, A. M. 1987a. Density-dependent dynamics of *Salicornia europaea* monocultures. *Ecology* 68: 737–741.

Ellison, A. M. 1987b. Effects of competition, disturbance, and herbivory on *Salicornia europaea*. *Ecology* 68: 576–586.

Ellison, A. M. 1989. Morphological determinants of self-thinning in plant monocultures and a proposal concerning the role of self-thinning in plant evolution. *Oikos* 54: 287–293.

Ellison, A. M., E. J. Farnsworth and R. R. Twilley. 1996. Facultative mutualism between red mangroves and root-fouling sponges in Belizean mangel. *Ecology* 77: 2431–2444.

Elner, R. W. and R. N. Hughes. 1978. Energy maximization in the diet of the shore crab *Carcinus maenas*. *Journal of Animal Ecology* 47: 103–116.

Elton, C. S. 1958. *The Ecology of Invasions by Plants and Animals*. Chapman and Hall, New York.

Estes, J. A. and J. Palmisano. 1974. Sea otters: Their role in structuring nearshore communities. *Science* 185: 1058–1060.

Estes, J. A., N. S. Smith and J. Palmisano. 1978. Sea otter predation and community organization in the Western Aleutian Islands, Alaska. *Ecology* 59: 822–833.

Etter, R. J. 1988a. Physiological stress and color polymorphisms in the intertidal snail *Nucella lapillus*. *Evolution* 42: 660–680.

Etter, R. J. 1988b. A symmetrical developmental plasticity in an intertidal snail. *Evolution* 42: 322–334.

Etter, R. J. 1989. Life history variation in the intertidal snail *Nucella lapillus* across a wave-exposure gradient. *Ecology* 70: 1857–1876.

Farrell, T. M. 1991. Models and mechanisms of succession: An example from a rocky intertidal community. *Ecological Monographs* 6: 95–113.

Fell, E. P., N. C. Olmstead, E. Carlson, W. Jacob, D. Hitchcock and G. Silber. 1982. Distribution and abundance of macroinvertebrates on certain Connecticut tidal marshes, with emphasis on dominant molluscs. *Estuaries* 5: 235–239.

Fell, E. P., K. A. Murphy, M. A. Peck and M. L. Recchia. 1991. Re-establishment of *Melampus bidentatus* and other macroinvertebrates on a restored impounded tidal marsh: Comparison of populations above and below the impoundment dike. *Journal of Experimental Marine Biology and Ecology* 15: 33–48.

Fenchel, T. 1975a. Character displacement and coexistence in mud snails (Hydrobiidae). *Oecologia* 20: 19–32.

Fenchel, T. 1975b. Factors determining the distribution patterns of mud snails (Hydrobiidae). *Oecologia* 20: 1–17.

Fenchel, T. 1976. Evidence for exploitative interspecific competition in mud snails (Hydrobiidae). *Oikos* 27: 367–376.

Fiorenza, M. 1997. Effects of predator foraging behavior on patterns of prey mortality in marine soft bottoms. *Ecological Monographs* 67: 203–224.

Flowers, T. J., P. F. Troke and A. R. Yeo. 1977. The mechanism of salt tolerance in halophytes. *Annual Reviews of Plant Physiology* 28: 89–121.

Flowers, T. J., M. A. Hajibagheri and N. J. W. Clipson. 1986. Halophytes. *Quarterly Review of Biology* 61: 313–337.

Forward, R. B. 1974. Negative phototaxis in crustacean larvae: Possible functional significance. *Journal of Experimental Marine Biology and Ecology* 16: 11–17.

Forward, R. B. 1976. Light and diurnal vertical migration: Photobehavior and photophysiology of plankton. *Photochemical and Photobiology Review* 1: 157–209.

Forward, R. B. 1988. Diel vertical migration: Zooplankton, photobiology and behavior. *Oceanography and Marine Biology Annual Review* 26: 361–393.

Foster, B. A. 1969. Tolerance of high temperatures by some intertidal barnacles. *Marine Biology* 4: 326–332.

Franz, D. R. and A. S. Merrill. 1980a. Distribution patterns of common seastars of the middle Atlantic continental shelf of the northwest Atlantic (Gulf of Maine to Cape Hatteras). *Biological Bulletin* 160: 394–418.

Franz, D. R. and A. S. Merrill. 1980b. The origins and determinants of distribution of molluscan faunal groups on the shallow continental shelf of the northwest Atlantic. *Malacologia* 19(2): 227–248.

Fréchette, M., C. A. Butman and W. R. Geyer. 1989. The importance of boundary-layer flows in supplying phytoplankton to the benthic suspension feeder, *Mytilus edulis* (L.). *Limnology and Oceanography* 34(1): 19–36.

Frey, R. W. and P. B. Basan. 1985. Coastal salt marshes. In *Coastal Sedimentary Environments*, R. A. Davis (ed.), 225–301. Springer-Verlag, New York.

Furbish, C. E. and M. Albano. 1994. Selective herbivory and plant community structure in a mid-Atlantic salt marsh. *Ecology* 75: 1015–1022.

Gagnon, P. S., R. L. Vadas, D. B. Burdick and B. P. May. 1980. Genetic identity of annual and perennial forms of *Zostera marina* L. *Aquatic Botany* 8: 157–162.

Gaines, S. D. 1985. Herbivory and between-habitat diversity: The differential effectiveness of defenses in a marine plant. *Ecology* 66(2): 473–485.

Gaines, S. D. and M. D. Bertness. 1992. Dispersal of juveniles and variable recruitment in sessile marine species. *Nature* 360: 579–580.

Gaines, S. D. and M. D. Bertness. 1993. Does variable transport generate variable settlement in coastal and estuarine species? In *Changes in Fluxes in Estuaries*, K. Dyer and R. Orth (eds.), 346–360. Olsen and Olsen Press, London.

Gaines, S. D. and M. D. Bertness. 1994. Measuring juvenile dispersal: Why field ecologists must learn to integrate. *Ecology* 74: 2430–2435.

Gaines, S. and J. Roughgarden. 1985. Larval settlement rate: A leading determinant of structure in an ecological community of the marine intertidal zone. *Proceedings of the National Academy of Science* 82: 3707–3711.

Gaines, S. D. and J. Roughgarden. 1987. Fish in offshore kelp forests affect recruitment to intertidal barnacle populations. *Science* 235: 479–481.

Gaines, S., S. Brown and J. Roughgarden. 1985. Spatial variation in larval concentrations as a cause of spatial variation in settlement for the barnacle, *Balanus glandula*. *Oecologia* 67: 267–272.

Gallagher, E. D., P. A. Jumars and D. D. Trueblood. 1983. Facilitation of soft-bottom benthic succession by tube builders. *Ecology* 64: 1200–1216.

Gallagher, J. L. 1975. Effect of an ammonium nitrate pulse on the growth and elemental composition of stands of *Spartina alterniflora* and *Juncus roemerianus*. *American Journal of Botany* 62: 644–648.

Gallagher, J. L., G. F. Somers, D. M. Grant and D. M. Seliskar. 1988. Persistent differences in two forms of *Spartina alterniflora*: A common garden experiment. *Ecology* 69: 1005–1008.

Gambi, M. C., A. R. Nowell and P. A. Jumars. 1990. Flume observations on flow dynamics in *Zostera marina* (eelgrass) beds. *Marine Ecology Progress Series* 61: 159–169.

Gartner-Kepkay, K. E., E. Zouros, L. M. Dickie and L. R. Freeman. 1983. Genetic differentiation in the face of gene flow: A study of mussel populations from a single Nova Scotian embayment. *Canadian Journal of Fisheries and Aquatic Sciences* 40: 443–451.

Gaylord, B. 1998. Implications of hydrodynamic length scales for the ability of fluid acceleration to limit the sizes of wave-swept organisms. *Western Society of Naturalists Meeting Abstracts* 29.

Geisselman, J. A. and O. J. McConnell. 1981. Polyphenols in brown algae *Fucus vesiculosus* and *Ascophyllum nodosum*: Chemical defenses against the marine herbivorous snail, *Littorina littorea*. *Journal of Chemical Ecology* 7: 1115–1133.

Gerard, V. A. 1987. In situ water motion and nutrient uptake by the giant kelp *Macrocystis pyrifera*. *Marine Biology* 69: 51–54.

Gerard, V. A. and K. H. Mann. 1979. Growth and production of *Laminaria longicruris* populations exposed to different intensities of water movement. *Journal of Phycology* 15: 33–41.

Gibbs, P. E., P. L. Pascoe and G. R. Burt. 1988. Sex change in the female dog-whelk, *Nucella lapillus*, induced by tributyltin from antifouling paints. *Journal of the Marine Biological Association* (UK) 68: 715–731.

Gillespie, M. C. 1971. Analysis and treatment of zooplankton of estaurine waters of Louisiana. *Cooperative Gulf of Mexico Inventory IV: Biology*, 108–175. Louisiana Wildlife Commission.

Gilwicz, M. Z. 1986. Predation and the evolution of vertical migration in zooplankton. *Nature* 320: 746–748.

Gleason, M. L. 1980. Influence of tidal inundation on internal oxygen supply of *Spartina alterniflora* and *Spartina patens*. Ph.D. dissertation, University of Virginia, Charlottesville.

Glude, J. B. 1955. The effects of temperature and predators on the abundance of the soft-shell clam, *Mya arenaria*, in New England. *Transactions of the American Fisheries Society* 84: 13–26.

Goldberg, D. E. and A. M. Barton. 1992. Patterns and consequences of interspecific competition in natural communities: A review of field experiments with plants. *American Naturalist* 139: 771–801.

Gosner, K. L. 1978. *A Field Guide to the Atlantic Seashore*. Houghton Mifflin, Boston.

Goss-Custard, J. D. 1969. The winter feeding ecology of the redshank, *Tringa totanus*. *Ibis* 111: 338–356.

Grace, J. B. 1987. The impact of preemption on the zonation of two *Typha* species along lakeshores. *Ecological Monographs* 57: 283–303.

Grant, J. 1981a. Dynamics of competition among estuarine sand-burrowing amphipods. *Journal of*

Experimental Marine Biology and Ecology 49: 255–265.

Grant, J. 1981b. Sediment transport and distribution on an intertidal sandflat: Infaunal distribution and recolonization. *Marine Ecology Progress Series* 6: 249–255.

Grant, J. 1983. The relative magnitude of biological and physical sediment reworking in an intertidal community. *Journal of Marine Research* 41: 673–689.

Grant, W. S. 1977. High intertidal community organization on a rocky headland in Maine, USA. *Marine Biology* 44: 15–25.

Gray, I. E . 1960. The seasonal occurrence of *Mytilus edulis* on the Carolina coast as a result of transport around Cape Hatteras. *Biological Bulletin* 119: 550–559.

Gray, J. 1974. Animal-sediment relationships. *Oceanography and Marine Biology Annual Review* 12: 223–261.

Green, G. 1977. Ecology of toxicity in marine sponges. *Marine Biology* 40: 207–215.

Grime, J. P. 1977. Evidence for the existence of three primary strategies in plants and its relevance to ecological and evolutionary theory. *American Naturalist* 111: 1169–1194.

Grosberg, R. K. 1981. Competitive ability influences habitat choice in marine invertebrates. *Nature* 290: 700–702.

Grosholz, E. D. 1992. Interactions among intraspecific, interspecific, and apparent competition with host-pathogen population dynamics. *Ecology* 73(2): 507–514.

Hacker, S. D. and M. D. Bertness. 1995a. A herbivore paradox: Why salt marsh aphids live on poor quality plants. *American Naturalist* 145: 192–210.

Hacker, S. D . and M. D. Bertness. 1995b. Morphological and physiological consequences of a positive plant interaction. *Ecology* 76: 2165–2175.

Hacker, S. D. and M. D. Bertness. 1996. The trophic consequences of a marsh plant positive interaction. *American Naturalist* 148: 559–575.

Hacker, S. D. and M. D. Bertness. 1999. The role of positive interactions in the plant species diversity of salt marsh plant communities. *Ecology*. In press.

Hacker, S. D. and S. D. Gaines. 1997. Some implications of direct positive interactions for community species diversity. *Ecology* 78: 1990–2003.

Haines, E. B. 1977. The origins of detritus in Georgia salt marsh estuaries. *Oikos* 29: 254–260.

Haines, E. B. and C. L. Montague. 1979. Food sources of estuarine invertebrates analyzed using $^{13}C/^{12}C$ ratios. *Ecology* 60: 48–56.

Hall, C. J., D. G. Raffaelli and W. R. Turrel. 1990. Predator caging experiments in marine systems: A reexamination of their value. *American Naturalist* 136: 657–672.

Hamilton, P. V. 1976. Predation on *Littorina irrorata* by *Callinectes sapidus*. *Bulletin of Marine Science* 26: 401–409.

Hamilton, W. D. 1971. Geometry for the selfish herd. *Journal of Theoretical Biology* 31: 295–311.

Hannan, C. A. 1981. Polychaete larval settlement: Correspondence of patterns in suspended jar collectors and in the adjacent natural habitat in Monterey Bay, CA. *Limnology and Oceanography* 26: 159–171.

Hansen, T. A. 1978. Larval dispersal and species longevity in lower Tertiary gastropods. *Science* 199: 885–887.

Hansen, T. A. 1980. Influence of larval dispersal and geographic distribution on species longevity in neogastropods. *Paleobiology* 6(2): 193–207.

Hare, J. A. and R. K. Cowan. 1991. Expatriation of *Xyrichytys novacula* (Pisecs: Labridae) larvae: Evidence of rapid cross-slope exchange. *Journal of Marine Research* 49: 801–823.

Harger, J. R. E. 1971. Competitive coexistence among intertidal invertebrates. *American Scientist* 60: 600–607.

Harley, C. D. G. and M. D. Bertness. 1996. Structural interdependence: An ecological consequence of morphological responses to crowding in marsh plants. *Functional Ecology* 10: 654–661.

Harlin, M. M. 1975. Epiphyte-host relations in seagrass communities. *Aquatic Botany* 1: 125–131.

Harlin, M. M. 1980. Seagrass epiphytes. In *Handbook of Seagrass Biology: An Ecosystem Perspective*, R. C. Phillips and C. P. McRoy (eds.), 117–151. Garland STPM Press, New York.

Harper, J. L. 1977. *Population Biology of Plants*. Academic Press. New York.

Hartman, J. M. and M. Engler. 1982. Density effects on the growth and survival of *Salicornia biglovii* and *Salicornia europaea*. *Biological Bulletin* 162: 365–366.

Hartman, J., H. Caswell and I. Valiela. 1983. Effects of wrack accumulation on salt marsh vegetation. *Proceedings of the 17th European Marine Biological Symposium* 99: 99–102.

Harvell, C. D. 1986. The ecology and evolution of inducible defenses in a marine bryozoan: Cues, costs, and consequences. *American Naturalist* 128: 810–823.

Harvell, C. D. 1992. Inducible defenses and allocation shifts in a marine bryozoan. *Ecology* 73(5): 1567–1576.

Hatcher, B. G., A. R. O. Chapman and K. Mann. 1977. An annual carbon budget for the kelp *Laminaria longicruris*. *Marine Biology* 44: 88–96.

Hawkins, S. J. and R. G. Hartnoll. 1982. Settlement patterns of *Semibalanus balanoides* (L.) on the Isle of Man (1977–1981). *Journal of Experimental Marine Biology and Ecology* 62: 271–283.

Hay, M. 1981. The functional morphology of turf forming seaweeds: Persistence in stressful marine habitats. *Ecology* 62(3): 739–750.

Hay, M. E. 1986. Associational plant defenses and the maintenance of species diversity. *American Naturalist* 128: 617–641.

Hay, M. E. 1996. Marine chemical ecology: What's known and what's next? *Journal of Experimental Marine Biology and Ecology* 200: 103–134.

Hay, M. E. and J. E. Duffy. 1990. Host-plant specialization decreases predation on a marine amphipod: An herbivore in plant's clothing. *Ecology* 7: 733–743.

Hay, M. E. and W. Fenical. 1988. Marine plant-herbivore interactions: The ecology of chemical defense. *Annual Review of Ecology and Systematics* 19: 111–145.

Hay, M. E., J. E. Duffy, C. A. Pfister and W. Fenical. 1987. Chemical defenses against different marine herbivores: Are amphipods insect equivalents? *Ecology* 68: 1567–1580.

Hay, M. A., V. J. Paul, S. M. Lewis, K. Gustafson, J. Tucker and R. N. Trindell. 1988. Can tropical seaweeds reduce herbivory by growing at night? Diel patterns of growth, nitrogen content, herbivory and chemical versus morphological defenses. *Oecologia* 75: 233–245.

Hay, M. E., J. R. Pawlik, J. E. Duffy and W. Fenical. 1989. Seaweed-herbivore-predator interactions: Host-plant specialization reduces predation on small herbivores. *Oecologia* 81: 418–427.

Heck, K. L. J. and T. A. Thoman. 1981. Experiments on predator-prey interactions in vegetated aquatic habitats. *Journal of Experimental Marine Biology and Ecology* 53: 125–134.

Hilbish, T. J. 1996. Population genetics of marine species: The interaction of natural selection and historically differentiated populations. *Journal of Experimental Marine Biology and Ecology* 200: 67–83.

Hilbish, T. J. and R. K. Koehn. 1985. The physiological basis of natural selection at the LAP locus. *Evolution* 39: 1302–1317.

Himmelman, J. H. 1975. Phytoplankton as a stimulus for spawning the three marine invertebrates. *Journal of Experimental Marine Biology and Ecology* 20: 199–214.

Himmelman, J. H. 1983. Community development following removal of urchins, *Strongylocentrotus drobachiensis*, from the rocky subtidal zone. *Oecologia* 59: 27–39.

Hines, A. H. and M. H. Posey. 1989. Complex trophic interactions in estuarine food webs: Jargon, concepts and examples. *American Zoologist* 29: 27A.

Hines, A. H., R. N. Lipcius and A. M. Haddon. 1987. Population dynamics and habitat partitioning by size, sex, and molt stage of blue crabs *Callinectes sapidus* in a subestuary of central Chesapeake Bay. *Marine Ecology Progress Series* 36: 55–64.

Hobson, E. S. and J. R. Chess. 1976. Trophic interactions among fishes and zooplankters near shore at Santa Catalina Island, California. *U.S. Fish and Wildlife Service Bulletin* 74: 567–598.

Hochachka, P. W. and G. N. Somero. 1973. *Strategies of Biochemical Adaptation*. Saunders, Philadelphia.

Hoffman, J. A., J. Katz and M. D. Bertness. 1984. Fiddler crab deposit-feeding and meiofaunal abundance in salt marsh habitats. *Journal of Experimental Marine Biology and Ecology* 82: 2–3.

Holbrook, N. M., M. Denny and M. A. R. Koehl. 1991. Intertidal trees: Consequences of aggregation on the mechanical and photosynthetic properties of sea palms. *Journal of Experimental Marine Biology and Ecology* 146: 39–67.

Holt, R. D. 1977. Predation, apparent competition and the structure of prey communities. *Theoretical Population Biology* 12: 197–229.

Hooper, R. 1981. Recovery of Newfoundland benthic marine communities from sea ice. *International Seaweed Symposium Proceedings* 8: 360–366.

Howarth, R. W. and J. M. Teal. 1979. Sulfate reduction in a New England salt marsh. *Limnology and Oceanography* 24: 999–1013.

Howarth, R. W. and J. M. Teal. 1980. Energy flow in a salt marsh ecosystem: The role of reduced inorganic sulfur compounds. *American Naturalist* 116: 862–872.

Howes, B. L. and J. M. Teal. 1994. Oxygen loss from *Spartina alterniflora* and its relationship to salt marsh oxygen balance. *Oecologia* 97: 431–438.

Howes, B. L., R. W. Howarth, J. M. Teal and I. Valiela. 1981. Oxidation-reduction potentials in a salt marsh: Spatial patterns and interactions with primary production. *Limnology and Oceanography* 26: 350–360.

Howes, B. L., J. W. H. Dacey and D. D. Goehringer. 1986. Factors controlling the growth form of *Spartina alterniflora*: Feedbacks between aboveground production, sediment oxidation, nitrogen and salinity. *Journal of Ecology* 74: 881–898.

Hughes, D. J. 1989. Variation in reproductive strategy among clones of the bryozoan *Celleporella hyalina*. *Ecological Monographs* 59(4): 387–403.

Hughes, R. N. 1980. Optimal foraging in a marine context. *Oceanography and Marine Biology Annual Review* 18: 423–481.

Hughes, R. N. and R. W. Elner. 1979. Tactics of a predator, *Carcinus maenas*, and morphological responses of the prey, *Nucella lapillus*. *Journal of Animal Ecology* 48: 65–78.

Hughes, R. N. and D. J. Roberts. 1981. Comparative demography of *Littorina rudis*, *L. nigrolineata* and *L. neritoides* on three contrasting shores in North Wales. *Journal of Animal Ecology* 50: 251–264.

Hulberg, L. W. and J. S. Oliver. 1980. Caging manipulations in marine soft-bottom communities: Importance of animal interactions or sedimentary habitat modifications. *Canadian Journal of Fisheries and Aquatic Sciences* 37: 1130–1139.

Hulbert, A. W. 1979. The ecological role of *Asterias vulgaris* in three subtidal communities. *Proceedings of the European Colloquim on Echinoderms: Echinoderms, past and present*, M. Jangoux (ed.), 191–196. Balkema, Rotterdam.

Huntley, M. and C. Boyd. 1984. Food-limited growth of marine zooplankton. *American Naturalist* 124: 455–478.

Huntley, M. E., and L. A. Hobson. 1978. Medusa predation and plankton dynamics in a temperate fjord, British Columbia. *Journal of the Fisheries Research Board of Canada* 35: 257–261.

Hurlbert, S. H. 1984. Pseudoreplication and the design of ecological field experiments. *Ecological Monographs* 54(2): 187–211.

Hylleberg, J. 1975. Selective feeding by *Arrenicola pacifica* with notes on *Arrenicola vagabunda* and a concept of gardening in lugworms. *Ophelia* 14: 113–137.

Irlandi, E. A. and C. H. Peterson. 1991. Modification of animal habitat by large plants: Mechanisms by which seagrasses influence clam growth. *Oecologia* 87: 307–318.

Irvine, G. V. 1973. The effect of selective feeding by two species of sea urchins on the structuring of algal communities. M.Sc. thesis, University of Washington.

Jablonski, D. 1986. Larval ecology and macoevolution in marine invertebrates. *Bulletin of Marine Science* 39: 565–587.

Jackson, J. B. C. 1974. Biogeographic consequences of eurytopy and stenotopy among marine bivalves and their evolutionary significance. *American Naturalist* 108: 541–559.

Jackson, J. B. C. 1977. Competition on marine hard substrata: The adaptive significance of solitary and colonial strategies. *American Naturalist* 111(980): 743–767.

Jackson, J. B. C. and L. W. Buss. 1975. Allelopathy and spatial competition among coral reef invertebrates. *Proceedings of the National Academy of Sciences USA* 72: 5160–5163.

Jackson, J. B. C. and T. P. Hughes. 1985. Adaptive strategies of coral reef invertebrates. *American Scientist* 73: 265–274.

Janson, K. 1983. Selection and migration in two distinct phenotypes of *Littorina saxatilis* in Sweden. *Oecologia* 59: 58–61.

Jefferies, R. L. and N. Perkins. 1977. The effects on the vegetation of the addition of inorganic nutrients to salt marsh soils at Stiffkey, Norfolk. *Journal of Ecology* 65: 867–882.

Johnson, D. J. 1920. *Shoreline Processes and Shoreline Development*. Columbia University Press, New York.

Jones, C. G., J. H. Lawton and M. Shachak. 1994. Organisms as ecosystem engineers. *Oikos* 69: 373–386.

Jordan, T. E. and I. Valiela. 1982. A nitrogen budget of the ribbed mussel, *Geukensia demissa*, and its significance in nitrogen flow in a New England salt-marsh. *Limnology and Oceanography* 27: 75–90.

Jorgensen, C. B. 1966. *Biology of Suspension Feeding*. Permamon Press, New York.

Jumars, P. A., A. R. Nowell and R. L. Self. 1981. A simple model of flow-sediment organism interaction. *Marine Geology* 42: 155–172.

Kadlec, J. A. and W. H. Drury. 1968. Structure of the New England herring gull population. *Ecology* 49: 644–676.

Kanwisher, J. 1957. Freezing and drying in intertidal algae. *Biological Bulletin* 113: 275–285.

Kanwisher, J. W. 1960a. Freezing in intertidal animals. *Biological Bulletin* 116: 258–264.

Kanwisher, J. W. 1960b. Histology and metabolism of frozen intertidal animals. *Biological Bulletin* 116: 264–269.

Katz, C. H. 1985. A nonequilibrium marine predator-prey interaction. *Ecology* 66: 1426–1438.

Keddy, J. and D. G. Patriquin. 1978. An annual form of eelgrass in Nova Scotia. *Aquatic Botany* 5: 163–170.

Keddy, P. 1989. *Competition*. Chapman and Hall, London.

Keefe, C. W. 1971. Marsh production: A summary of the literature. *Contributions to Marine Science* 16: 163–181.

Kemp, P. and M. D. Bertness. 1984. Snail shape and growth rates: Evidence for plastic shell allometry in *Littorina littorea*. *Proceedings of the National Academy of Sciences USA* 81: 811–813.

Kensler, C. B. 1967. Desiccation resistance of intertidal crevice species as a factor in their zonation. *Journal of Animal Ecology* 36: 391–406.

Keough, M. J. and B. J. Downes. 1982. Recruitment of marine invertebrates: The role of active larval choices and early mortality. *Oecologia* 54: 348–352.

Ketchum, B. H. 1954. Relation between circulation and plankton populations in estuaries. *Ecology* 35: 191–200.

King, G. M., M. J. Klug, R. G. Wiegert and A. G. Chalmers. 1982. Relation of soil water movement and sulfide concentration to *Spartina alterniflora* production in a Georgia salt marsh. *Science* 218: 61–63.

Kingsford, M. J., A. J. Underwood and S. J. Kennelly. 1991. Humans as predators on rocky reefs in New South Wales, Australia. *Marine Ecology Progress Series* 72: 2–14.

Kitching, J. A., L. Muntz and F. J. Ebling. 1966. The ecology of Lough Ine. XV. The ecological significance of shell and body form in *Nucella*. *Journal of Animal Ecology* 35: 113–125.

Kneib, R. T. 1981. Size-specific effects of density on the growth, fecundity and mortality of the fish *Fundulus heteroclitus* in an intertidal salt marsh. *Marine Ecology Progress Series* 6: 203–212.

Kneib, R. T. 1982. Habitat preference, predation, and the intertidal distribution of gammaridean amphipods in a North Carolina salt marsh. *Journal of Experimental Marine Biology and Ecology* 59: 219–230.

Kneib, R. T. 1985. Predation and disturbance by grass shrimp, *Palaemonetes pugio*, in soft-substratum benthic invertebrate assemblages. *Journal of Experimental Marine Biology and Ecology* 93: 91–102.

Kneib, R. T. 1986. The role of *Fundulus heteroclitus* in salt marsh trophic dynamics. *American Zoologist* 26: 259–269.

Kneib, R. T. 1988. Testing for indirect effects of predation in an intertidal soft-bottom community. *Ecology* 69: 1795–1805.

Kneib, R. T. 1991. Indirect effects in experimental studies of marine soft-sediment communities. *American Zoologist* 31: 874–885.

Kneib, R. T. 1997. Tidal marshes offer a different perspective on estuarine nekton. *Oceanography and Marine Biology Annual Review* 35: 1–120.

Kneib, R. T., S. Y. Newell and E. T. Hermeno. 1997. Survival, growth and reproduction of the salt-marsh amphipod *Uhlorchestia spartinophila* reared on natural diets of senescent and dead *Spartina alterniflora* leaves. *Marine Biology* 128: 423–431.

Koehl, M. A. R. 1984a. How do benthic organisms withstand moving water? *American Zoologist* 24: 57–70.

Koehl, M. A. R. 1984b. The interaction of moving water and sessile organisms. *Scientific American* 124–134.

Koehl, M. A. R. and S. A. Wainwright. 1977. Mechanical adaptations of a giant kelp. *Limnology and Oceanography* 22: 1067–1071.

Koehn, R. K. 1991. The genetics and taxonomy of the species in the genus *Mytilus*. *Aquaculture* 94: 125–145.

Koehn, R. K. and T. J. Hilbish. 1987. The adaptive importance of genetic variation. *American Scientist* 75: 182–189.

Koehn, R. K., J. G. Hall, D. J. Innes and A. J. Zera. 1984. Genetic differentiation of *Mytilus edulis* in eastern North America. *Marine Biology* 79: 117–126.

Kremer, J. N. and S. W. Nixon. 1978. *A Coastal Marine Ecosystem: Simulation and Analysis*. Springer-Verlag, New York.

Kuenzler, E. J. 1961. Structure and energy flow of a mussel population in a Georgia salt marsh. *Limnology and Oceanography* 6: 191–204.

Lehman, J. T. 1976. The filter feeder as an optimal forager, and the shape of predicted feeding curves. *Limnology and Oceanography* 22: 170–172.

Lent, C. M. 1969. Adaptations of the ribbed mussel, *Modiolus demissa*, to the intertidal habitat. *American Zoologist* 9: 283–292.

Leigh, E. G., R. T. Paine, J. F. Quinn and T. H. Suchanek. 1987. Wave energy and intertidal productivity. *Proceedings of the National Academy of Sciences USA* 84: 1314–1318.

Lenihan, H. S. and C. H. Peterson. 1997. How habitat degradation through fishery disturbance enhances impacts of hypoxia on oyster reefs. *Ecological Applications* 8: 128–140.

Leonard, G., J. M. Levine, P. Schmidt and M. D. Bertness. 1998a. Flow-generated bottom-up forcing of intertidal community structure in a Maine estuary. *Ecology* 79: 1395–1411.

Leonard, G., P. Yund and M. D. Bertness. 1998b. Predator induced structural defenses in the blue mussel. *Ecology*. In press.

Lessios, H. A., D. R. Robertson and J. D. Cubit. 1984. Spread of *Diadema* mass mortality through the Caribbean. *Science* 226: 335–337.

Levin, L. A. 1981. Dispersion, feeding behavior and competition in two spionid polychaetes. *Journal of Marine Research* 39: 99–117.

Levine, J., S. Brewer and M. D. Bertness. 1998. Nutrient availability and the zonation of marsh plant communities. *Journal of Ecology* 86: 285–292.

Levins, L. and T. S. Bridges. 1995. Pattern and diversity in reproduction and development. In *Ecology of Marine Invertebrate Larvae*, L. McEdwards (ed.), 1–48. CRC Press, New York.

Levington, J. S. 1981. Nutrition and food limitation of deposit feeders. I. The role of microbes in the growth of mud snails (Hydropbiidae). *Journal of Marine Research* 39: 531–545.

Levington, J. S. 1995. *Marine Biology: Function, Biodiversity, Ecology*. Oxford University Press, New York.

Levinton, J. S. and G. R. Lopez. 1977. A model of renewable resources and limitation of deposit-feeding benthic populations. *Oecologia* 31: 177–190.

Levinton, J. S., S. Stewart and T. H. Dewitt. 1985. Field and laboratory experiments on interference between *Hydrobia totteni* and *Ilyanassa obsoleta* (Gastropoda) and its possible relation to seasonal shifts in vertical mudflat zonation. *Marine Ecology Progress Series* 22: 53–58.

Levitan, D. R. 1995. The ecology of fertilization in free-spawning invertebrates. In *Ecology of Marine Invertebrate Larvae*, L. McEdwards (ed.), 123–156. CRC Press, New York.

Levitan, D. R., M. A. Sewell and F. S. Chia. 1992. How distribution and abundance influence fertilization success in the sea urchin *Strongylocentrotus franciscanus*. *Ecology* 73(1): 248–254.

Lewis, J. R. 1964. *The Ecology of Rocky Shores*. English University Press, London.

Lindeman, R. L. 1942. The trophic dynamic aspect of ecology. *Ecology* 23: 399–418.

Lindquist, N. and M. E. Hay. 1996. Palatability and chemical defense of marine invertebrate larvae. *Ecological Monographs* 66: 431–450.

Lipcius, R. N. and T. Hines. 1986. Variable functional responses of a marine predator in dissimilar homogeneous microhabitats. *Ecology* 67 (5): 1361–1371.

Little, C. and J. A. Kitching. 1996. *The Biology of Rocky Shores*. Oxford University Press, Oxford.

Littler, M. M. and D. S. Littler. 1980. The evolution of thallus form and survival strategies in benthic marine algae: Field and laboratory tests of a functional form model. *American Naturalist* 116(1): 25–44.

Littler, M. M. and S. N. Murray. 1975. Impact of sewage on the distribution, abundance and community structure of rocky intertidal macro-organisms. *Marine Biology* 30: 277–291.

Littler, M. M., D. Littler and P. R. Taylor. 1995. Selective herbivore increases biomass of its prey: A chiton-coralline reef building association. *Ecology* 76: 1666–1681.

Lively, C. M. 1986a. Competition, comparative life histories, and maintenance of shell dimorphism in a barnacle. *Ecology* 67: 858–864.

Lively, C. M. 1986b. Predator-induced shell dimorphism in the acorn barnacle *Chthamalus anisopoma*. *Evolution* 40: 232–242.

Lively, C. M. and P. T. Raimondi. 1987. Desiccation, predation, and mussel-barnacle interactions in the northern Gulf of California. *Oecologia* 74: 304–309.

Lobban, C. S. and P. J. Harrison. 1994. *Seaweed Ecology and Physiology*. Cambridge University Press, Cambridge.

Lopez, G. R. and J. S. Levinton. 1987. Ecology of deposit-feeding animals in marine sediments. *Quarterly Review of Biology* 2: 235–260.

Lubchenco, J. 1978. Plant species diversity in a marine intertidal community: Importance of herbivore food preference and algal competitive abilities. *American Naturalist* 112(983): 23–39.

Lubchenco, J. 1980. Algal zonation in the New England rocky intertidal community: An experimental analysis. *Ecology* 61(2): 333–344.

Lubchenco, J. 1982. Effects of grazers and algal competitors on fucoid colonization in tide-pools. *Journal of Phycology* 18: 544–550.

Lubchenco, J. 1983. *Littorina* and *Fucus*: Effects of herbivores, substratum heterogeneity and plant escapes during succession. *Ecology* 64: 1116–1123.

Lubchenco, J. and J. Cubit. 1980. Heteromorphic life histories of certain marine algae as adaptations to variations in herbivory. *Ecology* 61: 676–687.

Lubchenco, J. and S. D. Gaines. 1981. A unified approach to marine plant-herbivore interactions. I. Populations and communities. *Annual Review of Ecology and Systematics* 12: 405–437.

Lubchenco, J. and B. A. Menge. 1978. Community development and persistence in a low rocky intertidal zone. *Ecological Monographs* 48: 67–94.

Manahan, D. T. 1990. Adaptations by invertebrate larvae for nutrient acquisition from seawater. *American Zoologist* 30: 147–160.

Mann, K. H. 1972. Ecological energetics of the seaweed zone in a marine bay on the Atlantic coast of Canada. II. Productivity of seaweeds. *Marine Biology* 14: 199–209.

Mann, K. H. 1982. *The Ecology of Coastal Waters: A Systems Approach*. Blackwell Scientific, New York.

Mann, K. H. and J. R. N. Lazier. 1991. *Dynamics of Marine Ecosystems: Biological-Physical Interactions in the Ocean*. Blackwell Scientific, Boston.

Marsh, C. P. 1986. Impact of avian predators on high intertidal limpet populations. *Journal of Experimental Marine Biology and Ecology* 104: 185–201.

Mathieson, A. C. and R. L. Burns. 1975. Ecological studies of economic red algae. V. Growth and reproduction of natural and harvested populations of *Chondrus crispus* (Stackhouse) in New Hampshire. *Journal of Experimental Marine Biology and Ecology* 17: 137–156.

Mathieson, A. C., C. A. Penniman, P. K. Busse and E. Tveter-Gallagher. 1982. Effects of ice on *Ascophyllum nodosum* within the Great Bay Estuary System of New Hampshire-Maine. *Journal of Phycology* 18: 331–336.

Mathieson, A. C., C. A. Penniman and L. Harris. 1991. Northwest Atlantic rocky shore ecology. In *Ecosystems of the World: Intertidal and Littoral Ecosystems*, A. C. Mathieson and P. H. Nienhuis (eds.), 109–192. Elsevier, Amsterdam.

McCall, P. 1977. Community patterns and adaptive strategies of the infaunal benthos of Long Island Sound. *Journal of Marine Research* 35: 221–261.

McClintock, J. B. 1990. Chemical defense in the eggs and embryo antarctic sea stars (Echinodermata). *Marine Biology* 105: 491–495.

McDonald, J. H., Seed, R. and R. K. Koehn 1991. Allozymes and morphometric characteristics of three species of *Mytilus* in the Northern and Southern Hemispheres. *Marine Biology* 111: 323–333.

McKee, K. L., I. M. Mendelssohn and M. W. Hester. 1988. Reexamination of pore water sulfide concentrations and redox potentials near the aerial roots of mangroves. *American Journal of Botany* 75: 1352–1359.

McRoy, C. P. and C. McMillan. 1977. Production ecology and physiology of seagrasses. In *Seagrass Ecosystems: A Scientific Perspective*, C. P. McRoy and C. Helfferich (eds.), 53–88. Marcel Dekker, New York.

Meadows, P. S. and J. I. Campbell. 1971. Habitat selection by aquatic invertebrates. *Advances in Marine Biology* 19: 271–282.

Mendelssohn, I. A. 1979a. The influence of nitrogen level, form and application method on the growth response of *Spartina alterniflora* in North Carolina. *Estuaries* 2: 106–112.

Mendelssohn, I. A. 1979b. Nitrogen metabolism in the height forms of *Spartina alterniflora* in North Carolina. *Ecology* 60: 106–112.

Mendelssohn, L. A., K. L. McKee and W. H. Patrick. 1981. Oxygen deficiency in *Spartina alterniflora* roots: Metabolic adaptation to anoxia. *Science* 214: 439–441.

Menge, B. 1975. Brood or broadcast? The adaptive significance of different reproductive strategies in two intertidal sea stars *Leptasterias hexactis* and *Pisaster ochraceus*. *Marine Biology* 31: 87–100.

Menge, B. A. 1976. Organization of the New England rocky intertidal community: Role of predation, competition, and environmental heterogeneity. *Ecological Monographs* 46: 355–393.

Menge, B. A. 1978. Predation intensity in a rocky intertidal community. I. Relation between predator foraging activity and environmental harshness. II. Effect of wave action and desiccation on predator feeding rates. *Oecologia* 34: 1–35.

Menge, B. A. 1991. Relative importance of recruitment and other causes of variation in rocky intertidal community structure. *Journal of Experimental Marine Biology and Ecology* 146: 69–100.

Menge, B. A. 1992. Community regulation: Under what conditions are bottom-up factors important on rocky shores? *Ecology* 73(3): 755–765.

Menge, B. A. 1995. Indirect effects on marine rocky intertidal interaction webs: Patterns and importance. *Ecological Monographs* 65: 21–74.

Menge, B. A. and J. P. Sutherland. 1976. Species diversity gradients: Synthesis of the roles of predation, competition, and temporal heterogeneity. *American Naturalist* 110: 351–369.

Menge, B. A. and J. P. Sutherland. 1987. Community regulation: Variation in disturbance, competition, and predation in relation to environmental stress. *American Naturalist* 130: 730–757.

Micheli, F. 1997. Effects of predator foraging behavior on patterns of prey mortality in marine soft bottoms. *Ecological Monographs* 67: 203–224.

Miller, D. C. 1960. The feeding mechanisms of fiddler crabs with ecological considerations of feeding adaptations. *American Zoologist* 46: 89–101.

Miller, W. B. and F. E. Egler. 1950. Vegetation of the Wequetequock-Pawcatuck tidal marshes, Connecticut. *Ecological Monographs* 20: 143–172.

Mills, L. S., M. E. Soulé and D. F. Doak. 1993. The keystone-species concept in ecology and conservation. *BioScience* 43: 213–224.

Milne, L. J. and M. J. Milne. 1951. The eelgrass catastrophe. *Scientific American* 184(1): 52–55.

Minchinton, T. E., R. E. Scheibling and H. L. Hunt. 1997. Recovery of an intertidal assemblage following a rare occurrence of scouring by sea ice in Nova Scotia, Canada. *Botanica Marina* 40: 139–148.

Montague, C. L. 1982. The influence of fiddler crab burrows and burrowing on metabolic processes in salt marsh sediments. In *Estaurine Comparisons*, V. S. Kennedy (ed.), 283–301. Academic Press, New York.

Moore, P. G. and R. Seed (eds). 1995. *The Ecology of Rocky Coasts*. Hodder and Stoughton, London

Morgan, S. G. 1989. Adaptive significance of spination in estuarine crab larvae. *Ecology* 70: 464–482.

Morgan, S. G. 1992. Impact of planktivorous fishes on the dispersal, hatching and morphology of estuarine crab larvae. *Ecology* 71: 1639–1652.

Morgan, S. G. 1995. Life and death in the plankton: Larval mortality and adaptation. In *Ecology of Marine Invertebrate Larvae*, L. McEdwards (ed.), 279–322. CRC Press, New York.

Morgan, S. G. and J. H. Christy. 1995. Adaptive significance of the timing of larval release by crabs. *American Naturalist* 145: 457–479.

Mullineaux, L. S. and C. A. Butman. 1991. Initial contact, exploration and attachment of barnacle cyprids settling in flow. *Marine Biology* 110: 93–103.

Nagle, J. S. 1968. Distribution of the epibiota of macroepibenthic plants. *Contributions to Marine Science* 13: 105–114.

Nelson, W. G. 1980. A comparative study of amphipods in seagrasses from Florida to Nova Scotia. *Bulletin of Marine Science* 30(1): 80–89.

Nelson, W. G. 1981. Experimental studies of decapod and fish predation on seagrass macrobenthos. *Marine Ecology Progress Series* 5: 141–149.

Neushul, M. 1960. Observation on sublittoral Antarctic marine algae. *British Phycological Bulletin* 2: 19.

Newell, R. C. 1969. Effects of fluctuations in temperature on the metabolism of intertidal invertebrates. *American Zoologist* 9: 293–307.

Newell, R. C. 1979. *The Biology of Intertidal Animals*. Marine Ecological Surveys. Faversham, Kent, England.

Newell, R. I. E. 1988. Ecological changes in Chesapeake Bay: Are they the result of overharvesting the American Oyster? In *Understanding the Estuary: Advances in Chesapeake Bay Research*, M. P. Lynch and E. C. Krome (eds.), 536–546. Chesapeake Research Consortium, Solomons, MD.

Newell, S. Y. 1993. Decomposition of shoots of a salt-marsh grass. *Advances in Microbial Ecology* 13: 301–326.

Newell, S. Y. 1996. Established and potential impacts of eukaryotic mycelia decomposers in marine/terrestrial ecotones. *Journal of Experimental Marine Biology and Ecology* 200: 187–206.

Newell, S. Y., T. L. Arsuffi, P. F. Kemp and L. A. Scott. 1991. Water potential of standing-dead shoots of an intertidal grass. *Oecologia* 85: 321–326.

Niering, W. A. and R. S. Warren. 1980. Vegetation patterns and processes in New England salt marshes. *BioScience* 30: 301–307.

Nixon, S. W. 1980. Between coastal marshes and waters: A review of twenty years of speculation and research on the role of salt marshes in estuarine productivity and water chemistry. In *Estuarine and Coastal Processes*, P. Hamilton and K. B. Macdonald (eds.), 437–525. Plenum Press, New York.

Nixon, S. W. 1982. *The Ecology of New England High Salt Marshes: A Community Profile*. Biological Report 81/55. U.S. Fish and Wildlife Service.

Nixon, S. W. and C. A. Oviatt. 1973. Ecology of a New England salt marsh. *Ecological Monographs* 43: 463–498.

Nixon, S. W. and C. A. Oviatt. 1978. Analysis of local standing crop of *Spartina alterniflora*. *Botanica Marina* 26: 103–109.

Nowell, A. R. M. and P. Jumars. 1984. Flow environments of aquatic benthos. *Annual Review of Ecology and Systematics* 15: 303–328.

Odum, E. P. 1969. The strategy of ecosystem development. *Science* 164: 262–270.

Odum, E. P. 1970. *Fundamentals of Ecology*. Saunders, Philadelphia.

Odum, E. P. and A. de la Cruz. 1967. Particulate organic detritus in a Georgia salt marsh-estuarine system. In *Estuaries*, G. H. Lauff (ed.), 383–385. AAAS Publication 83. Washington, DC.

Odum, W. E. 1978. The importance of tidal freshwater wetlands in coastal zone management. In *Coastal Zone '78: Symposium on Technical, Environmental, Socioeconomic and Regulatory Aspects of Coastal Zone Management*, 1196–1203. American Society of Civil Engineers, New York.

Officer, C. B., T. J. Smayda and R. Mann. 1982. Benthic filter feeding: A natural eutrophication control. *Marine Ecology Progress Series* 9: 203–210.

Okamura, B. 1986. Group living and the effects of spatial position in aggregates of *Mytilus edulis*. *Oecologia* 69: 341–347.

Olafsson, E. B. 1986. Density dependence in suspension-feeding and deposit-feeding populations of the bivalve *Macoma balthica*: A field experiment. *Journal of Animal Ecology* 55: 517–526.

Olafsson, E. B., C. W. Peterson and W. G. Ambrose. 1994. Does recruitment limitation structure populations and communities of macro-invertebrates in marine soft sediments? The relative significance of pre- and post-settlement processes. *Oceanography and Marine Biology Annual Review* 32: 65–109.

Oliver, J. S., R. G. Kvitek and P. N. Slattery. 1985. Walrus feeding disturbance: Scavenging habits and recolonization of the Bering Sea benthos. *Journal of Experimental Marine Biology and Ecology* 91: 233–246.

Olney, P. J. S. 1963. The food and feeding habits of teal *Anas crecca* L. *Proceedings of the Zoological Society of London* 140: 169–210.

Olson, R. R. and M. H. Olson. 1989. Food limitation of planktotrophic marine invertebrate larvae: Does it control recruitment success? *Annual Review of Ecology and Systematics* 20: 225–247.

Oney, J. 1954. *Final Report: Clapper Rail Survey and Investigation Study*. Georgia Game and Fish Commission.

Orr, P. R. 1955. Heat death. I. Time-temperature relationships in marine animals. *Physiological Zoology* 28: 290–294.

Orth, R. J. 1975. Destruction of eelgrass, *Zostera marina*, by the cownose ray, *Rhinoptera bonasus*, in the Chesapeake Bay. *Chesapeake Science* 16: 205–208.

Orth, R. J. 1977. The importance of sediment stability in seagrass communities. In *Ecology of marine benthos*, B. C. Coull (ed.), 281–300. University of South Carolina Press, Columbia.

Orth, R. J. and D. A. Moore. 1981. Submerged aquatic vegetation of the Chesapeake Bay: Past, present and future. *Transactions of the North American Wildlife and Natural Resources Conference* 46: 271–283.

Orth, R. J. and K. A. Moore. 1983. Chesapeake Bay: An unprecedented decline in submerged aquatic vegetation. *Science* 222: 51–53.

Orth, R. J. and J. van Montfrans. 1984. Epiphyte-seagrass relationships with an emphasis on the role of micrograzing: A review. *Aquatic Botany* 18: 43–69.

Orth, R. J., M. Luckenbach and K. A. Moore. 1994. Seed dispersal in a marine macrophyte: Implications for colonization and restoration. *Ecology* 75(7): 1927–1939.

Orton, J. H. 1912. An account of the natural history of the slipper limpet. *Journal of the Marine Biological Association* 9: 437–443.

Osman, R. W. 1977. The establishment and development of a marine epifaunal community. *Ecological Monographs* 47(1): 37–63.

Oviatt, C. A., S. W. Nixon and J. Garber. 1977. Variation and evaluation of coastal salt marshes. *Environmental Management* 1: 201–211.

Pace, M. L. 1979. The effect of grazing by a gastropod, *Nassarius obsoletus*, on the benthic microbial community of a salt marsh mudflat. *Estuarine and Coastal Marine Science* 9: 121–134.

Paine, R. T. 1966. Food web complexity and species diversity. *American Naturalist* 100: 65–75.

Paine, R. T. 1974. Intertidal community structure: Experimental studies on the relationship between a dominant competitor and its principal predator. *Oecologia* 15: 93–120.

Paine, R. T. 1976. Size-limited predation: An observational and experimental approach with the *Mytilus-Pisaster* interaction. *Ecology* 57: 858–873.

Paine, R. T. 1979. Disaster, catastrophe, and local persistence of the sea palm *Postelsia palmaeformis*. *Science* 205: 685–686.

Paine, R. T. 1994. *Marine Rocky Shores and Community Ecology: An Experimentalist's Perspective*. Ecology Institute, Oldendorf/Luhe, Germany.

Paine, R. T. and S. A. Levin. 1981. Intertidal landscapes. *Ecological Monographs* 51: 145–178.

Paine, R. T. and R. L. Vadas. 1969. The effects of grazing by sea urchins, *Strongylocentrotus* spp., on benthic algal populations. *Limnology and Oceanography* 14: 710–719.

Palmer, A. R. 1979. Fish predation and the evolution of gastropod shell sculpture. *Evolution* 33: 697–713.

Palmer, A. R. 1982. Predation and parallel and evolution: Recurrent parietal plate reduction in balanomorph barnacles. *Paleobiology* 8: 31–41.

Palmer, A. R. 1991. Effect of crab effluent and scent of damaged conspecifics on feeding, growth and shell morphology of the Atlantic dogwhelk *Nucella lapillus*. *Hydrobiologia* 193: 155–182.

Palmer, A. R. and R. R. Strathmann. 1989. Scale of dispersal in varying environments and its implications for life histories of marine invertebrates. *Oecologia* 48: 308–318.

Palumbi, S. R. 1984. Tactics of acclimation: Morphological changes in sponges in an unpredictable environment. *Science* 225: 1478–1480.

Palumbi, S. R. 1986. How body plans limit acclimation: Responses of a demosponge to wave force. *Ecology* 67: 208–214.

Parrish, R. H., C. S. Nelson and A. Bakun. 1981. Transport mechanisms and reproductive success of fishes in the California current. *Biological Oceanography* 1: 175–201.

Paulay, G., L. Boring and R. R. Strathmann. 1985. Food limited growth and development of larvae: Experiments with natural seawater. *Journal of Experimental Marine Biology and Ecology* 93: 1–10.

Pawlik, J. R. 1992. Chemical ecology of the settlement of benthic marine invertebrates. *Oceanography and Marine Biology Annual Review* 30: 273–335.

Pawlik, J. R., C. A. Butman and V. R. Starczak. 1991. Hydrodynamic facilitation of a reef-building tube worm. *Science* 251: 421–424.

Pechenik, J. A. 1979. Role of encapsulation in invertebrate life histories. *American Naturalist* 114: 859–870.

Penhale, P. A. and G. W. Thayer. 1980. Uptake and transfer of carbon and phosphorus by eelgrass (*Zostera marina*) and its epiphytes. *Journal of Experimental Marine Biology and Ecology* 42: 113–123.

Penhale, P. A. and R. G. Wetzel. 1983. Structural and functional adaptations of eelgrass (*Zostera marina* L.) to the anaerobic sediment environment. *Canadian Journal of Botany* 61: 1421–1428.

Pennings, S. C. and M. D. Bertness. 1998. Using latitudinal variation to examine effects of climate on coastal salt marsh pattern and process. *Proceedings of the OECD Workshop on Global Change and Wetlands. Journal of Wetland Geochemistry*, in press.

Pennings, S. C. and R. M. Callaway. 1992. Salt marsh plant zonation: The relative importance of competition and physical factors. *Ecology* 73: 681–690.

Pennings, S. C. and R. M. Callaway. 1996. Impact of a parasitic plant on the structure and dynamics of salt marsh vegetation. *Ecology* 77: 1410–1419.

Pennings, S. C. and V. J. Paul. 1992. Effect of plant toughness, calcification, and chemistry on herbivory by *Dolabella auricularia*. *Ecology* 73(5): 1606–1619.

Pennings, S. C. and V. J. Paul. 1993. Sequestration of algal secondary metabolites by sea hares: Location, specificity and dynamics. *Marine Biology* 117: 535–546.

Pennings, S. C., T. H. Carefoot, E. L. Siska, M. E. Chase and T. A. Page. 1998. Feeding preferences of a generalist salt-marsh crab: Relative importance of multiple plant traits. *Ecology*. In press.

Pennington, J. T. 1985. The ecology of fertilization of echinoid eggs: The consequences of sperm dilution, adult aggregation, and synchronous spawning. *Biological Bulletin* 169: 417–430.

Pennington, J. T. and F. S. Chia. 1984. Morphological and behavioral defenses of trochophore larvae of *Sabellaria* against four planktonic predators. *Biological Bulletin* 167: 168–175.

Peterson, C. H. 1977. Competitive organization of the soft-bottom communities of southern California lagoons. *Marine Biology* 43: 343–359.

Peterson, C. H. 1979. Predation, competitive exclusion, and diversity in the soft-sediment benthic communities of estuaries and lagoons. In *Ecological Processes in Coastal and Marine Systems*, R. J. Livingston (ed.), 233–263. Plenum Press, New York.

Peterson, C. H. 1982a. Clam predation by whelks (*Busycon* spp.): Experimental tests of the importance of prey size, prey density, and seagrass cover. *Marine Biology* 66: 159–170.

Peterson, C. H. 1982b. The importance of predation and intra- and interspecific competition in the population biology of two infaunal suspension-feeding bivalves, *Protothaca staminea* and *Chione undatella*. *Ecological Monographs* 52: 437–475.

Peterson, C. H. 1986. Enhancement of *Mercenaria mercenaria* densities in seagrass beds: Is pattern fixed during settlement season or altered by subsequent differential survival? *Limnology and Oceanography* 31: 200–205.

Peterson, C. H. 1992. Intertidal zonation of marine invertebrates in sand and mud. *American Scientist* 70: 236–249.

Peterson, C. H. and R. Black. 1987. Resource depletion by active suspension feeders on tidal flats: Influence of local density and tidal elevation. *Limnology and Oceanography* 32: 143–166.

Peterson, C. H. and N. M. Peterson. 1979. *The Ecology of Intertidal Flats of North Carolina: A Community Profile*. National Coastal Ecosystems Team, U.S. Fish and Wildlife Service, 48.

Peterson, C. H. and G. Skilleter. 1994. Control of foraging behavior of individuals within an ecosystem context: The clam, *Macoma balthica*, flow environment, and siphon-cropping fishes. *Oecologia* 100: 256–267.

Peterson, C. H., H. C. Summerson and P. B. Duncan. 1984. The influence of seagrass cover on population structure and individual growth rate of a suspension-feeding bivalve, *Mercenaria mercenaria*. *Journal of Marine Research* 42: 123–142.

Pethick, J. S. 1974. The distribution of salt pannes on tidal salt marshes. *Journal of Biogeography* 7: 57–62.

Pethick, J. S. 1984. *An Introduction to Coastal Geomorphology*. Arnold Press, London.

Petraitis, P. S. 1987. Factors organizing rocky intertidal communities of New England: Herbivory and predation in sheltered bays. *Journal of Experimental Marine Biology and Ecology* 109: 117–136.

Petraitis, P. S. 1991. Recruitment of the mussel *Mytilus edulis* on sheltered and exposed shores. *Journal of Experimental Marine Biology and Ecology* 147: 65–80.

Petraitis, P. S., R. E. Latham and R. A. Niesenbaum. 1989. The maintenance of species diversity by disturbance. *Quarterly Review of Biology* 64: 393–418.

Phillips, R. C. 1971. Ecological life history of *Zostera marina* L. (eelgrass) in Puget Sound, Washington. Ph.D. dissertation, University of Washington, Seattle.

Pickett, S. T. A. and P. S. White. 1985. *The Ecology of Natural Disturbance and Patch Dynamics*. Academic Press, New York.

Pineda, J. 1991. Predictable upwelling and the shoreward transport of planktonic larvae by internal tidal bores. *Science* 253: 548–551.

Pineda, J. 1994. Spatial and temporal patterns in barnacle settlement along a southern California rocky shore. *Marine Ecology Progress Series* 107: 125–138.

Platt, T., A. Prakash and B. Irwin. 1972. Phytoplankton, nutrients and flushing of inlets on the coast of Nova Scotia. *Nature Canada* 99: 253–261.

Porter, D. 1967. Observations on the cytology and motility of *Labyrinthula*. Ph.D. dissertation, University of Washington, Seattle.

Power, M. E. 1992. Top-down and bottom-up forces in food webs: Do plants have primacy? *Ecology* 73(3): 733–746.

Power, M. E., D. Tilman, J. A. Estes, B. A. Menge, W. J. Bond, L. S. Miles, G. Daily, J. C. Castilla, J. Lubchenco and R. T. Paine. 1996. Challenges in the quest for keystones. *BioScience* 46: 609–620.

Pullen, J. and M. LaBarbera. 1991. Modes of feeding in aggregations of barnacles and the shape of aggregations. *Biological Bulletin* 181: 442–452.

Purcell, J. E., F. P. Cresswell, D. G. Cargy and V. S. Kennedy. 1991. Differential ingestion and digestion of bivalve larvae by the scyphozoan *Chrysaora quinquecirrha* and the ctenophore *Mnemiopis leidyi*. *Biological Bulletin* 180: 103–111.

Quinn, J. F. 1982. Competitive hierarchies in marine benthic communities. *Oecologia* 54: 129–135.

Quinn, J. F., S. R. Wing and L. W. Botsford. 1993. Harvest refugia in marine invertebrate fisheries: Models and applications to the red sea urchin, *Strongylocentrotus franciscanus*. *American Zoologist* 33: 537–550.

Raffaelli, D. and S. Hawkins. 1996. *Intertidal Ecology*. Chapman and Hall, London.

Raffaelli, D., I. Karakassis and A. Galloway. 1991. Zonation schemes on sandy shores: A multivariate approach. *Journal of Experimental Marine Biology and Ecology* 148: 241–253.

Raimondi, P. T. 1988a. Rock type effects on barnacle zonation. *Journal of Experimental Marine Biology and Ecology* 123: 253–267.

Raimondi, P. T. 1988b. Settlement cues and determination of the vertical limit of an intertidal barnacle. *Ecology* 69(2): 400–407.

Rampino, M. R. and J. Sanders. 1981. Episodic growth of Holocene tidal marshes in the northeastern United States: A possible indicator of eustatic sea level fluctuations. *Geology* 9: 63–67.

Ramus, J. 1983. A physiological test of the theory of complementary chromatic adaptation. II. Brown, red and green seaweeds. *Journal of Phycology* 19: 173–178.

Rand, T. 1999. Seed supply, habitat suitability and the distribution of halophytes across a New England salt marsh. *Ecology*, in press.

Randell, J. E. 1967. Food habits of reef fishes of the West Indies. *Studies in Tropical Biology, Oceanography Institute, University of Miami* 5: 665–847.

Ranwell, D. S. 1961. *Spartina* salt marshes in southern England. I. The effects of sheep grazing at the upper limits of *Spartina* marsh in Bridgewater Bay. *Journal of Ecology* 49: 325–340.

Ranwell, D. S. 1971. *Ecology of Salt Marshes and Sand Dunes*. Chapman and Hall, London.

Rasmussen, E. 1977. The wasting disease of eelgrass (*Zostera marina*) and its effect on environmental factors and fauna. In *Seagrass Ecosystems: A Scientific Perspective*, C. P. McRoy and C.

Helfferich (eds.), 1–51. Marcel Dekker, New York.

Redfield, A. C. 1965. Ontogeny of a salt marsh. *Science* 147: 50–55.

Redfield, A. C. 1972. Development of a New England salt marsh. *Ecological Monographs* 42: 201–237.

Reed, D. C. 1990. The effects of variable settlement and early competition on patterns of kelp recruitment. *Ecology* 71: 776–787.

Reid, D. G. 1990a. A cladistic study of the genus *Littorina* (Gastropoda): Implications for evolution of reproductive strategies and for classification. *Hydrobiologia* 193: 1–20.

Reid, D. G. 1990b. Trans-Arctic migration and speciation induced by climate change: The biogeography of *Littorina* (Mollusca: Gastropoda). *Bulletin of Marine Science* 47: 35–49.

Reidenbaugh, T. G. and W. C. Banta. 1982. Origin and effects of tidal wrack in a Virginia salt marsh. *Gulf Research Reports* 6: 393–401.

Reimold, R. J. 1977. Mangals and salt marshes of the eastern United States. In *Wet Coastal Ecosystems*, V. J. Chapman (ed.), 157–166. Elsevier, Amsterdam.

Reise, K. 1983. Biotic enrichment of intertidal sediments by experimental aggregates of the deposit-feeding bivalve *Macoma balthica*. *Marine Ecology Progress Series* 12: 229–236.

Reise, K. 1985. *Tide Flat Ecology: An Experimental Approach to Species Interactions*. Springer-Verlag, Berlin.

Rhoads, D. C. 1973. The influence of deposit-feeding benthos on water turbidity and nutrient recycling. *American Journal of Science* 273: 1–22.

Rhoads, D. C. 1974. Organism-sediment relations on the muddy sea floor. *Oceanography and Marine Biology Annual Review* 12: 263–300.

Rhoads, D. C. and D. K. Young. 1970. The influences of deposit-feeding organisms on sediment stability and community trophic structure. *Journal of Marine Research* 28: 150–178.

Rhoads, D. C. and D. K. Young. 1971. Animal-sediment relations in Cape Cod Bay, Massachusetts. II. Reworking by *Molpadia oolitica* (Holothuroidea). *Marine Biology* 11: 255–261.

Ringold, P. 1979. Burrowing, root mat density, and the distribution of fiddler crabs in the eastern United states. *Journal of Experimental Marine Biology and Ecology* 36: 11–21.

Rivest, B. R. 1983. Development and the influence of nurse-egg allotment on hatching size in *Searlesia dira* (Reeve, 1846) (Prosobrachia: Buccinidae). *Journal of Experimental Marine Biology and Ecology* 69: 217–241.

Rivest, B. R. 1986. Extra-embryonic nutrition in the prosobranch gastropod *Urosalpinx cinera* (Say, 1882). *Bulletin of Marine Science* 39: 498–505.

Robertson, J. R. and S. Y. Newell. 1982. Experimental studies of particle ingestion by the sand fiddler crab *Uca pugilator* (Bose). *Journal of Experimental Marine Biology and Ecology* 52: 1–21.

Roughgarden, J. S., Y. Iwasa and C. Baxter. 1985. Demographic theory for an open marine population with space-limited recruitment. *Ecology* 66: 54–67.

Roughgarden, J. S., S. D. Gaines and S. W. Pacala. 1987. Supply side ecology: The role of physical transport. In *Organization of Communities Past and Present*, J. Gee and P. S. Giller (eds.), 491–518. Blackwell Scientific, Oxford.

Roughgarden, J. S., S. D. Gaines and H. Possingham. 1988. Recruitment dynamics in complex life cycles. *Science* 241: 1460–1466.

Rozas, L. P. and W. F. Odum. 1987. Use of tidal marshes by fishes and macrofaunal crustaceans along a marsh stream-order gradient. *Estuaries* 10: 36–43.

Ryland, J. S. 1970. *Bryozoans*. Hutchinson University Library, London.

Ryther, J. H. and W. M. Dunstan. 1971. Nitrogen, phosphorus and eutrophication in the coastal marine environment. *Science* 171: 1008–1013.

Saffo, M. B. 1987. New light on seaweeds. *BioScience* 37: 654–664.

Sale, P. F. 1977. Maintenance of high diversity in coral reef fish communities. *American Naturalist* 111: 337–359.

Sanders, H. L. 1956. The biology of marine bottom communities. *Bulletin of the Binghamton Oceanographic Collection of Yale University* 15: 344–414.

Sanford, E., D. Bermudez, M. D. Bertness and S. D. Gaines. 1994. Flow, food supply and acorn barnacle population dynamics. *Marine Ecology Progress Series* 104: 49–62.

Schaffner, L. C. 1990. Small-scale organism distributions and patterns of species diversity: Evidence for positive interactions in an estuarine benthic community. *Marine Ecology Progress Series* 61: 107–117.

Scheibling, R. 1986. Increased macroalgal abundance following mass mortalities of sea urchins (*Strongylocentrotus droebachiensis*) along the Atlantic coast of Nova Scotia. *Oecologia* 68: 186–198.

Scheibling, R. E. and R. L. Stephenson. 1984. Mass mortality of *Strongylocentrotus droebachiensis* (Echinodermata: Echinoidea) off Nova Scotia, Canada. *Marine Biology* 78: 153–164.

Scheltema, R. S. 1971a. The dispersal of larvae of shoal-water benthic invertebrate species over long distances by ocean currents. In *Fourth European Marine Biology Symposium*, D. Crisp (ed.), 7–28. Cambridge University Press, Cambridge.

Scheltema, R. S. 1971b. Larval dispersal as a means of genetic exchange between geographically separated populations of shallow-water benthic marine gastropods. *Biological Bulletin* 140: 284–322.

Scheltema, R. S. 1986. On dispersal and planktonic larvae of benthic invertebrates: An eclectic overview and summary of problems. *Bulletin of Marine Science* 39(2): 290–322.

Schindler, D. E., B. M. Johnson, N. A. MacKay, N. Bouwes and J. F. Kitchell. 1994. Crab/snail size-structured interactions and salt marsh predation gradients. *Oecologia* 97: 49–61.

Schmidt, P. R. and D. M. Rand. 1999. Thermal microhabitat and patterns of genetic variation in the northern acorn barnacle *Semibalanus balanoides*. *Evolution*. In press.

Schmidt, R. J. 1982. Consequences of dissimilar defenses against predation in a subtidal marine community. *Ecology* 63: 1588–1601.

Schmidt, R. J. 1987. Indirect interactions between prey: Apparent competition, predator aggregation, and habitat segregation. *Ecology* 68: 1887–1897.

Schneider, D. C. 1978. Equalization of prey numbers by migratory shorebirds. *Nature* 271: 353–354.

Schneider, D. C. and B. A. Harrington. 1981. Timing of shorebird migration in relation to prey depletion. *Auk* 98: 801–811.

Sebens, K. P. 1982. The limits to indeterminate growth: An optimal size model applied to passive suspension feeders. *Ecology* 63: 209–222.

Sebens, K. P. 1985. The ecology of the rocky subtidal zone. *American Scientist* 73: 548–557.

Sebens, K. P. and M. A. R. Koehl. 1984. Predation on zooplankton by the benthic anthozoans *Alcyonium siderium* (Alcyonacea) and *Metriduim senile* (Actinaria) in the New England subtidal. *Marine Biology* 18: 225–271.

Seed, R. 1969. The ecology of *Mytilus edulis* L. (Lamellibranchiata) on exposed rocky shores. 2. Growth and mortality. *Oecologia* 3: 317–350.

Seeley, R. H. 1986. Intense natural selection caused a rapid morphological transition in a living marine snail. *Proceedings of the National Academy of Sciences USA* 83: 6897–6901.

Seliskar, D. M. 1983. Root and rhizome distribution as an indictor of upper salt marsh wetland limits. *Hydrobiologia* 107: 231–236.

Seliskar, D. M. 1985. Morphometric variations of five tidal marsh halophytes along environmental gradients. *American Journal of Botany* 72: 1340–1352.

Shanks, A. L. 1983. Surface slicks associated with tidally forced internal waves may transport pelagic larvae of benthic invertebrates and fishes shoreward. *Marine Ecology Progress Series* 13: 311–315.

Shanks, A. L. 1985. Behavioral basis of internal-wave-induced shoreward transport of megalopae of the crab *Pachygrapsus crassipes*. *Marine Ecology Progress Series* 24: 289–295.

Shanks, A. L. and W. G. Wright. 1986. Adding teeth to wave action: The destructive effects of waveborne rocks on intertidal organisms. *Oecologia* 69: 420–428.

Shanks, A. L. and W. G. Wright. 1987. Internal-wave mediated shoreward transport of cyprids, megalopae, and gammarids and correlated longshore differences in the settling rate of intertidal barnacles. *Journal of Experimental Marine Biology and Ecology* 114: 1–13.

Shat, H. 1984. A comparative ecophysiology study of the effects of waterlogging and submergence of dune slack plants. *Oecologia* 62: 279–286.

Shea, M. L., R. S. Warren and W. A. Niering. 1975. Biochemical and transplantational studies of the growth form of *Spartina alterniflora* on Connecticut salt marshes. *Ecology* 56: 461–466.

Shimeta, J. and P. A. Jumars. 1991. Physical mechanisms and rates of particle capture by suspension feeders. *Oceanography and Marine Biology Annual Review* 29: 191–257.

Shumway, S. W. 1995. Physiological integration among clonal ramets during invasion of disturbance patches in a New England salt marsh. *Annals of Botany* 76: 225–233.

Shumway, S. W. and M. D. Bertness. 1992. Salt stress limitation of seedling recruitment in a salt marsh plant community. *Oecologia* 92: 490–497.

Silliman, B. R. 1998. Top-down control of *Spartina alterniflora* growth: Salt marsh periwinkles eat more than just detritus. Master's thesis, University of Virginia, Charlottesville.

Skilleter, G. A. 1994. Refuges from predation and the persistence of estuarine clam populations. *Marine Ecology Progress Series* 109: 29–42.

Skilleter, G. A. and C. H. Peterson. 1994. Control of foraging behavior of individuals within an ecosystem context: The clam, *Macoma balthica*, and interactions between competition and siphon cropping. *Oecologia* 100: 268–278.

Slocum, C. J. 1980. Differential susceptibility to grazers in two phases of an intertidal alga: Advantages of heteromorphic generations. *Journal of Experimental Marine Biology and Ecology* 46: 99–110.

Smalley, A. E. 1960. Energy flow of a salt marsh grasshopper population. *Ecology* 41: 672–677.

Smith, L. D. 1990. Patterns of limb loss in the blue crab, *Callinectes sapidus*, and the effects of autotomy on growth. *Bulletin of Marine Science* 46: 23–36.

Smith, T. J. and W. E. Odum. 1981. The effects of grazing by snow geese on coastal salt marshes. *Ecology* 62: 98–106.

Smith, T. J., K. G. Bnoto, S. D. Frusher and R. L. Giddins. 1991. Keystone species and mangrove forest dynamics: The influence of burrowing by crabs on soil nutrient status and forest productivity. *Estuarine, Coastal and Shelf Science* 33: 419–432.

Snelgrove, R. V. R., C. A. Butman and J. P. Grassle. 1993. Hydrodynamic enhancement of larval settlement in the bivalve *Mulinia lateralis* (Say) and the polychaete *Capitella* sp. I in microdepositional environments. *Journal of Experimental Marine Biology and Ecology* 168: 71–109.

Snow, A. and S. Vince. 1984. Plant zonation in an Alaskan salt marsh. II. An experimental study of the role of edaphic conditions. *Journal of Ecology* 72: 669–684.

Sousa, W. P. 1979a. Disturbance in marine intertidal boulder fields: The nonequilibrium maintenance of species diversity. *Ecology* 60(6): 1225–1239.

Sousa, W. P. 1979b. Experimental investigations of disturbance and ecological succession in a rocky intertidal algal community. *Ecological Monographs* 49(3): 227–254.

Sousa, W. P. 1985. Disturbance and patch dynamics on rocky intertidal shores. In *The Ecology of Natural Disturbance and Patch Dynamics*, S. T. A. Pickett and P. S. White (eds.), 101–124. Academic Press, New York.

Sponaugle, S. and P. Lawton. 1990. Portunid crab predation on juvenile hard clams: Effects of substrate type and prey density. *Journal of Experimental Marine Biology and Ecology* 67: 43–53.

Stanley, S. M. 1970. Relation of shell form to life habits of the bivalve molluscs. *Geological Society of America Monographs* 125: 1–296.

Stanley, S. M. 1975. Why clams have the shape they have: An experimental analysis of burrowing. *Paleobiology* 1: 1–48.

Stanley, S. M. 1986. Anatomy of a regional mass extinction: Plio-Pleistocene decimation of the western Atlantic bivalve fauna. *Palios* 1: 17–36.

Stanley, S. M. and W. A. Newman. 1980. Competitive exclusion in evolutionary time: The case of the acorn barnacles. *Paleobiology* 6: 173–183.

Steneck, R. S. 1982. A limpet-coralline algal association: Adaptations and defenses between a selective herbivore and its prey. *Ecology* 63: 507–522.

Steneck, R. S. 1983. Escalating herbivory and resulting adaptive trends in calcareous algal crusts. *Paleobiology* 9(1): 44–61.

Steneck, R. S. and M. N. Dethier. 1994. A functional group approach to the structure of algal-dominated communities. *Oikos* 69: 476–498.

Steneck, R. S. and L. Watling. 1982. Feeding capabilities and limitation of herbivorous molluscs: A functional group approach. *Marine Biology* 68: 299–319.

Stephens, E. G. and M. D. Bertness. 1991. Mussel facilitation of barnacle survival in a sheltered bay habitat. *Journal of Experimental Marine Biology and Ecology* 145: 33–48.

Stephenson, T. A. and A. Stephenson. 1949. The universal features of zonation between tidemarks on rocky coasts. *Journal of Ecology* 38: 289–305.

Stephenson, T. A. and A. Stephenson. 1971. *Life Between Tidemarks on Rocky Shores*. W. H. Freeman, San Francisco.

Stiven, A. E. and S. A. Gardner. 1992. Population processes in the ribbed mussel *Geukensia demissa* in a North Carolina salt marsh. *Journal of Experimental Marine Biology and Ecology* 160: 81–102.

Stiven, A. E. and E. J. Kuenzler. 1982. The response of two salt marsh molluscs, *Littorina irrorata* and *Geukensia demissa*, to field manipulations of density and *Spartina* litter. *Ecological Monographs* 49: 151–171.

Stout, J. P. 1984. *The Ecology of Irregularly Flooded Salt Marshes of the Northeastern Gulf of Mexico: A Community Profile*. Biological Reports 85(7.1): 1–98. U.S. Fish and Wildlife Service, Washington, D.C.

Strathmann, R. R. 1990. Why life histories evolve differently in the sea. *American Zoologist* 30: 197–207.

Strathmann, R. R. 1993. Hypotheses on the origins of marine larvae. *Annual Review of Ecology and Systematics* 24: 89–117.

Strathmann, R. R. and M. F. Strathmann. 1982. The relation between adult size and brooding in marine invertebrates. *American Naturalist* 119: 91–101.

Strathmann, R. R., E. S. Branscomb and K. Vedder. 1981. Fatal errors in set as a cost of dispersal and the influence of intertidal flora on set of barnacles. *Oecologia* 48: 13–18.

Strong, D. R. 1992. Are trophic cascades all wet? Differentiation and donor-control in speciose ecosystems. *Ecology* 73(3): 747–754.

Suchanek, T. H. 1981. The role of disturbance in the evolution of life history strategies in the intertidal mussels *Mytilus edulis* and *Mytilus californianus. Oecologia* 50: 143–152.

Suchanek, T. H. 1986. Mussels and their role in structuring rocky shore communities. In *The Ecology of Rocky Shores*, P. G. Moore and R. Seed (eds.), 70–96. Hodder and Stoughton, Sevenoaks, England.

Suchanek, T. H. 1993. Oil impacts on marine invertebrate populations and communities. *American Zoologist* 33: 510–523.

Sulkin, S. D. 1984. Behavioral basis of depth regulation in the larvae of brachyuran crabs. *Marine Ecology Progress Series* 15: 181–205.

Sullivan, M. and F. C. Daiber. 1974. Response in production of the cordgrass *Spartina alterniflora* to inorganic nitrogen and phosphorus fertilization. *Chesapeake Science* 15: 121–123.

Summerson, H. C. and C. H. Peterson. 1984. Role of predation in organizing benthic communities of a temperate zone seagrass bed. *Marine Ecology Progress Series* 15: 63–77.

Sutherland, J. P. 1974. Multiple stable points in natural communities. *American Naturalist* 108: 859–873.

Sutherland, J. P. and R. H. Karlson. 1977. Development and stability of the fouling community at Beaufort, North Carolina. *Ecological Monographs* 47: 425–446.

Taghon, G. L. and P. A. Jumars. 1984. Variable ingestion rate and its role in optimal foraging behavior of marine deposit feeders. *Ecology* 65: 549–558.

Taghon, G. L., A. R. M. Nowell and P. J. Jumars. 1980. Induction of suspension feeding in spionid polychaetes by high particulate fluxes. *Science* 210: 562–564.

Teal, J. M. 1962. Energy flow in the saltmarsh ecosystem of Georgia. *Ecology* 43: 614–624.

Teal, J. M. 1985. *The Ecology of New England Low Marsh Habitats: A Community Profile*. U.S. Fish and Wildlife Service, Washington, D.C.

Teal, J. M. and J. W. Kanwisher. 1966. Gas transport in a marsh grass *Spartina alterniflora. Journal of Experimental Botany* 17: 355–361.

Thayer, G. W. and H. H. Stuart. 1974. The bay scallop makes its bed of eelgrass. *U.S. Fish and Wildlife Service Biological Reports* 36(7): 27–39.

Thayer, G. W., D. A. Wolfe and R. B. Williams. 1975. The impact of man on seagrass systems. *American Scientist* 63: 288–296.

Thayer, G. W., H. H. Stuart, W. J. Kenworthy, J. F. Ustach and A. B. Hall. 1979. Habitat values of salt marshes, mangroves, and seagrasses for aquatic organisms. In *Wetland Functions and Values: The State of Our Understanding*, P. E. Greeson, J. R. Clark and J. E. Clark (eds.), 235–247. American Water Resources Association, Minneapolis, MN.

Thorson, G. 1950. Reproductive and larval ecology of marine bottom invertebrates. *Biological Review* 25: 1–45.

Tomlinson, P. B. 1974. Vegetative morphology and meristem dependence: The foundation of productivity in seagrasses. *Aquaculture* 4: 107–130.

Trager, G. C., J. S. Hwang and J. R. Strickler. 1990. Barnacle suspension feeding in variable flow. *Marine Biology* 105: 117–127.

Trussell, G. C. 1996. Phenotypic plasticity in an intertidal snail: The role of a common crab predator. *Evolution* 50: 448–454.

Trussell, G. C., A. S. Johnson, S. G. Rudolph and E. S. Gilfillan. 1993. Resistance to dislodgement: Habitat and size-specific differences in morphology and tenacity in an intertidal snail. *Marine Ecology Progress Series* 100: 135–144.

Turner, R. E. 1976. Geographic variations in salt marsh macrophyte production: A review. *Contributions in Marine Science* 20: 47–68.

Turner, R. E. 1977. Intertidal vegetation and commercial yields of Paneid shrimp. *Transactions of the American Fisheries Society* 106: 411.

Tutlin, T. G. 1938. The autecology of *Zostera marina* in relation to its wasting disease. *New Phytologist* 37: 50–71.

Ulanowicz, R. E. and J. H. Tuttle. 1992. The trophic consequences of oyster stock rehabilitation in Chesapeake Bay. *Estuaries* 15: 298–306.

Underwood, A. J. and E. J. Denley. 1984. Paradigms, explanations, and generalizations in models for the structure of intertidal communities on rocky shores. In *Ecological Communities*, D. Strong, D. Simberloff, L. Abele and A. Thistle (eds.), 151–180. Princeton University Press, Princeton, NJ.

Vadas, R. L., W. A. Wright and S. L. Miller. 1990. Recruitment of *Ascophyllum nodosum*: Wave action as a source of mortality. *Marine Ecology Progress Series* 61: 263–272.

Vadas, R. L., S. Johnson and T. A. Norton. 1992. Recruitment and mortality of early post-settlement stages of benthic algae. *British Journal of Phycology* 27: 331–351.

Valentine, J. F. and K. L. Heck. 1993. Mussels in seagrass meadows: Their influence on macroinvertebrate abundance and secondary production in the Gulf of Mexico. *Marine Ecology Progress Series* 96: 63–74.

Valiela, I. 1995. *Marine Ecological Processes*. Springer-Verlag, New York.

Valiela, I. and C. S. Riestma. 1995. Disturbance of salt marsh vegetation by wrack mats in Great Sippewissett Marsh. *Oecologia* 102: 106–112.

Valiela, I. and J. M. Teal. 1974. Nutrient limitation of salt marsh vegetation. In *Ecology of Halophytes*, R. J. Reimold and W. H. Queen (eds.), 547–563. Academic Press, New York.

Valiela, I. and J. M. Teal. 1979. The nitrogen budget of a salt marsh ecosystem. *Nature* 280: 652–656.

Valiela, I., D. F. Babiec, W. Atherton, S. Setzinger and C. Krebs. 1974. Some consequences of sexual dimorphism: Feeding in male and female fiddler crabs *Uca pugnax* (Smith). *Biological Bulletin* 147: 652–660.

Valiela, I., J. M. Teal and N. Y. Persson. 1976. Production and dynamics of experimentally enriched salt marsh vegetation: Belowground biomass. *Limnology and Oceanography* 21: 245–252.

Valiela, I., J. M. Teal and W. G. Deuser. 1978. The nature of growth forms in the salt marsh grass *Spartina alterniflora*. *American Naturalist* 112: 461–470.

Valiela, I., J. M. Teal, C. Cogswell, J. Hartman, S. Allen, R. V. Etten and D. Goehringer. 1985. Some long-term consequences of sewage contamination in salt marsh ecosystems. In *Ecological Considerations in Wetlands Treatment of Municipal Wastewaters*, P. J. Godfrey, E. R. Kaynor, S. Pelezarski and J. Benforado (eds.), 301–316. Van Nostrand Reinhold, New York.

Valiela, I., P. Peckol, C. D'Avanzo, C.–H. Sham and K. Lajtha. 1992. Couplings of watersheds and coastal waters: Sources and consequences of nutrient enrichment in Waquoit Bay, Massachusetts. *Estuaries* 15: 443–457.

Van Alstyne, K. L. 1988. Herbivore grazing increases polyphenolic defenses in the intertidal brown alga, *Fucus distichus*. *Ecology* 69: 655–633.

Van Blaricom, G. E. 1982. Experimental analysis of structural regulation in marine sand communities. *Ecological Monographs* 52: 283–305.

Vance, R. R. 1973. On reproductive strategies in marine benthic invertebrates. *American Naturalist* 107: 339–352.

Van Dolah, R. F. 1978. Factors regulating the distribution and population dynamics of the amphipod *Gammarus palustris* in an intertidal salt marsh community. *Ecological Monographs* 48: 191–217.

Vermeij, G. J. 1971. Temperature relations of some tropical Pacific intertidal gastropods. *Marine Biology* 10: 308–314.

Vermeij, G. J. 1972. Intraspecific shore-level size gradient in intertidal molluscs. *Ecology* 53: 693–700.

Vermeij, G. J. 1973. Morphological patterns in high intertidal gastropods: Adaptive strategies and their limitation. *Marine Biology* 20: 319–346.

Vermeij, G. J. 1978. *Biogeography and Adaptation: Patterns of Marine Life*. Harvard University Press, Cambridge, MA.

Vermeij, G. J. 1982a. Environmental change and the evolutionary history of the periwinkle *Littorina littorea* in North America. *Evolution* 36: 561–580.

Vermeij, G. J. 1982b. Phenotypic evolution in a poorly dispersing snail after arrival of a predator. *Nature* 299: 349–350.

Vermeij, G. J. 1987. *Evolution and Escalation: An Ecological History of Life*. Princeton University Press, Princeton, NJ.

Vermeij, G. J. 1989. Geographical restriction as a guide to the causes of extinction: The case of the cold northern oceans during the Neogene. *Paleobiology* 15(4): 335–356.

Vermeij, G. J. 1991. Anatomy of an invasion: The trans-Arctic interchange. *Paleobiology* 17(3): 281–307.

Vermeij, G. J. and J. Vail. 1978. A latitudinal pattern in bivalve shell gaping. *Malacologia* 17: 57–61.

Vince, S., I. Valiela, N. Backus and J. M. Teal. 1976. Predation by the salt marsh killifish *Fundulus heteroclitus* (L.) in relation to prey size and habitat structure: Consequences for prey distribution and abundance. *Journal of Experimental Marine Biology and Ecology* 23: 255–266.

Vince, S. W., I. Valiela and J. M. Teal. 1981. An experimental study of the structure of herbivorous insect communities in a salt marsh. *Ecology* 62(6): 1662–1678.

Virnstein, R. W. 1977. The importance of predation by crabs and fishes on benthic infauna in Chesapeake Bay. *Ecology* 58: 1199–1217.

Vogel, S. 1974. Current-induced flow through sponges in situ. *Proceedings of the National Academy of Sciences USA* 74: 2069–2071.

Vogel, S. 1977. Current-induced flow through the sponge *Halichondria*. *Biological Bulletin* 147: 443–456.

Vogel, S. 1981. *Life in Moving Fluids*. Princeton University Press, Princeton, NJ.

Vogel, S. 1988. *Life's Devices*. Princeton University Press, Princeton, NJ.

Walne, P. R. 1971. The influence of current speed, body size and water temperature on the filtration rate in five species of bivalves. *Journal of the Marine Biology Association* (UK) 52: 345–374.

Warren, R. S. and W. A. Niering. 1993. Vegetation change on a northeast tidal marsh: Interaction of sea-level rise and marsh accretion. *Ecology* 74: 96–103.

Watzin, M. C. 1983. The effects of meiofauna on settling macrofauna: Meiofauna may structure macrofaunal communities. *Oecologia* 59: 163–166.

Watzin, M. C. 1986. Larval settlement into marine soft-sediment systems: Interactions with the meiofauna. *Journal of Experimental Marine Biology and Ecology* 98: 65–113.

Weissburg, M. 1992. Functional analysis of fiddler crab foraging: Sex specific mechanics and constraints. *Journal of Experimental Marine Biology and Ecology* 156: 105–124.

Weissburg, M. 1993. Sex and the single forager: Gender specific energy maximization strategies in fiddler crabs. *Ecology* 74: 279–291.

Weissburg, M. J. and R. K. Zimmer-Faust. 1993. Life and death in moving fluids: Hydrodynamic effects on chemosensory-mediated predation. *Ecology* 74: 1428–1443.

Wethey, D. S. 1983a. Geographical limits and local zonation: The barnacles *Semibalanus* and *Chthamalus* in New England. *Biological Bulletin* 165: 330–341.

Wethey, D. S. 1983b. Intrapopulation variation in growth of sessile organisms: Natural populations of the intertidal barnacle *Balanus balanoides*. *Oikos* 40: 14–23.

Wethey, D. S. 1984a. Effects of crowding on fecundity in barnacles: *Semibalanus* (*Balanus*) *balanoides*, *Balanus glandula* and *Chthamalus dalli*. *Canadian Journal of Zoology* 62: 1788–1795.

Wethey, D. 1984b. Thermal effects on the distribution of barnacle populations in New England. *Biological Bulletin* 59: 160–169.

Wethey, D. S. 1985. Catastrophe, extinction and species diversity: A rocky intertidal example. *Ecology* 66: 445–456.

Wheeler, W. N. 1980. Effect of boundary layer transport on the fixation of carbon by the giant kelp *Macrocytis pyrifera*. *Marine Biology* 56: 103–110.

Whitlatch, R. B. 1981. Animal-sediment relationships in intertidal marine benthic habitats: Some determinants of deposit-feeding species diversity. *Journal of Experimental Marine Biology and Ecology* 53: 31–45.

Whitlatch, R. B. and R. N. Zajac. 1985. Biotic interactions among estuarine infaunal opportunistic species. *Marine Ecology Progress Series* 21: 299–311.

Whittaker, R. H. 1975. *Communities and Ecosystems*. Macmillan, New York.

Wicksten, M. K. 1980. Decorator crabs. *Scientific American* 242: 146–242.

Wiegel, R. L. 1964. *Oceanographic Engineering*. Prentice-Hall, New York.

Wiegert, R. G. and B. J. Freeman. 1990. *Tidal Salt Marshes of the Southeast Atlantic Coast: A Community Profile*. Biological Report 85(7.29). U.S. Fish and Wildlife Service, Washington, DC.

Wiegert, R. G. and L. R. Pomeroy. 1981. *The Ecology of a Salt Marsh*. Springer-Verlag, New York.

Wiegert, R. G., A. G. Chalmers and P. F. Randerson. 1983. Productivity gradients in salt marshes: The response of *Spartina alterniflora* to experimentally manipulated soil water movement. *Oikos* 41: 1–6.

Wilber, T. P. and W. Herrnkind. 1984. Predaceous gastropods regulate new-shell supply to salt marsh hermit crabs. *Marine Biology* 79: 145–150.

Wilce, R. T. 1959. The marine algae of the Labrador Peninsula and northwest Newfoundland (ecology and distribution). *Bulletin of the National Museum of Canada* 158: 1–103.

Wildish, D. J. and D. D. Kristmanson. 1979. Tidal energy and sublittoral macrobenthic animals in estuaries. *Journal of the Fisheries Research Board of Canada* 36: 1197–1206.

Williams, G. A. 1992. The effect of predation on the life histories of *Littorina obtusata* and *Littorina*

mariae. *Journal of the Marine Biological Association* (UK) 72: 403–416.

Williams, S. L. 1980. Experimental studies of Caribbean seagrass bed development. *Ecological Monographs* 60: 449–469.

Williams, S. L. and M. H. Ruchelshaus. 1993. Effects of nitrogen availability and herbivory on eelgrass and epiphytes. *Ecology* 74: 904–918.

Wilson, J. B. and D. Agnew. 1992. Positive-feedback switches in plant communities. *Advances in Ecological Research* 23: 263–333.

Wilson, W. H. 1983. The role of density dependence in a marine infaunal community. *Ecology* 64: 295–306.

Wilson, W. H. 1984. Non-overlapping distributions of spionid polychaetes: The relative importance of habitat and competition. *Journal of Experimental Marine Biology and Ecology* 75: 119–127.

Wilson, W. H. 1989. Predation and the mediation of intraspecific competition in an infaunal community in the Bay of Fundy. *Journal of Experimental Marine Biology and Ecology* 132: 221–245.

Wilson, W. H. 1991. Competition and predation in marine soft sediment communities. *Annual Review of Ecology and Systematics* 21: 221–241.

Wiltse, W. I. 1980. Effects of *Polinices duplicatus* (Gastropoda: Naticidae) on infaunal community structure at Barnstable Harbor, Massachusetts. *Marine Biology* 56: 301–310.

Witman, J. D. 1985. Refuges, biological disturbance, and rocky subtidal community structure in New England. *Ecological Monographs* 55(4): 421–445.

Witman, J. D. 1987. Subtidal coexistence: Storms, grazing, mutualisms, and the zonation of kelps and mussels. *Ecological Monographs* 57: 167–187.

Witman, J. D. and K. Sebens. 1992. Regional variation in fish predation intensity: A historical perspective in the Gulf of Maine. *Oecologia* 90: 305–315.

Witman, J. D. and T. Suchanek. 1984. Mussels in flow: Drag and dislodgment by epizoans. *Marine Ecology Progress Series* 16: 167–187.

Woodin, S. A. 1976. Adult-larval interactions in dense infaunal assemblages: Patterns of abundance. *Journal of Marine Research* 34: 25–41.

Woodin, S. A. 1977. Algal gardening behavior by nereid polychaetes: Effects on soft-bottom community structure. *Marine Biology* 44: 39–42.

Woodin, S. A. 1978. Refuges, disturbance and community structure: A marine soft-bottom example. *Ecology* 59(2): 274–284.

Woodin, S. A. 1981. Disturbance and community structure in a shallow water sand flat. *Ecology* 62: 1052–1066.

Woodwell, G. M., D. E. Whitney, C. A. S. Hall and R. A. Houghton. 1977. The Flax Pond ecosystem study: Exchanges of carbon in water between a salt marsh and Long Island Sound. *Limnology and Oceanography* 22: 833–838.

Wootton, T. J. 1992. Indirect effects, prey susceptibility, and habitat selection: Impacts of birds on limpets and algae. *Ecology* 73(3): 981–991.

Wootton, T. J. 1993. Indirect effects and habitat use in an intertidal community: Interaction chains and interaction modifications. *American Naturalist* 141: 71–89.

Wu, R. S. S. 1980. Effects of crowding on the energetics of the barnacle *Balanus glandula* Darwin. *Canadian Journal of Zoology* 58: 559–566.

Wu, R. S. S., C. D. Levings and D. J. Randell. 1977. Differences in energy partitioning between crowded and uncrowded individual barnacles. *Canadian Journal of Zoology* 55: 643–647.

Yates, J. L. and P. Peckol. 1993. Effects of nutrient availability and herbivory on polyphenolics in the seaweed *Fucus vesiculosus*. *Ecology* 74: 1757–1766.

Yoshioka, P. M. 1982. The role of planktonic and benthic factors in the population dynamics. *Ecology* 63: 457–468.

Young, C. M. and B. L. Bingham. 1987. Chemical defense and aposematic coloration in larvae of the ascidian *Ecteinascidia turbinata*. *Marine Biology* 96: 539–544.

Yund, P. O. 1990. An in situ measurement of sperm dispersal in a colonial marine hydroid. *Journal of Experimental Zoology* 253: 102–106.

Zach, R. 1978. Selection and dropping of whelks by north-western crows. *Behaviour* 67: 134–148.

Zajac, R. N. and R. B. Whitlatch. 1982. Responses of estuarine infauna to disturbance. II. Spatial and temporal variation of succession. *Marine Ecology Progress Series* 10: 15–27.

Zajac, R. N. and R. B. Whitlatch. 1985. A hierarchical approach to modeling soft-bottom sucessional dynamics. In *Proceedings of the 19th European Marine Biology Symposium*, P. E. Gibbs (ed.), 265–276. Cambridge University Press, Cambridge.

Zaneveld, J. S. 1966. The occurrence of benthic marine algae under shore fast-ice in the western Ross Sea Antarctic. *International Seaweed Symposium Proceedings* 5: 217–31.

Zieman, J. C. 1981. *The Food Webs within Seagrass Beds and Their Relationship to Adjacent Systems*. Special Report FWS/OBS-80/59. U.S. Fish and Wildlife Service, Washington, DC.

Index